Introduction to Liquid Crystals

Chemistry and Physics

Second Edition

Introduction to Liquid Crystals
Chemistry and Physics
Second Edition

Peter J. Collings
John W. Goodby

CRC Press
Taylor & Francis Group
Boca Raton London New York

CRC Press is an imprint of the
Taylor & Francis Group, an **informa** business

CRC Press
Taylor & Francis Group
6000 Broken Sound Parkway NW, Suite 300
Boca Raton, FL 33487-2742

© 2020 by Taylor & Francis Group, LLC
CRC Press is an imprint of Taylor & Francis Group, an Informa business

No claim to original U.S. Government works

Printed on acid-free paper

International Standard Book Number-13: 978-1-138-29885-9 (Hardback)
International Standard Book Number-13: 978-1-138-29876-7 (Paperback)

Library of Congress Cataloging-in-Publication Data

Names: Collings, Peter J., 1947- author. | Goodby, J. W. G., author.
Title: Introduction to liquid crystals : chemistry and physics / Peter J.
  Collings and John W. Goodby.
Other titles: Introduction to liquid crystals chemistry and physics
Description: [Second edition]. | [Boca Raton, Florida] : [CRC Press],
  [2019] | Revised edition of: Introduction to liquid crystals chemistry
  and physics / by Peter J. Collings and Michael Hird. [1997] | Includes
  bibliographical references and index. | Summary: "Introduction to Liquid
  Crystals: Chemistry and Physics, Second Edition, relies on only
  introductory level chemistry and physics as the foundation for
  understanding liquid crystal science. Liquid crystals combine the
  material properties of solids with the flow properties of fluids. As
  such they have provided the foundation for a revolution in low-power,
  flat-panel display technology (LCDs). In this book, the essential
  elements of liquid crystal science are introduced and explained from the
  perspectives of both the chemist and physicist"-- Provided by publisher.

Identifiers: LCCN 2019031550 (print) | LCCN 2019031551 (ebook) | ISBN
  9781138298767 (paperback) | ISBN 9781138298859 (hardback) | ISBN
  9781315098340 (ebook)
Subjects: LCSH: Liquid crystals.
Classification: LCC QD923 .C635 2019 (print) | LCC QD923 (ebook) | DDC
  530.4/29--dc23
LC record available at https://lccn.loc.gov/2019031550
LC ebook record available at https://lccn.loc.gov/2019031551

Visit the Taylor & Francis Web site at
http://www.taylorandfrancis.com

and the CRC Press Web site at
http://www.crcpress.com

To our wives, Diane and Ann,
who through their love and patience
have supported us in our lifelong adventures
in the science of liquid crystals.

Contents

Preface

This is the second edition of a textbook originally written by Peter J. Collings and Michael Hird. The first edition was designed to be the first book in a series on liquid crystals, and in order for it to appear before the other books in the series, it was written on a very tight schedule. As a result, the topics were selected extremely carefully and the coverage was at times narrow. Aspects of a textbook that would make it more useful to teachers and students, like end-of-chapter exercises, were absent. Still, the goal was to provide an introduction to the chemistry and physics of liquid crystals at a fundamental level for students with little or no knowledge of the field.

The authors of the second edition are Peter J. Collings and John W. Goodby. First and foremost, it must be stated that the second edition is very different from the first. The number of topics is greatly expanded, and scientific findings over the past 20 years have caused some old topics to be revised and many new topics to be added. The most significant change is in the scope and perspective of the chemistry discussions, which is not surprising given the change in author. The physics coverage is also expanded, with the discussion reflecting a more modern analytical approach. The result is a much deeper and broader introduction into the science of liquid crystals. Of almost equal importance are the changes designed to make the textbook more useful for students and teachers. Figures are used extensively throughout the book in hopes of making concepts clear and examples plentiful. In addition, exercises have been added at the end of every chapter so that students can test their understanding and review the material in the chapter.

While the limited coverage of the first edition might have made it appropriate to discuss the entire book in a course on liquid crystals, the expanded coverage of the second edition may make it necessary for a course to omit some topics. It is hoped that this option and other refinements will make the second edition an even more successful resource for the learning of liquid crystal science.

CHAPTER 1

Welcome to a Delicate and Special Phase of Matter

1.1 PHASES OF MATTER

The difference between solids and liquids, the two most common condensed matter phases, is that the atoms, molecules, or molecular complexes in a solid are ordered whereas in a liquid they are not. The order in a solid is usually both positional and orientational, in that the molecules are constrained both to occupy specific positions and to point their molecular axes in specific directions. The molecules in liquids, on the other hand, diffuse randomly throughout the sample container with the molecular axes rotating, tumbling, and oscillating wildly. Interestingly enough, many phases with more order than is present in liquids but less order than is typical of solids also exist in nature. These phases are grouped together and called by the oxymoron "liquid crystals", or in German, *flüssige kristalle*, since they share properties normally associated with both liquids and solids.

The molecules in all liquid crystal phases diffuse about much like the molecules of a liquid, but as they do so they maintain some degree of fluctuating orientational order and sometimes some local positional order. The degree of order in a liquid crystal is often quite small relative to a solid, although there are liquid crystals with considerable order. It is important to remember that the order in a liquid crystal is dynamic. This means that molecules move and tumble fairly freely, with the order that is present only observable on average. For most liquid crystals, there is only a slight tendency for the molecules to point more in one direction than other directions or to spend more time in some positions than other positions. The fact that most of the order of a solid is lost when it transforms to a liquid crystal is revealed by the value of the latent heat. Values are around 250 J/g, which is very typical of a solid to liquid transition. When a liquid crystal transforms to a liquid, however, the latent heat is much smaller, typically about 5 J/g. Yet this small amount of order in a liquid crystal reveals itself with mechanical, electrical, magnetic, and optical properties typical of solids.

1.2   ORIENTATIONAL AND POSITIONAL ORDER

In the simplest liquid crystal phase, called the nematic phase, one molecular axis tends to point along a preferred direction as the molecules undergo diffusion. This preferred direction is called the director and is denoted by the unit vector $\hat{n}$, where $\hat{n} = -\hat{n}$ because the preferred direction can be defined to be in either of two opposite directions. Examples of this type of order are how logs tend to align when flowing down a river or how pencils in a box tend to align if the box is shaken. An orientational distribution function $f(\theta)d\theta$ can be defined by either taking a snapshot of the molecules at any one time and noting the fraction of molecular axes making an angle between θ and $(\theta + d\theta)$ with the director, or by following a single molecule, noting the fraction of time the molecular axis makes an angle between θ and $(\theta + d\theta)$ with the director. Since all directions perpendicular to the director are equivalent in the nematic liquid crystal phase, the orientational distribution function does not depend on the azimuthal angle ϕ. Since $f(\theta)$ represents a fraction, the integral of $f(\theta)$ over all solid angle Ω ($d\Omega = \sin\theta d\theta d\phi$) must equal one. A snapshot of elongated molecules in such a phase is shown in Figure 1.1, which also presents the definition of θ.

To specify the amount of orientational order in such a liquid crystal phase, an order parameter S is defined. This can be done in many ways, but the most useful formulation is to find the average of a specific function of θ called the second Legendre polynomial, $P_2(\cos\theta)$,

$$S = \langle P_2(\cos\theta) \rangle = \left\langle \frac{3}{2}\cos^2\theta - \frac{1}{2} \right\rangle, \tag{1.1}$$

where the brackets denote an average over many molecules at the same

Figure 1.1 Molecular order in a nematic liquid crystal. The vertical arrow represents the director and the arrow at an angle θ to the director represents the long axis of one of the "molecules".

time or an average over time for a single molecule. The second Legendre polynomial is encountered in many physical and technical situations, and is chosen here for reasons that will become clear in the next paragraph.

To understand how the order parameter S represents the degree of orientational order, let us calculate S when there is no orientational order, that is, when the molecular axes point in all directions with equal probability. This means that $f(\theta)$ is equal to $1/4\pi$, so

$$
\begin{aligned}
S &= \int_0^\pi f(\theta) \left(\frac{3}{2} \cos^2 \theta - \frac{1}{2} \right) \sin \theta d\theta \int_0^{2\pi} d\phi \\
&= \frac{1}{2} \int_{-1}^1 \left(\frac{3}{2} x^2 - \frac{1}{2} \right) dx = 0,
\end{aligned}
\tag{1.2}
$$

where $x = \cos \theta$. Thus S is zero if there is no orientational order. Now let us calculate the value of S if the molecules are perfectly oriented. In this case $\theta = 0$ for all molecules, so the average of the second Legendre polynomial is just its value with $\theta = 0$. Thus $S = 1$ for a perfectly oriented system. In a typical liquid crystal, S decreases as the temperature is raised, usually taking on values between 0.3 and 0.8. This temperature variation is illustrated in Figure 1.2.

The order parameter can be measured in a number of ways. Usu-

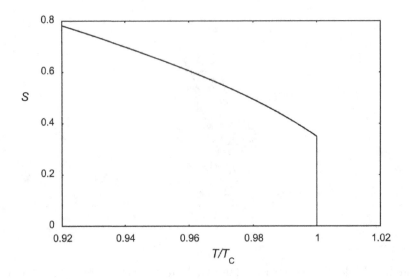

Figure 1.2 Typical temperature dependence of the nematic order parameter S. T_C is the temperature of the nematic liquid crystal to liquid phase transition.

ally a macroscopic property of the liquid crystal phase is measured, which then can be used to determine S if the molecular parameters that produce the macroscopic property are known. Diamagnetism, optical birefringence, nuclear magnetic resonance, and Raman scattering measurements are examples.

As an example of just how this works, let us assume the orientational distribution function varies as $\cos^2\theta$, that is $f(\theta) = 3\cos^2\theta/(4\pi)$. Such a function is shown in Figure 1.3. Evaluation of the first integral in Eq.(1.2) yields a value of S of 0.4 for this distribution of molecular axes.

Additional orientational order parameters can also be defined. For example, the average of the fourth Legendre polynomial could be used, $\langle P_4(\cos\theta) \rangle$, where

$$\langle P_4(\cos\theta) \rangle = \left\langle \frac{1}{8} \left(35\cos^4\theta - 30\cos^2\theta + 3 \right) \right\rangle. \qquad (1.3)$$

Measuring such an order parameter is more difficult than measuring S, but it can be done using Raman scattering, X-ray scattering, neutron scattering, or electron spin resonance. Since the director can be defined in either of two opposite directions, there is no need to define order

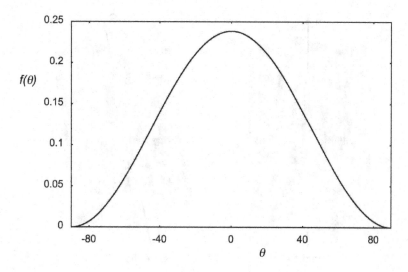

Figure 1.3 Example of an orientational distribution function $f(\theta) = 3\cos^2\theta/(4\pi)$.

parameters based on the odd Legendre polynomials, because all of these averages are zero.

Describing the positional order in a liquid crystal phase is a little different. The most straightforward case is when there is a tendency for the centers of mass of the molecules to spend more time in layers than between layers as they diffuse throughout the sample. Figure 1.4 depicts two such phases, the smectic A and smectic C phases.

Let us denote the normal to the layers as the z-axis. The density of the centers of mass oscillates up and down along the z-axis in going from one layer to the next. The most useful way to define a positional order parameter is to describe this oscillation as a sinusoidal function,

$$\rho(z) = \rho_0\left[1 + \psi\cos(2\pi z/d)\right], \tag{1.4}$$

where ρ_0 is the average density and d is the distance between layers. ψ is the order parameter and it is complex. The modulus of ψ, $|\psi|$, represents the amplitude of the density oscillation and is the more important part. The phase of ψ merely describes where along the z-axis the layers are located relative to a fixed coordinate system. Typical values for $|\psi|$ are much less than one, decreasing as the temperature is increased. Additional order parameters involving the higher harmonics of $\sin(2\pi z/d)$ can also be defined. These usually are very small relative to $|\psi|$, but can be significant when the positional order is quite

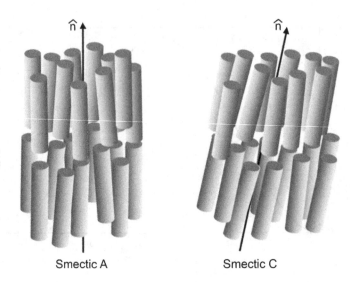

Figure 1.4 Two liquid crystal phases with layer order.

high. But for the two phases depicted in Figure 1.4, a simple sinusoidal description for the density of the centers of mass is very close to the actual case.

In contrast to liquid crystals that always possess orientational order and sometimes have positional order, there is another phase of matter that possesses positional order but no orientational order. Materials in this phase are called plastic crystals, and are said to exhibit rotator phases, because the molecules freely rotate around one or more of their molecular axes, even though their centers of mass are fixed.

The nematic, smectic A, and smectic C phases are only three of many different liquid crystal phases. What they all have in common, with only a couple of exceptions, is that they are anisotropic. In an isotropic phase, all spatial directions are equivalent. In an anisotropic phase, this is not the case. For example, in the nematic phase, a property along the director can have a different value compared to its value along a direction perpendicular to the director. An electric field applied parallel to the director produces a different electric polarization (dipole moment per unit volume) than the electric polarization produced by an electric field applied perpendicular to the director. In general, either the molecular shape is such that one molecular axis is very different from the other two, or in some cases different parts of the molecules have very different solubility properties. In either case, the interactions between these anisotropic molecules promote orientational and sometimes positional order in an otherwise fluid phase.

A basic question that will be addressed many times in many different ways is why some molecules form liquid crystal phases, *i.e.*, ordered fluid phases. How molecules pack together without leaving much free volume (the space in between the molecules) is a central criterion. Molecular interactions are also important, with the polarity of the molecules important to the nature of these interactions. The molecular entities that form liquid crystals can be quite complex, in which case the packing considerations become even more important, especially since complex molecular architectures allow for different and sometimes intricate packing structures. An example of extreme complexity is when assemblies, made by weakly bonding molecules together, form the entities that possess orientational and/or positional order. It must be mentioned here also that such systems are extremely dynamic. Not only do the assemblies diffuse with only averages showing order, the molecules within the assemblies also diffuse both within the assembly and between assemblies.

1.3 TYPES OF LIQUID CRYSTALS

The molecules of liquid crystal phases are usually rod-like, (*i.e.*, one molecular axis is much longer than the other two). Such systems are called rod-like or lath-like liquid crystals, but today are commonly called calamitic, where calamitic is Greek for *kalamos*. In Greek myth, Kalamos and Karpos were two friends who competed in a swimming contest; unfortunately Karpos drowned. In his grief, Kalamos drowned himself. The Gods transformed him into a water reed. Thus, Calamus means "reed" or "pen", and is also the root of the word calamari. Apart from the nematic and smectic A and C phases, many other phases are also possible. It is important that the molecule is fairly rigid for at least some portion of its length, since it must maintain an elongated shape in order to produce interactions that favor alignment and pack together in an ordered way. In fact, the architecture of such a molecule is two or more rigid ring structures, linked together directly or via linking groups, plus hydrocarbon chains at each end. As is discussed at length later, there are numerous variations of this recipe, for example, the replacement of one hydrocarbon chain with a group containing a permanent dipole. A molecule that forms several liquid crystal phases is shown in Figure 1.5.

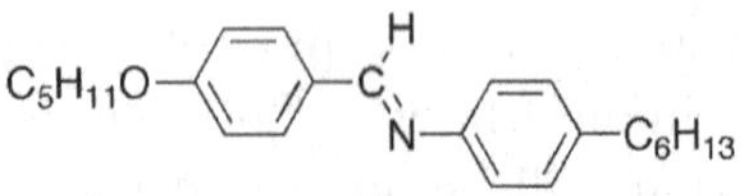

Figure 1.5 Typical molecule that forms a calamitic liquid crystal phase.

Disk-like molecules, namely molecules with one molecular axis much shorter than the other two, also form liquid crystal phases. Such compounds are called discotic liquid crystals, and again, rigidity in the central part of the molecule is generally essential. The cores of molecules that form discotic liquid crystal phases are usually based on benzene, triphenylene, or truxene, with six or eight side chains. Figure 1.6 is an example of a molecule that forms a discotic liquid crystal phase.

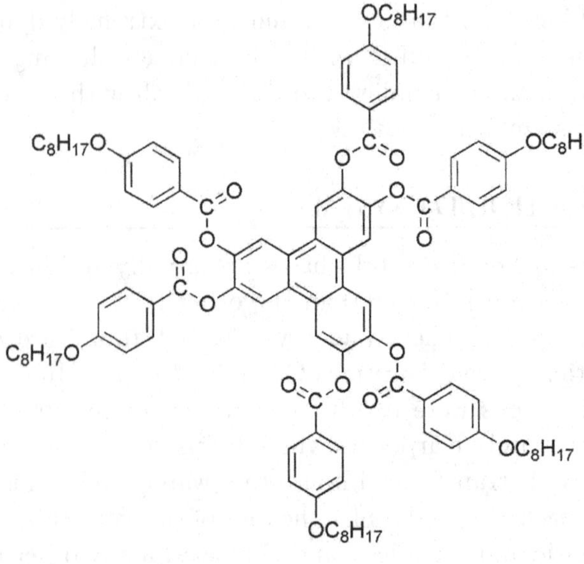

Figure 1.6 Typical molecule that forms a discotic liquid crystal phase.

Both calamitic and discotic liquid crystals are also called thermotropic liquid crystals because the liquid crystal phase is stable for a certain temperature interval. Pure compounds or mixtures of compounds fall into this category. There is another type of molecule, however, which forms liquid crystal phases only when mixed with a solvent of some kind. For these compounds, the concentration of the solution is just as important, if not more important, than the temperature in determining whether a liquid crystal phase is stable. To differentiate these substances from thermotropic liquid crystals (which need no solvent), these compounds have been given the name lyotropic liquid crystals. Indeed the first observation of a lyotropic liquid crystal was made by Virchow in 1854 when studying biological systems composed of myelin in water.

A recipe for a molecule that forms a lyotropic liquid crystal phase is one which combines a hydrophobic group at one end with a hydrophilic group at the other end, *e.g.*, soaps. Such amphiphilic molecules associate to form ordered structures in both polar and non-polar solvents. Good examples are glycolipids and phospholipids found in nearly all living organisms, including us! As can be seen from Figure 1.7, such compounds have a polar head group attached to a hydrocarbon tail group.

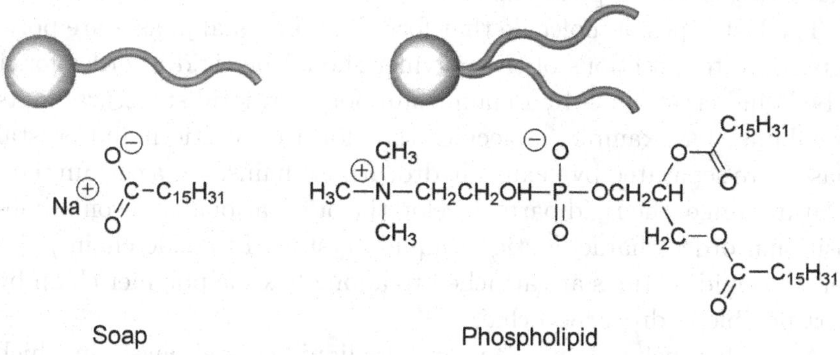

Figure 1.7 Two molecules that form lyotropic liquid crystal phases.

When dissolved in a polar solvent such as water, the hydrophobic tails associate together and present the hydrophilic heads to the solvent. The typical resulting structure for soap molecules is called a micelle and for phospholipids is called a vesicle. These are shown in Figure 1.8. Both soaps and phospholipid molecules also associate to form a bilayer structure, with the hydrocarbon chains separated from the water by the head groups. These lamellar phases are of extreme importance in the case of phospholipids, since such a lipid bilayer is the structural unit for biological membranes.

If these amphiphilic molecules are mixed with a non-polar solvent such as hexane, similar structures form but now the polar heads associate together with the non-polar tail groups in contact with the

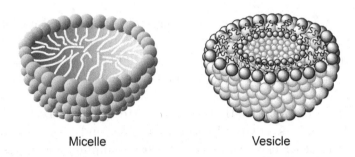

Micelle   Vesicle

Figure 1.8 Structures formed by amphiphilic molecules in a polar solvent.

solvent. These are called reverse phases to distinguish them from the phases that occur in polar solvents.

The last types of molecule that form liquid crystal phases are polymers. Here too, sections of the polymer should be rigid in order for it to be liquid crystalline. In a main chain polymer, rigid structural units resembling, for example, molecules that form calamitic liquid crystal phases, are separated by flexible hydrocarbon chains. For a certain temperature range, the rigid parts develop the orientational and sometimes positional order characteristic of liquid crystals. In a side chain polymer, the rigid sections are attached to a long flexible polymer chain by short flexible hydrocarbon chains.

Again, for such polymers one or more liquid crystal phases in which the rigid parts possess orientational and sometimes positional order (with the long flexible chains unordered) are stable for certain intervals of temperature. Figure 1.9 depicts both a main chain and side chain polymer. Main chain polymers are often found in applications that require strong fibers or the capacity to be engineered, such as Vectran®.

Certain long, rigid molecules develop liquid crystalline order when dissolved in a solvent at high concentrations. Many of the examples of the molecules that fall in this category are biologically important. Two examples are poly(γ-benzyl L-glutamate) (PBLG) and tobacco mosaic virus (TMV). These types of phases are normally called polymer solutions to differentiate them from liquid crystalline polymers and lyotropic liquid crystals. Sometimes, polymer solutions can be used in the processing of materials such as Kevlar®, which is found, for example, in protective vests and tires.

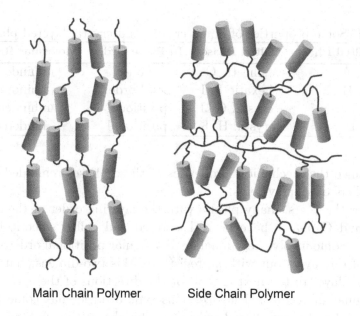

Main Chain Polymer   Side Chain Polymer

Figure 1.9 Two types of polymer liquid crystals.

1.4 LIQUID CRYSTAL PHASES

As already mentioned, the nematic phase of calamitic liquid crystals is the simplest liquid crystal phase. In this phase the molecules maintain a preferred orientational direction as they diffuse throughout the sample. There is no positional order in this phase, and this is the phase depicted in Figure 1.1. The name comes from the Greek word *nematos* for thread, since in a polarizing microscope, there are often many dark lines visible in thick film samples. These lines are defects in the orientational order and are called disclinations, and are unique to liquid crystals because they are not found in solids. The preferred direction is undefined at these disclinations.

Two other phases common to calamitic liquid crystals are shown in Figure 1.4. In addition to the orientational order of nematics, there is positional order since the centers of mass are arranged in layers. If the director is perpendicular to the layers in which molecules are more likely to be, then it is called a smectic A liquid crystal. If the director makes an angle other than 90° to these layers, it is a smectic C phase. In these two phases, there is no positional order of the molecules within the layers. The name smectic comes from the Greek word *smectos* for

Table 1.1 Some properties of liquid crystal and crystal layered phases.

Non-Tilted Phases	Tilted Phases	In Plane Order	Molecular Rotation
smectic A	smectic C	none	unhindered
hexatic B	hexatic F, I	bond order	unhindered
crystal B	crystal G, J	positional	unhindered
crystal E	crystal H, K	positional	hindered

soap, since the mechanical properties of these phases reminded early researchers of soap systems.

Over the years, many phases with the layering order of the smectic A and C phases have been discovered and called smectic liquid crystals. Some involve hexagonal or rectangular positional order in the plane of the layers but with no registry of this two-dimensional order from one layer to the next except for the directions of the axes of the rectangular or hexagonal lattice. This type of order has come to be known as bond orientational order, and the phases that possess it are called hexatic phases. Although all of these phases are proper liquid crystal phases, there are other phases in which the positional order is three-dimensional. In these phases, the two-dimensional positional order carries from one layer to the next with respect to both the orientation and position of the lattice. Although such phases were originally labeled as smectic liquid crystals, they are now more properly called crystal mesophases. These layered phases also differ in the nature of the rotation of the molecules about their long axes. In many phases there is unhindered thermal rotation about the long axes of the molecules; in other phases the rotation is hindered with only flips of 180° taking place.

Just as with the smectic A and C phases, examples of each of these more ordered phases with a director both perpendicular to the layers and at an angle to the layer normal have been observed. For the tilted phases, the director can point toward one of the nearest neighbor molecules or toward the midpoint of the line joining two nearest neighbors.

The best way to keep track of all these layered smectic and crystal phases is to use a table listing the various phases along with a few of their most important properties. Table 1.1 is such a chart. A more detailed discussion of these phases can be found in Chapter 2.

If the molecules that form a liquid crystal phase are chiral (lack inversion or reflection symmetry like a human hand), then chiral phases sometimes exist in place of non-chiral phases. In calamitic liquid crystals, the nematic phase is replaced by the chiral nematic phase, in which the director rotates in helical fashion about an axis perpendicular to the director. Such a phase is illustrated in Figure 1.10.

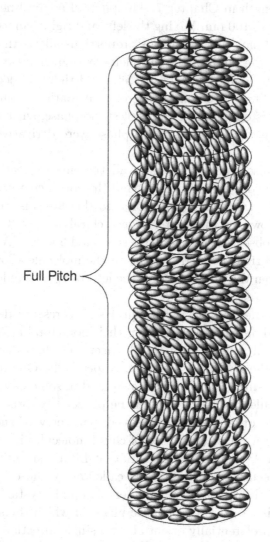

Full Pitch

Figure 1.10 Structure of the chiral nematic phase. The director is always perpendicular to the helical axis (denoted by the arrow), and the distance along the helical axis for the director to rotate by 360° is called the pitch.

The pitch of a chiral nematic phase is the distance along the helical axis over which the director rotates by 360°. It should be noted, however, that the structure repeats itself every half pitch due to the equivalency of $\hat{n}$ and $-\hat{n}$. Interesting optical effects occur when the wavelength of light in the liquid crystal is equal to the pitch. These are described at length in Chapter 7. The pitch of a chiral nematic phase can be as short as 100 nm. Mixing the left- and right-handed versions of the same chiral molecule in various proportions allows the pitch to be increased from the pitch of the pure left- or right-handed molecule. A racemic mixture (equal parts of the left- and right-handed molecules) possesses an infinite pitch and is therefore nematic. Finally, the chiral nematic phase is often called the cholesteric phase, since many of the first compounds that possessed this phase were derivatives of cholesterol.

There are also chiral versions of all the tilted smectic phases. In these phases the director rotates around the cone generated by the tilt angle as the position along the normal to the layers is varied. Such a structure is shown in Figure 1.11 for a chiral smectic C phase. The shortest pitch observed in this phase is around 250 nm. Again, mixing of the left- and right-handed versions of the molecule allows the pitch to be varied from the pitch of either the left- or right-handed molecules to infinity for a racemic mixture.

Sometimes these phases with chiral structures are designated by an asterisk to differentiate them from their non-chiral analogs. For example, the chiral nematic and chiral smectic C phases are frequently designated as the N* and C* phases, respectively. One must be cautious, however, since the asterisk is also used in some cases to indicate that the molecules are chiral even if the phase they form possesses no chiral structure other than the reduced symmetry of the molecules. Thus a smectic A phase composed of chiral molecules is sometimes referred to as an A* phase to differentiate it from a smectic A phase of non-chiral molecules, which is simply called an A phase.

Disc-like molecules form a number of liquid crystal phases. The simplest type is the discotic nematic phase in which the short axes of the molecules preferentially orient along a single direction again called the director. This phase is depicted in Figure 1.12, where it is clear that there is no positional order in this phase. Positional order in discotic liquid crystals displays itself by the tendency of the molecules to arrange themselves in columns. This means that in the plane perpendicular to the columns, the molecules tend to arrange themselves

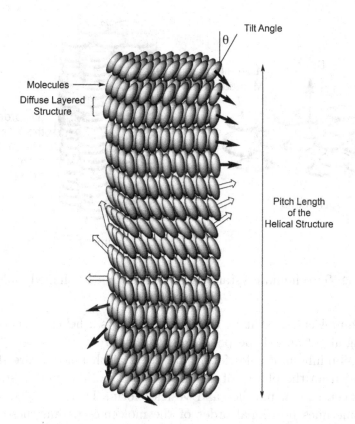

Figure 1.11 Structure of the chiral smectic C phase. The director makes an angle θ with the helical axis and the distance along the helical axis for the director to rotate around the helical axis by 360° is called the pitch. The electric polarization is perpendicular to the plane containing the helical axis and the director, and is depicted by several small arrows.

in a two-dimensional lattice, either rectangular or hexagonal, as they diffuse throughout the sample. Such a columnar phase is also shown in Figure 1.12. Keep in mind that the stacking along the columns is sometimes not regular. In some columnar phases, the molecules are tilted with their short axes not parallel to the column axes. The tilt direction alternates from one column to the next.

Chiral disk-like molecules also form chiral phases, the best example being the chiral nematic phase. From the standpoint of the director, the structure of this phase is identical to the chiral nematic phase in

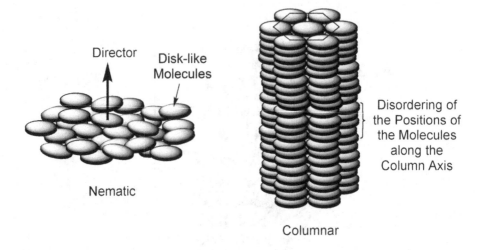

Figure 1.12 Two liquid crystal phases formed by disc-shaped molecules.

rod-like molecules. That is, the director adopts a helical structure by rotating about an axis perpendicular to the director.

Amphiphilic molecules form liquid crystal phases that are slightly different from the phases of calamitic and discotic liquid crystals. At low concentration, micelles and vesicles form. There is orientational and sometimes positional order of the molecules within these structures, but there is no ordering of the micelles or vesicles themselves. At higher concentrations, the structure changes to one in which the micelles or vesicles themselves are also ordered. One example is the hexagonal phase (or middle soap phase), which possesses a hexagonal arrangement of long cylindrical rods of amphiphilic molecules. At some concentrations the lamellar phase (sometimes called the neat soap phase) forms with a uniform amount of water separating the bilayers. Cross-sections of the hexagonal and lamellar phases are illustrated in Figures 1.13 and 1.14. Sometimes a cubic phase (or viscous isotropic phase) forms between the hexagonal and lamellar phases. In this phase the amphiphilic molecules arrange themselves in spheres, which in turn form a cubic lattice. These spheres may be closed or they may be connected to one another, thus forming a bicontinuous phase.

If the amphiphilic character of a molecule is between its center and outside, a nematic phase of assemblies sometimes forms. By stacking one molecule on top of the other, the more polar parts of the molecule are on the outside exposed to the polar solvent (usually water) and the

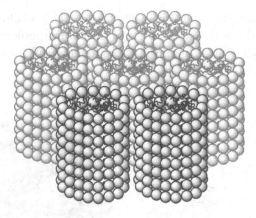

Figure 1.13 Structure of the lyotropic hexagonal liquid crystal phase.

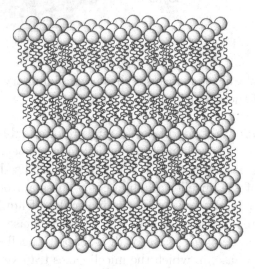

Figure 1.14 Structure of the lyotropic lamellar liquid crystal phase.

more non-polar aromatic molecular centers are shielded from the polar solvent and experience attractive interactions due to overlap of the π-orbitals of the aromatic groups (see Figure 1.15). These lyotropic phases of rod-like associations are called chromonic liquid crystals, since many of the molecules forming them are dyes. If the associations possess orientational and perhaps positional order, the nematic and columnar phases are usually the result, and as is the case for most lyotropic liquid crystals, the liquid crystal phase exists for a range of both concentration and temperature. Besides dyes, certain drugs and biologically important molecules such as nucleic acids form chromonic liquid crystals. If the molecule is chiral or if chiral molecules are added to the solvent, the chiral nematic phase forms in place of the nematic phase.

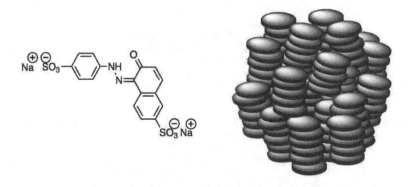

Figure 1.15 A molecule that forms a chromonic liquid crystal phase and the structure of the nematic phase it forms.

Under just the right conditions, a mixture of a highly polar liquid, a slightly polar liquid, and amphiphilic molecules associate to form micelles that are not spherical. They can be rod-like, disc-like, or biaxial (all three axes of the micelles are different). These anisotropic micelles sometimes order in the solvent as the molecules of liquid crystals order in thermotropic phases. There is a nematic phase of rod-shaped micelles, another nematic phase of disk-shaped micelles, and even a biaxial nematic phase, in which the micelle axes transverse to the long micelle axis partially order. Chiral versions of these phases with the same structure as the chiral nematic phase also form.

The most common liquid crystal phases of polymers are the nematic and smectic phases. Here it is the rigid segments of the polymer, whether they are in main chains or on side chains, that order. The ordering of these rigid mesogenic units is identical to the order in the nematic and smectic phases of small molecules. Chiral nematic and chiral smectic phases also form if some component of the polymer is chiral. It should also be noted that the solid phase in many polymers is a glass phase and not a crystalline phase. The order in a glass phase is very similar to crystalline order, in that the molecules are more or less locked into place, but instead of a regular lattice of molecules there is a random network of molecules.

Rigid macromolecules form nematic and smectic phases when their concentration in certain solvents is high enough. There are also chiral phases if the molecular entities are chiral. The best examples of these are the rod-like macromolecules with structures based on a single or double helix, which form chiral nematic and chiral smectic phases in solution.

1.5   COMPLEX MOLECULAR ARCHITECTURES

The liquid crystal material shown in Figure 1.5 has a fairly simple molecular structure. However, it must be pointed out right from the beginning that complex molecular structures also form liquid crystal phases and sometimes exhibit unexpected structures. The simple structure of Fig. 1.5 can be made a bit more complex by (1) adding another ring structure, (2) adding a lateral group to one or more of the ring structures, (3) changing the linkage group between the ring structures, and (4) changing the length, bonds, and/or branching of the end chains. By making such changes, materials can be found that exhibit a wide range of properties, a result that has been crucial for the development of liquid crystal devices. But even with these changes, the molecular architecture does not approach the complexity possible for liquid crystalline materials.

For example, the molecules of two liquid crystals can be bound to a small central unit such that the two rod-like parts are not collinear. These bent-core molecules resemble boomerangs or bananas more than rods, and they order into liquid crystal phases with novel properties. Layered structures are the norm, but within the layers the points of the boomerangs are aligned. There is still plenty of room for nature to be creative with such building blocks; two examples are a twisted structure even though the molecule is non-chiral and an ordering of the direction of the points from one layer to the next.

Adding even more complexity, multiple molecules that form liquid crystal phases can be bound together with fairly long chains, creating molecules called bi-mesogens, tri-mesogens, tetra-mesogens, *etc.* Or multiple molecules of liquid crystals can be bound to a central core group, producing trimers, tetramers, pentamers, *etc.* One might think such supermolecules would never form a liquid crystal phase, but in some cases the attached molecules line up on either side of the core to produce something that is rod-shaped. Or the attached molecular units line up in a plane around the core, adopting a disk shape that forms discotic liquid crystal phases. Disk-like and rod-like molecules can even be bound together by fairly long chains to produce a super-molecule that orders into a liquid crystal phase!

This huge amount of richness in the structures that form liquid crystal phases is explored in depth in later chapters. Figure 1.16 displays the structure of a complicated molecule and the liquid crystal phase it forms. As a rule, looking into how these supermolecules with such complicated structures can pack together leaving as little free volume as possible explains why a particular liquid crystal structure forms.

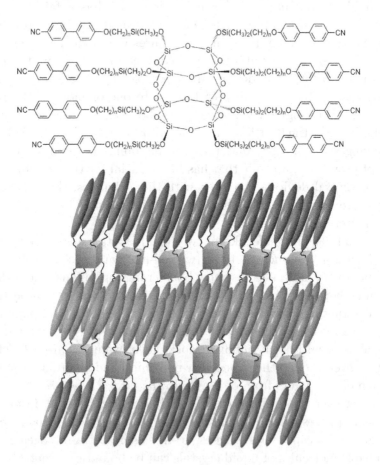

Figure 1.16 The molecular structure of an octamer and the structure of the layered liquid crystal phase it forms.

1.6 LIQUID CRYSTAL DISPLAYS

There are more liquid crystals displays (LCDs) in the world than people, so why do so many of today's displays use liquid crystals? The answer is quite simple and involves two very basic attributes of liquid crystals. First, because liquid crystals are ordered, their optical properties can be varied by changing how they are ordered. Second, because liquid crystals are fluid, a weak external influence is all that is necessary to change how they are ordered. Thus the typical liquid crystal display is ordered in one way when no electric field is applied to it, and ordered in another way when an electric field is applied to it. Often there are polarizers on either side of the liquid crystal, so for example, no light gets through the polarizers and liquid crystal when no electric field is applied, and lots of light gets through the polarizers and liquid crystal when an electric field is applied. In short, an LCD typically acts as a light valve, letting light from a source behind it get through or not. Because it does not take much electrical power to apply the necessary electric field, LCDs were first used as displays in battery powered devices. Amazingly, it was found that a watch display could be operated using a miniature battery for over ten years. But as the quality of the display improved, LCDs found their way into all kinds of devices, even when power consumption is not a concern.

There are many different designs for LCDs. A simple example is a vertically aligned nematic (VAN) LCD found in many TVs. In Figure 1.17, a thin cell is made to contain the liquid crystal using two pieces of glass with a transparent electrode and a surfactant on their inside surfaces. The transparent electrode is a conducting film, thick enough to spread electric charge around but thin enough so most of the light passes through it. The surfactant interacts with the molecules of the liquid crystal so the director is perpendicular to both surfaces. In thin cells, aligning the director at the two surfaces causes the director to point in the same direction throughout the liquid crystal. This is why it is called a vertically aligned LCD. If polarizers with their transmission axes perpendicular are added to the outside surface of the two pieces of glass, then no light is transmitted through the polarizers and liquid crystal. However, if the molecules of the liquid crystal tend to align perpendicular to an electric field, application of a voltage between the two transparent electrodes produces an electric field in the liquid crystal perpendicular to the glass surfaces. The director reacts by tilting away from the direction perpendicular to the glass surfaces, allowing light to be transmitted through the polarizers and liquid crystal. Thus this LCD has two states, an off-state that is dark when no electrical voltage is applied, and an on-state that is bright when an electrical voltage is applied.

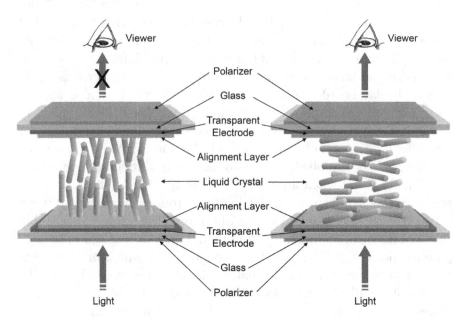

Figure 1.17 A vertically aligned liquid crystal display. The transmission axes of the polarizers are perpendicular to one another. Left: zero applied voltage. Right: non-zero applied voltage.

What has just been described is one pixel of an LCD with many, many pixels. If the light behind the liquid crystal and polarizers is white, then color filters on the outside surface of the glass produce red, green, or blue pixels. By addressing all of the pixels appropriately, a wonderfully colored image appears on the LCD. While this description of an LCD makes everything sound straightforward, it should be kept in mind that it took an increasing knowledge of molecular properties, molecular interactions, and molecular packing to engineer the materials necessary for the evolution of the LCDs, from the early ones that barely worked to the absolutely gorgeous, high-definition, high-speed LCDs we routinely use today.

Figure 1.18 traces the history of LCD technology. The first displays (upper right) were based on a dynamic scattering mode (DSM), but quickly these were replaced by twisted nematic (TN) displays and super twisted nematic (STN) displays. By putting the two electrodes on the same substrate, the in-plane switching (IPS) display achieved a much wider viewing angle. Then vertical aligned nematic (VAN) displays were important as the area of displays increased substantially. A more recent display, called the zenithal bistable nematic display (ZBD) is based on grating surfaces, which allow the display to operate in a bistable mode and provide an image without the application of an electric field other than to switch the image.

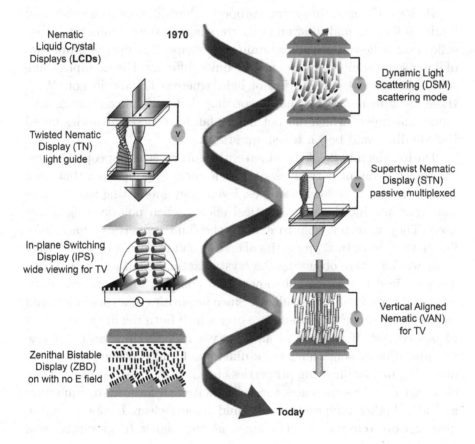

Figure 1.18 Historical evolution of liquid crystal displays from the dynamic scattering mode (DSM) display shown in the upper right to the zenithal bistable display (ZBD) shown in the lower left.

1.7   HISTORICAL PERSPECTIVE

Liquid crystals are currently an important phase of matter both scientifically and technologically. They are recognized as a stable phase for many compounds, thus putting them on equal footing with the solid, liquid, and gas phases. This situation is quite a recent development, existing for about the last 50 years. It is both interesting and informative to review quickly how the present understanding of this phase of matter developed.

Perhaps the most important property that differentiates solids and liquids is flow. Liquids flow and take the shape of the container, whereas solids do not flow and tend to retain their shape. The optical properties of liquids and some solids can also be quite different. For example, some solids change the polarization of light whereas liquids do not. With these ideas in mind, it is not surprising that early investigators who found substances which did not neatly fall into these categories noted their findings and began to ask questions.

During the middle of the nineteenth century, several people observing the covering of nerve fibers under a microscope noticed that they formed flexible and flowing shapes. Even more interesting was the fact that they produced unusual optical effects when polarized light was used. They were not able to explain the fluid properties typical of a liquid with the optical properties of certain solids. About this same time other workers were observing the crystallization of various substances. They noticed that some substances first changed to a non-crystalline form and then finally crystallized. Since impurities in a substance tend to produce a range of temperature over which both the liquid and solid phases coexist, it was not at all clear whether this was an effect due to impurities or something quite different. Finally, other people were observing unusual melting properties of some natural substances. They observed that in some cases the solid melted to form an opaque fluid and at a higher temperature the fluid turned clear. In one case, the investigators referred to the change at the higher temperature as a second melting point.

The discovery of liquid crystals is usually attributed to an Austrian botanist by the name of Friedrich Reinitzer. In 1888, he experimented with a substance related to cholesterol and noted that it had two melting points. At 145.5°C it melted from a solid to a cloudy liquid and at 178.5°C it turned into a clear liquid. He also observed some unusual color behavior upon cooling; first a pale blue color appeared as the clear liquid turned cloudy and second a bright blue-violet color was present as the cloudy liquid crystallized. Reinitzer sent samples of this substance to Otto Lehmann, a professor of natural philosophy (physics) in Germany. Lehmann was one of the people studying the crystallization properties of various substances and Reinitzer wondered whether what he observed was related to what Lehmann was reporting. Lehmann had constructed a polarizing microscope with a stage on which he could precisely control the temperature of his samples. This instrument allowed him to observe the crystallization of his sample under controlled conditions. Lehmann observed Reinitzer's substance with his microscope and noted its similarity to some of his own samples. He first referred to them as soft crystals; later he used the term crystalline fluids. As he became more convinced that the opaque phase was a uniform phase of matter sharing properties of both liquids and solids, he began to call them *flüssige kristalle* or liquid crystals.

Lehmann was quite influential in getting a group of French workers interested in liquid crystals. The culmination of this effort was the classification scheme introduced by Georges Freidel in 1922. This consisted of three phases, the nematic, smectic, and cholesteric phases, with the smectic phase having a layered structure. There was some experimentation on liquid crystals during the war years of the 1930s and 1940s, mainly concerning their elastic properties, X-ray diffractive characteristics, and the effects of electric and magnetic fields. The first use of the order parameter occurred during this time. After the war, work on liquid crystals slowed significantly, probably due to the fact that textbooks did not discuss them and because no one saw an application for them.

Shortly before 1960, interest in liquid crystals awakened in the United States, Great Britain, and the Soviet Union. This resurgence was probably due to the lure of applications, but there was also a feeling that these phases were scientifically significant. An American chemist, Glenn Brown, published a lengthy review article on liquid crystals and founded the Liquid Crystal Institute at Kent State University. George Gray, a British chemist, published a full-length book on liquid crystals, and I. G. Chystyakov started a group working on liquid crystals in Moscow. Research on liquid crystals also began in Germany and France. Progress was both swift and substantial. A theory for the nematic liquid crystal phase added an important foundation to the research and the first demonstration of a liquid crystal display gave the whole enterprise a practical purpose. Also during this time, the first room temperature, moderately stable liquid crystal was discovered.

Research on liquid crystals exploded during the 1970s and 1980s, and became part of a growing scientific field investigating soft matter in the years since. Scientifically, liquid crystals are an important phase for the investigation of cooperative phenomena, and liquid crystal synthesis is a field in its own right, especially in investigating structure-property relationships. Technologically, liquid crystals have become a part of our daily lives, first showing up in wristwatches and pocket calculators, but now being used for displays in all sorts of instrumentation, including tablets, smartphones, and televisions. In fact, the effect that LCDs have had on communications across the world must be recognized, including the crucial role it has had on social media such as Facebook, Skype, *etc.* At first their advantage was low power consumption and small size; now liquid crystal displays compete with other technologies for attractiveness, ease of viewing, cost, and durability.

EXERCISES

1.1 As a rule with only a few exceptions, the amount of order decreases as the temperature increases. With this in mind, determine which of the two phases given occurs at the higher temperature: (a) smectic A and nematic, (b) discotic nematic and columnar, (c) smectic A and hexatic B.

1.2 With $f(\theta, \phi) = 1/(4\pi)$, show that $\langle P_3(\cos\theta)\rangle = \langle(5\cos^3\theta - 3\cos\theta)/2\rangle = 0$.

1.3 Check to see if the distribution function $f(\theta, \phi) = (3/(4\pi))\cos^2\theta$ is properly normalized. That is, does the integral of this $f(\theta, \phi)$ over all solid angle equal one?

1.4 With $f(\theta, \phi) = (3/(4\pi))\cos^2\theta$, show that $\langle P_1(\cos\theta)\rangle = \langle\cos\theta\rangle = 0$.

1.5 Two order parameters are necessary to describe the order in a smectic A liquid crystal: S to characterize the orientational order and ψ to specify the density variation due to the layering. Is another order parameter necessary to describe the order in a smectic C liquid crystal? If so, what might it be?

1.6 Figure 1.8 shows the mesophase structures amphiphilic molecules form in a polar solvent. Sketch the corresponding structures for the same molecules in a non-polar solvent, $i.e.$, the reverse phase structures.

1.7 Assume the helical axis of a chiral nematic phase is along the z-axis (see Figure 1.10) and that at $z = 0$ the director $\hat{n}$ is along the x-axis $(n_x(0) = 1, n_y(0) = 0, n_z(0) = 0)$. If the pitch is given by P, find expressions for the components of $\hat{n}$ at any z ($i.e.$, find expressions for $n_x(z), n_y(z), n_z(z)$) for both a right-handed and left-handed chiral nematic phase. For a right-handed chiral nematic phase, the director rotates counterclockwise in the xy plane (looking from $z = +\infty$) as z increases.

CHAPTER 2

A Liquid Crystal – What Is It?

2.1 FACE-TO-FACE WITH LIQUID CRYSTALS

The oxymoron "liquid crystal" suggests that this unique state of matter has properties of both liquids and crystals, and therefore has structures derived from both. It should "flow" like a liquid, with ease like water or with a high viscosity like molasses, and it should have local order making it quasi-rigid like a crystal. However, this does not mean this phase of matter is composed of mixtures of liquids and solids; rather it is homogeneous and the constituents are uniformly dispersed in the bulk. What liquid crystals are **NOT** can be found on the internet, and the following description is similar to what is there. *Liquid crystals are minerals with amazing properties. Their geometric structure and irides-cent color exude energy that can be harnessed by humans. When placed in water, the liquid crystal imparts its vibrational energy to the water, using its unique natural geometry to imbue the water with exceptional healing power.* In fact, the closest you can come to liquid crystals in your everyday life is through your mobile telephones, flat screen tele-visions, soaps and detergents, and high-performance polymers such as Kevlar®.

As liquid crystals behave like fluids, their constituents (molecules, complexes, supramolecular assemblies, *etc.*) must tumble, rotate, and oscillate relatively freely. They are in dynamic motion. Thus when a nematic phase is viewed under a magnification of around 100 times, "Brownian" motion is clearly apparent and can be disturbed by changes in temperature, pressure, and mechanical stress.

The dynamic motion of the constituents can affect the physical properties of a material. Take for example a perfect crystal that has a cubic lattice, where the constituents have individual properties that are not the same in 3D space. In the solid state, although the components vibrate on their lattice sites, their optical properties are different along the three main axes of the cubic lattice, as shown in Figure 2.1. Thus, the material is anisotropic. If the constituents are allowed to rotate diffusely on their lattice sites, then the directional properties become the same in 3D space, and the materials are isotropic as shown in the figure. Such changes can be achieved by heating the material, thereby inducing a displacive transformation. Displacive is a word that describes a number of phase transitions, an example of which is a structural change caused by a shuffle of components, *i.e.*, they are driven by displacements of atoms or ions. For an opaque material the phase transition occurs with the material becoming transparent. These phase transitions often correspond to the formation of plastic crystals. There are many examples of materials that exhibit plastic crystals, such as adamantane, camphor, *etc.*, for which the constituents typically have either spherical or near-spherical shapes, as shown in Figure 2.2. Generally for such materials, the rotational diffusion occurs at rates of 10^{10} - 10^{12} times per second.

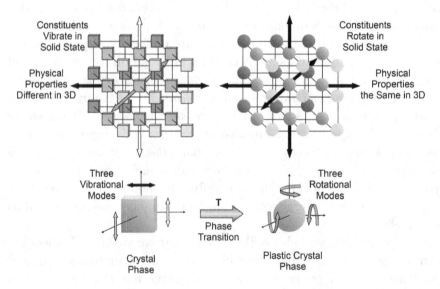

Figure 2.1 Directional properties in a cubic lattice. The constituents are allowed to vibrate (left) or rotate diffusely (right) on their lattice sites.

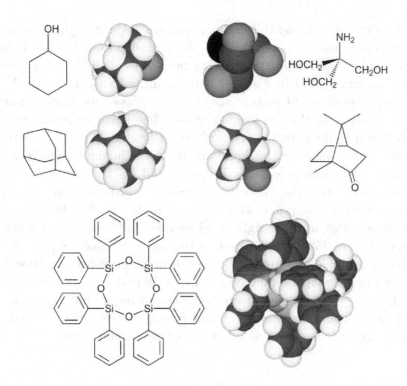

Figure 2.2 Structures of compounds that exhibit plastic crystal phases.

The transition from a solid to a liquid can also show similar changes because the crystal structure of the solid can be anisotropic, but the structure of the liquid is amorphous and therefore its properties are isotropic. Given this information for solids and liquids, an important question to ask is, "What happens at the phase transition from a liquid crystal to a liquid?". First, we need to examine the bulk appearance of a typical liquid crystal, and in this example the nematic phase is selected, in particular the relatively fluid nematic phase of 4-pentyl-4'-cyanobiphenyl, commonly known as 5CB.

A photograph of 5CB in a 250 ml conical flask is shown in Figure 2.3. It is an opaque white liquid that flows like water and has a similar density. The chemical structure of 5CB, shown in Figure 2.4, is composed of a biphenyl moiety, a nitrile (or cyano) functional group, and an aliphatic chain. The chemical structure contains only one chromophoric group, the biphenyl unit, which does not support the material being white in color. Moreover, if the nematic phase were really a liquid, then it would appear colorless and transparent. Thus the nematic phase cannot be a true liquid, and the white appearance is due to light refraction or scattering, which means that the phase must possess local ordering of the molecules. The light scattering properties of the nematic phase can be demonstrated in a simple experiment, which is shown in Figure 2.4. The left-hand picture shows a long glass tube that contains about 0.5 kg of 5CB in its nematic phase. The liquid crystal is gently heated near to the nematic-air interface. At around 30 to 40°C there is a phase transition from the liquid crystal to the liquid, at which point the light scattering properties disappear and 5CB becomes isotropic in its liquid phase. This transition is usually called the clearing point, or the isotropization point. The meniscus at the interface between the two phases, shown in the right-hand picture, will be at the transition temperature of 35°C. Thus, this experiment, which may be performed on a smaller scale and used in demonstrations, provides a method for determining the local temperature of a body that has been heated to $(35 \pm 2)°C$. For other liquid crystals this technique could be used as a sensor to determine whether or not packaged items such as food or pharmaceuticals are maintained at an optimum temperature.

Figure 2.3 A 250 ml flask containing 4-pentyl-4′-cyanobiphenyl, 5CB.

We can further investigate the liquid to nematic phase transition by viewing it between polarizers where the vibration axes are perpendicular to one another. This is usually done using a polarizing, transmission light microscope at a magnification of 100 to 200 times, as shown in Figure 2.5. The use of the microscope to determine transition temperatures is akin to using a melting point apparatus for determining a melting point of a material, except it can be done more accurately and with expanded observation of the process. Hence microscopic studies of liquid crystals are usually the first experimental methods used

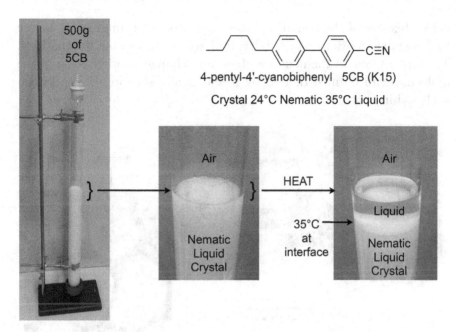

Figure 2.4 The bulk clearing point transition at 35°C from the nematic phase to the isotropic liquid for 5CB.

to investigate mesogenic materials. In the figure shown, the left-hand photograph and drawing show a typical instrumental arrangement for the study of liquid crystals. A polarizing transmission light microscope, with a rotatable sample stage and rotatable upper (analyzer) and lower polarizers, is fitted with a heating/cooling unit and microfurnace and a camera. The heating/cooling unit is usually accurate to 0.1°C, and the camera preferentially is digital with a 10 to 20 megapixel capacity. Sample specimens can be inserted into the microfurnace, which has built-in portals for observations. The specimens are usually sandwiched between a glass slide and a cover slip and are usually between 10 to 50 μm in thickness. The image to the right of the figure is of the phase transition from the nematic phase of 5CB to the liquid. The liquid appears extinct when placed between the polarizers because the liquid has isotropic optical properties. Conversely, the nematic appears bright with lines and shadows in the phase. These patterns are called textures, and overall the pattern is called a defect texture. The patterns are related to the macroscopic structure of the nematic phase, and although the picture is static, direct observation shows that the defect texture of the nematic phase shimmers. The shimmering is due to Brownian motion, which is related to the molecular fluctuations. The ease of flow of the nematic phase and change in orientation of the molecules can be investigated by mechanically sheering the specimen with a thin spatula.

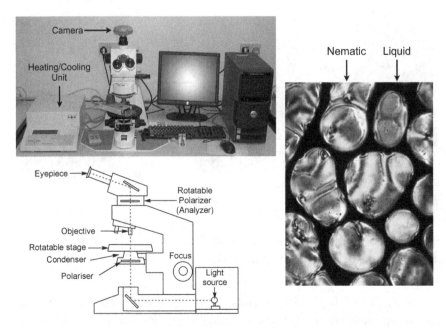

Figure 2.5 A polarizing transmission light microscope (left) and a photomicrograph of the liquid to nematic phase transition when viewed in transmission between crossed polarizers (x100). The bright areas are the nematic phase and the black areas are the liquid phase (right).

This simple experiment demonstrates that the nematic phase is liquidlike, is composed of dynamically fluctuating clusters or domains, has local anisotropic short-range ordering of the molecules, and has a transition temperature to and from the liquid that is reproducible on both heating and cooling (unlike that of a solid which can be subject to supercooling). These are properties of both solids and liquids, but as 5CB is a single component system, the phase must be homogeneous.

As with the demonstration using a bulk sample of 5CB, the microscopy experiment, coupled with a projector, can be used to show images of the defect texture on a much larger scale where flow and Brownian motion can be clearly visualized.

Optical microscopy, which has always been at the heart of studies on liquid crystals, was the first technique used to demonstrate definitively that liquid crystals constituted a new state of matter. In the 1880s it was known that cholesterols could be obtained from animals, and that they were chemically related to carotene, and possibly chlorophyll. There was also interest in determining if all cholesterols were the same or if they were made up of a number of different, but closely related compounds. Reinitzer of the German University in Prague was at the time investigating the physico-chemical properties of various derivatives of carrot cholesterol, and to his surprise he found that cholesteryl benzoate did not melt like other compounds; instead it appeared to have two melting points. Reinitzer sought the help of Lehmann at the Polytechnical School in Aachen in his studies because Lehmann possessed a polarizing light microscope that was not commercially available, and which he jealously guarded. Moreover Lehmann's microscope was also fitted with a heating stage enabling high temperature *in-situ* observations to be made. By August of 1889, Lehmann had his first article ready for publication in *Zeitschrift für Physikalische Chemie*, titled "On Flowing Crystals". Lehmann went on to study a variety of other materials with double melting points, including some that exhibited three. Thus he found one phase, which he called *fliessende Kristalle* (flowing crystals) or *schleimig flüssige Kristalle* (slimy liquid crystals), and a second that he called *kristalline Flüssigkeit* (crystalline fluid) or *tropfbar flüssige Kristalle* (liquid crystals that form drops). Thus, there began the concept of polymorphism in this new state of matter.

Flow within a nematic liquid crystal can also be demonstrated using the apparatus shown in Figure 2.4. Instead of heating the glass tube in the region of the air to liquid crystal interface, the nematic phase can be heated much further down the tube in the region of the bulk phase, as shown in Figure 2.6. When the bulk of 5CB is heated to a temperature above the clearing point, an interesting question to ask is: will the bulk become liquid bounded by the nematic phase above and below? The answer to the experiment, as shown in the figure, is that the liquid phase appears at the air interface leaving the region heated as nematic. Thermal conduction has occurred in the nematic phase with transfer of energy to the interface with the air, leaving a liquid to nematic boundary at 35°C above the point where the 5CB was heated, with the original area that was heated still remaining below 35°C. One of the reasons for the transfer of energy is that the hotter isotropic phase is less dense than the cooler nematic phase.

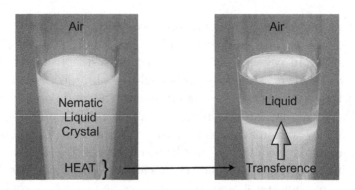

Figure 2.6 Thermal transference occurring in the nematic phase of 5CB.

2.2   POLYMORPHISM AND HEATS OF TRANSITION

For many materials there is the possibility of phase transitions occurring within the solid state of matter. For example, iron has α-, γ-, and δ-polymorphs, with transition temperatures on heating of 910°C and 1400°C, respectively. Transitions in the solid state can also be subject to supercooling and glassification, and the structures of the phases can have the same lattices, for instance α- and δ- are both body-centered cubic whereas γ- is face-centered cubic. Conversely for all materials there are no phase transitions in the liquid state, and thus there is only one type of liquid. Not surprisingly, therefore, there is the possibility of numerous phases being formed in the liquid crystal state. There can also be combinations of polymorphisms in which there are multiple solid forms but only one liquid crystal phase, and vice versa.

Transitions in the solid state, and from the solid to the liquid or liquid crystal, involve differences in the lattice energies, which are usually large, whereas the converse is the case for transitions between liquid crystal polymorphs and between liquid crystals and the liquid. The heats of transition, in terms of enthalpies and entropies, can be determined using differential scanning calorimetry. A typical instrument for this is shown in Figure 2.7.

A differential scanning calorimeter, which has been standardized with respect to the melting of a known material (usually indium), measures the heat capacity of a material as a function of temperature. The calorimeter determines the relative uptake or output of energy from a specimen with reference to the empty sample holder that does not experience any change, via a reference platinum sensor as shown in the

Figure 2.7 Typical microprocessor controlled differential scanning calorimeter with autosampler.

schematic in Figure 2.8. The calorimeter reports energy flow in milliwatts as a function of temperature. The output shows peaks for the phase transitions, the areas under which are the latent heats of transition. The DSC thermogram of dodecyl glucopyranoside, which exhibits polymorphic behavior in the solid state and one liquid crystal phase, is shown as an example in Figure 2.9.

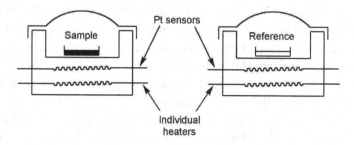

Figure 2.8 A schematic of the sample holders in a typical differential scanning calorimeter.

The first heating cycle for dodecyl glucopyranoside shows two peaks in the thermogram before the melting transition to the liquid crystal phase. These peaks correspond to phase transitions between three different crystal polymorphs, and the associated latent heats are due to changes in the lattice structures. The transition to the liquid crystal

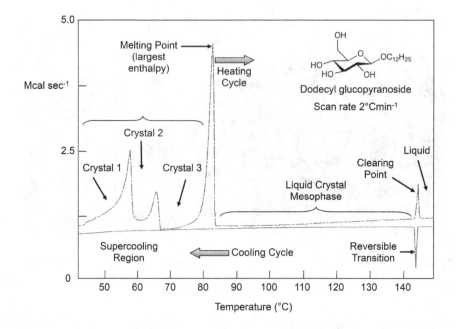

Figure 2.9 A DSC thermogram of dodecyl β-D-glucopyranoside.

phase has the largest enthalpy, because at this point the crystal lattice collapses and the associated lattice energy is released. As the compound is heated further, there is a transition to the liquid from the liquid crystal. The enthalpy for this transition is much smaller than for the melting point because there is no associated lattice energy involved. The cooling cycle, which is always required of DSC experiments, shows that the transition from the liquid back to the liquid crystal is reversible with respect to temperature and enthalpy, whereas the transition back to the solid state at the same temperature as the melting point is not seen in the scan due to supercooling of the sample. This simple experiment demonstrates that the liquid crystal phase does not possess long-range ordering, as in a lattice of its constituents (molecules), and therefore is not a solid or solid particles floating in a liquid phase. Thus the liquid crystal phase of dodecyl glucopyranoside is nearer in structure to that of a liquid than a solid. Structural studies show that the liquid crystal phase of dodecyl glucopyranoside is a disordered lamellar phase, which is sometimes referred to as a smectic A phase. In practice, the transition temperatures for the phase transitions are read by determining the temperature for the maximum gradient in the heat flow from the peaks in the DSC output, as shown in Figure 2.10 for a material that has one crystal form and one liquid crystal phase. The scan rates are usually between 2 and 10°C/min, with faster scan rates producing sharper peaks.

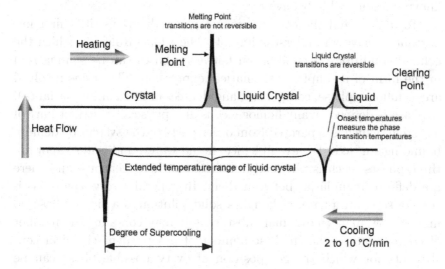

Figure 2.10 A schematic of a classical DSC thermogram for a material that has one solid phase and one liquid crystal phase. The horizontal axis is temperature, and the vertical axis is heat flow.

2.3 MIXING SOLIDS AND LIQUIDS

One of the most important aspects of the applications of materials is their formulation in mixtures. For most people, mixtures are just throwing a few different constituents into a mixing pot and hoping the resultant product has properties that are suitable and reproducible for the targeted application. Formulation, on the other hand, is the selection of constituents for use in a desirable concentration range and for a specific task in an application. For example, the liquid crystals used in display device formulations are particularly important for the creation of mixtures that exhibit nematic phases over wide temperature ranges. The mixing of liquid crystals also gives further insights into the properties and nature of this unusual state of matter. It is important to understand, first, that the formulation of mixtures requires an understanding of phase diagrams, second, that liquids are either continuously miscible or segregate, and third, that solids tend to exhibit eutectic phase diagrams. Therefore the mixing behavior of liquid crystal phases can indicate if a mesophase is closely related to a liquid or to a solid. Moreover, co-miscibility in phase diagrams can be used to identify mesophase types relative to standard materials of known mesomorphic behavior.

To understand the construction of phase diagrams, it is first important to have an understanding of "The Phase Rule", in which the general conditions of equilibrium between phases can be summarized in the form of a simple mathematical expression. The terms involved are as follows: phase, component, and degrees of freedom (or variance). A phase is defined as any homogeneous and physically distinct part of a system, which is separated from other parts of the system by definite bounding surfaces. For example, ice, liquid water, and water vapor are three phases; each is physically distinct and homogeneous, and there are definite boundaries between them. In general, every solid constitutes a separate phase, although a solid solution is a single phase no matter how many individual substances it may contain. The number of components is the smallest number of independent chemical constituents for which the composition of every possible phase can be expressed. Thus, the water system consists of one component.

The number of degrees of freedom of a system is the number of variable factors, such as temperature, pressure, and concentration, that need to be fixed so that the condition of a system at equilibrium can be completely defined. Systems possessing one, two, three, *etc.*, degrees of freedom are said to be invariant, bi-variant, tri-variant, and so on, respectively. Provided the equilibrium between the phases is not influenced by gravity, by electrical or magnetic forces, or by surface action, and only by temperature, pressure, and concentration, then the phase rule states that the number of degrees of freedom (F) of the system is related to the number of components (C) and number of phases (P) present at equilibrium by the equation

$$F = C - P + 2. \tag{2.1}$$

The simplest application of the phase rule is for a one-component system. In this case, C is one, and so the phase rule takes the form

$$F = 3 - P \quad \text{or} \quad F + P = 3. \tag{2.2}$$

If only one phase is present, $P = 1$, so the number of degrees of freedom $F = 2$. It is necessary to state both the temperature and pressure of the phase. If two phases are present, $P = 2$, then the number of degrees of freedom $F = 1$, meaning either the temperature or pressure can be specified, but there is no choice of the other variable in order for the system to be at the transition between both phases. At the triple point, $P = 3$ so $F = 0$ (therefore there are no degrees of freedom) since both the temperature and pressure must be specified to be at the point where all three phases coexist. On the other hand, since the maximum number of degrees of freedom in a two-component system $(C = 2)$ is three, fixing the pressure leaves two adjustable variables, temperature and composition, which is important in the formulation of mixtures of liquid crystals for applications to display devices.

In a two-component system, when a liquid mixture of components A and B is cooled, the solid phase commences to separate out at a definite temperature, termed the freezing point. The actual value of the freezing point depends on the composition of the liquid mixture, and if the results for a series of such mixtures of components varying from pure A to pure B are plotted against the corresponding compositions, two curves, AC and BC, are found as shown in Figure 2.11. The points A and B are the freezing points of the pure components. The curves AC and BC may be regarded as representing the temperature below which liquid phases of various compositions are in equilibrium with the solid phase A or the solid phase B, respectively. At the point C, where the curves AC and BC meet, both solids A and B are in equilibrium with the liquid phase. There are consequently three phases present, and since the system involves two components, there can be only one degree of freedom, according to the phase rule,

$$P = 3, C = 2, \quad \text{and hence} \quad F = 2 - 3 + 2 = 1. \qquad (2.3)$$

Since the pressure is arbitrarily fixed at 1 atm, this reduces the system by one-degree of freedom, meaning there are no degrees of freedom if three phases are present. This means that the existence of the two solid phases A and B in equilibrium with the liquid, at 1 atm pressure, completely defines the point C; therefore, there is only one temperature at which this equilibrium is possible, as the phase diagram shown in Figure 2.11 indicates. The point C, where the freezing point curves AC and BC meet, is the lowest temperature at which any liquid mixture can be in equilibrium with solid A or B; consequently, it is the lowest temperature at which any mixture of solid A and B will melt. For this reason C is called the eutectic point.

The inclusion of a liquid crystal phase in the polymorphism of the

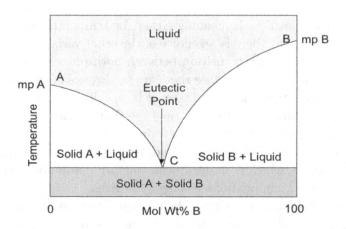

Figure 2.11 A schematic of a classical phase diagram as a function of temperature at constant pressure for a mixture of two components A and B (mol wt%) that possess solid and liquid phases.

two components in the phase diagram complicates matters. If the components exhibit liquid, nematic, and solid phases, then the classical phase diagram is reproduced, but with the addition of a liquid to the nematic phase transition curve. No eutectic point is found for the liquid to nematic phase transitions, and the nematic phase behaves like a liquid with continuous miscibility across the phase diagram, as shown in Figure 2.12. In comparison, the nematic to solid transitions appear to be similar to those found in the classical phase diagram shown in the preceding figure. The eutectic point in this case appears to be the lowest temperature melting point, and because the liquid to nematic phase transition curve is ideally linear with respect to concentration, the maximum temperature range of the nematic phase also occurs at the concentration of the eutectic point. This type of study therefore confirms two points: (1) the nematic phase is very similar in structure to the liquid phase, and (2) the continuous miscibility of the nematic phases across the phase diagram indicates that the nematic phases of components A and B have the same structures and have the same phase identity.

Figure 2.13 shows a phase diagram determined by experiment (microscopy and calorimetry) between the two compounds: 4-pentyl-4′-cyanobiphenyl (5CB) and 4-octyloxy-4′-cyanobiphenyl (8OCB). The temperature range of the nematic phase for 5CB is 11°C because it melts at 24°C and transforms to the liquid at 35°C. The melting point of the eutectic mixture is 5°C and the transition to the liquid is 50°C. It therefore has a nematic range of 45°C, which is very advantageous for applications.

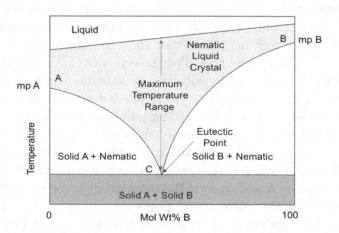

Figure 2.12 A schematic of a classical phase diagram as a function of temperature at constant pressure for a mixture of two components A and B (mol wt%) that possess solid, nematic, and liquid phases.

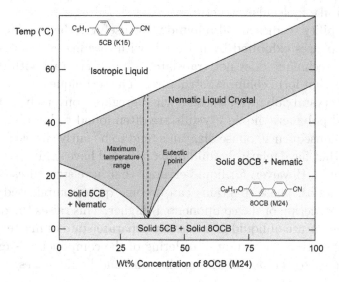

Figure 2.13 The phase diagram as a function of temperature (°C) at constant pressure for a mixture of 4-pentyl-4′-cyanobiphenyl, 5CB, and 4-octyloxy-4′-cyanobiphenyl, 8OCB (wt%).

As noted above, two materials that have liquid crystal phases of the same type tend to show miscibility of those phases across the entire concentration range. This simple observation provides a method by which liquid crystal phases can be classified. Starting with a standard material that exhibits previously classified phases, a test material can be mixed with it, and the phases that show continuous miscibility are deemed to be of the same type. For those phases that are immiscible, classification is said to be inconclusive because de-mixing is always a possibility. Figure 2.14 shows a typical phase diagram that is used for the classification of smectic phases. Smectics can show polymorphism, meaning a material can exhibit numerous smectic phases. The standard material in the miscibility diagram, compound A, exhibits smectic A, smectic B, and crystal E phases, whereas the test material, B, is shown to exhibit smectic A and smectic B phases via their co-miscibility. The other phase that the test material possesses is found to be a nematic phase, but the classification of this phase would have to be determined through another phase diagram with a third material that was standardized and shown to exhibit a nematic phase.

Through these combinations of phase diagrams, co-miscibility versus immiscibility is used in creating a definitive family of smectic and soft crystal phases (A to K) formed by materials possessing molecules with rod-like molecular structures.

Miscibility studies are also found to be useful in the classification of nematic phases exhibited by materials with disc-like molecules. However, the technique does not translate to liquid crystals with disc-like materials that form columnar structures. The technique does not apply to liquid crystal polymers, as these will segregate; conversely, low molar mass and polymeric liquid crystals are often found to be miscible.

As a consequence, phase diagrams tend to be fairly simple for liquid crystals that have relatively fluid properties and have similar structures to the liquid. However, for liquid crystals that have structures closer to those of solids, phase diagrams can become quite complicated, as the degree of ordering of the components is higher. This raises the question about the nature of liquid crystals as a separate state of matter. In the following section, the extent of ordering of the components is explored with respect to the boundaries of the solid and liquid states.

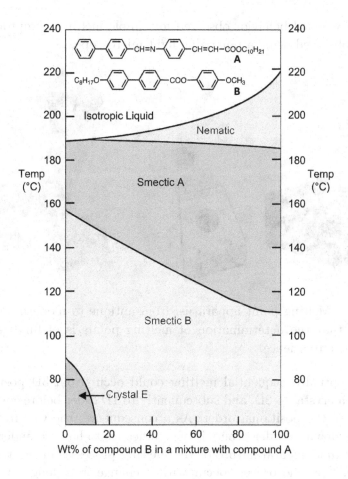

Figure 2.14 A miscibility phase diagram as a function of temperature (°C) at constant pressure for a mixture of a standard material A, and a test material B (wt%).

2.4 PERIODIC ORDER AND LENGTH SCALES IN 1D, 2D, AND 3D SYSTEMS

At the melting point from a solid to a liquid, there is a collapse in the positional order of the constituent atoms or molecules to complete disorder. For molecular materials this event is usually observed through the lens of a melting point apparatus, from antique to modern, as shown in Figure 2.15. However for liquid crystals, since the process occurs in a discrete step-wise fashion over a temperature range rather than the type of extended behavior observed for simple melting, the process is usually observed using a polarizing light microscope.

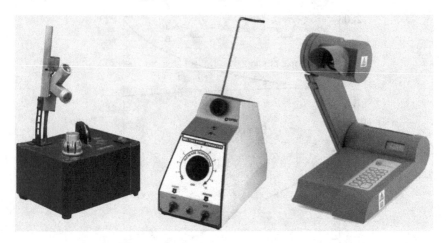

Figure 2.15 Melting point apparatus, from antique to modern, demonstrating that the determination of melting points is of fundamental importance to science.

Conceptually, sequential melting could occur from 3D positional order in a crystal to 2D, and subsequently to 1D order before complete collapse of the positional order. As a consequence, the way in which the breakdown in order occurs is of significance to how we understand the very fundamental nature of the melting processes of matter. For example, does the process occur with a change from long- to short-range order, and what do we mean by long- and short-range order, or might there be intermediate pathways and stop-off states from a solid to a liquid?

To answer this question we have two issues to consider. First, positional order, sometimes called translational order, is related to the distances between adjacent atoms or molecules in a condensed phase. If the distances between the constituents repeat over long distances, this is said to be long-range positional order, as illustrated by the black dotted arrow in Figure 2.16(a). If on the other hand the repeat distance lasts over only a few atoms or molecules, the positional order is said to be short-range as shown by the light-gray arrow in Figure 2.16(c). The second issue is concerned with the orientation of the packing array of the atoms and molecules, sometimes called bond orientational ordering, and the distance over which this persists. It can be long-range as shown in Figure 2.16(a) or short-range as illustrated by the two domains in Figure 2.16(c).

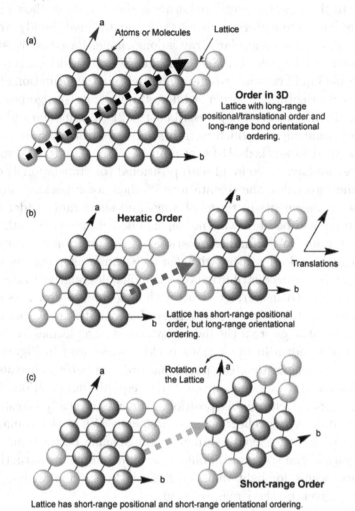

Figure 2.16 Positional and bond-orientational order: (a) long-range positional and bond orientational order; (b) short-range positional order and long-range orientational order; (c) short-range positional and orientational order. The dotted arrows show the direction of progression across a sample; the darker the arrow the more certainty in the depiction of the second domain.

For the crystalline organization shown in Figure 2.16(a), the positional and orientational order are both long-range, whereas for Figure

2.16(c), both are short-range. However, there is the possibility in which one is short-range and the other is long-range, as depicted in Figure 2.16(b). In this case the positional order is short-range as shown by the two domains of the material, whereas the orientational (bond) ordering of the atomistic or molecular array is long-range. Based on Kosterlitz and Thouless (KT) theory, Halperin and Nelson theorized that two-dimensional liquid crystal systems could melt via the formation of long-range bond orientational order, while retaining short-range positional order of the molecules. They described the 2D intermediary phase as "hexatic", analogies of which were first observed in three dimensions and described as stacked-3D hexatic phases. Thus, in general melting processes, we have to contend with positional (or translational) ordering of the molecules, the orientations of their local packing, and the stacking arrangements in terms of long- and short-range order where clearly there can be both with no dependence of one on the other.

This picture raises further questions. If the positional order is not long-range, how does it decay with distance from a reference point, and secondly, how do the distributions of the molecules relative to the packing array vary? To answer, let's start with long-range order. This means that there is no decay in the positional order with distance away from a reference point, *e.g.,* from the light gray atoms or molecules in the left-hand corners shown in the various packing structures in Figure 2.16. This does not mean that the positional order is perfect, because the molecules are always moving around their equilibrium positions due to thermal energy. Determining positional order is typically investigated by performing X-ray reflection measurements. When electromagnetic radiation with a wavelength similar to the spacing between molecules (X-ray part of the spectrum) is incident on a sample, constructive interference occurs for light that is diffracted by specific angles. These angles are given by the Bragg relation,

$$2d \sin \theta = n\lambda, \tag{2.4}$$

where d is the spacing between molecules, θ is the diffraction angle, λ is the wavelength of the X-rays, and n is an integer giving the order of the diffraction ($n = 1, 2, 3, \cdots$). Figure 2.17 shows what the X-ray diffraction pattern might look like for classical and soft crystal phases (top two diagrams). There is no decay in the positional order with distance for either, but for the classical crystalline phase, the thermal motions are related to lattice vibrations. For the soft crystal phase, rotations and oscillations are also possible. In both cases for long-range positional order, the X-ray diffraction pattern exhibits sharp peaks, but due to thermal motion, the amplitude of the peaks decreases as the order of the peak increases. Thus the crystal phase shows more peaks than the soft crystal phase.

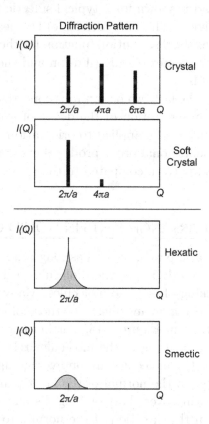

Figure 2.17 The X-ray diffraction patterns for different types of positional order; (top) long-range order for crystals and soft crystal phases; (middle) quasi-long-range order for hexatic phases; (bottom) short-range order for smectic phases.

Conversely, if the positional order decays with distance away from a reference point, the diffraction pattern depends on exactly how the positional order decays. If the positional order decays algebraically, meaning it is proportional to $r^{-\eta}$, where r is the distance from the reference point and η is a temperature-dependent constant, then the diffraction pattern contains broadened, cusp-like peaks instead of very narrow ones. This is known as quasi-long-range order, and this is the order found in hexatic phases (next to lowest diagram of the figure). Finally, if the positional order decays exponentially, meaning it is proportional to $e^{-r/\xi}$, where ξ is a temperature-dependent constant, then the diffraction pattern contains even broader peaks. This represents short-range order and is shown for a typical smectic phase in the lowest diagram of the figure. Thus the breadth of the peaks in the diffraction pattern indicates the distribution function for the components; the broader the peak the greater the distribution and the less positionally ordered the material is.

The discussion of the last two paragraphs concerns positional order, but exactly the same description of the decay of order with distance from a reference point can be applied to orientational order. The only difference is that orientational order produces even more diffuse maxima in the diffraction pattern compared to the peaks due to positional order.

2.5 STRUCTURAL ARRANGEMENTS IN LIQUID CRYSTALS

Apart from the nematic phase, which has long-range orientational order but no positional ordering of the molecules, other structures are possible for the packing of rod-like molecules. Many of these are similar in their local packing arrangements to those of crystals, as shown for the Bravais lattices in Figure 2.18. Furthermore, two groups can be differentiated in the tilting of the molecules relative to the formation of layers. Either the molecules are on average upright or tilted at an angle with respect to the normal of the layer planes, with the layers themselves repeating over short or long distances, *i.e.*, soft layers versus hard layers, in the direction of the normal to the layer planes. Again there are issues of the nature of correlation lengths between the layers. For instance, in some cases the layers are soft and almost do not exist except for a sinusoidal density wave of the centers of mass of the molecules, as shown in Figure 2.19 for two smectic phases, one in which the molecules are on average upright and the other in which they are tilted. Conversely, the layers can be better defined, repeating over a number of layers (>10).

Given the various aspects concerning the structures of liquid crystals, the phases close to the solid state are usually more ordered and are effectively disordered or soft crystals, whereas those nearer to the liquid state are relatively disordered. In this context, if the varieties of structural features discussed above are combined in various ways,

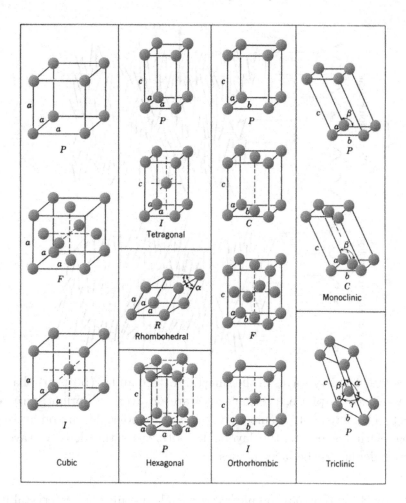

Figure 2.18 Bravais lattices of the solid state.

a unified picture arises with respect to the stepwise breakdown in order on passing from the solid to the liquid. The stepwise breakdown is commensurate with the formation of liquid crystal phases, particularly for liquid crystals formed by rod-like molecules. Figure 2.20 shows a schematic of this unified picture, in which the plane and side elevations of the packing of rod-like molecules in the possible phase are shown. At the top of the figure the phases are more disorganized, and nearer to the liquid state, and to the bottom nearer to the solid. To the left of the figure, molecules that are upright in layers are shown to form disordered layers (smectic A), hexagonal layers with geometric decay

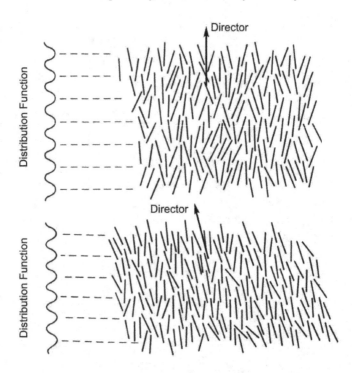

Figure 2.19 Density waves in the smectic A (top) and C (bottom) phases. In a smectic A phase, the molecules are on average upright with respect to the layers. In a smectic C phase, the molecules are on average tilted with respect to the layers. In both cases the density wave is perpendicular to the layer planes.

(hexatic B), and orthorhombic layers with long-range order (crystal E). An almost identical situation applies to phases in which the molecules are tilted in layers, except for the hexatic and crystal phases where the tilt can be towards the apexes or edges of the packing arrangements, and for cases where the tilt can flip backwards and forwards between layers (anticlinic phases).

Additional items given in the figure include the freedom of the molecules to rotate or oscillate around their long molecular axes. Rotation nearer to the liquid state is freer than nearer to the solid state, where oscillations as opposed to rotations are more likely to be found. The symmetry elements for each mesophase are given at the top-right, and lattices are depicted with straight lines. This beautiful picture is consistent and almost complete. There are arguments for AAA, ABAB,

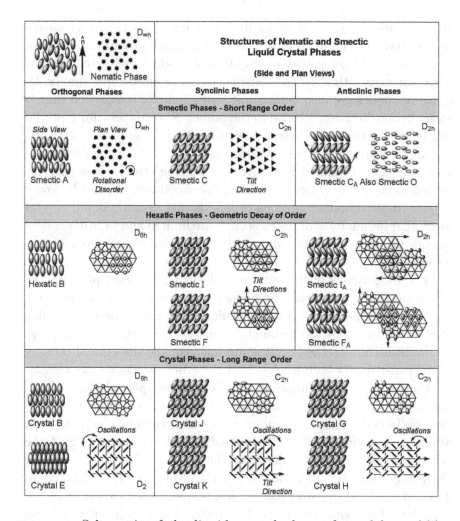

Figure 2.20 Schematic of the liquid crystal phases formed by rod-like molecules.

and ABC stacking arrangements for the molecules, for anticlinic structures in the crystal phases, and for interdigitation and bilayers in hexatic and smectic phases, but these are difficult to observe in terms of the thermodynamics of separate mesophases.

2.6 DYNAMICS AND FLUCTUATIONS IN LIQUID CRYSTALS

It has already been hinted that the constituents of liquids are organized in a random way such that every orientation is represented. The constituents are also in constant motion, rotating about their various axes on relatively short time scales, undergoing translation in various directions, and undergoing various vibrational modes. The nematic phase is similar except that, as the constituents have anisotropic shapes, the dynamic modes are also anisotropic. For example, translation may be easier in one direction relative to the others. Nevertheless, for materials like 5CB, the rotational modes are still very fast: approximately 10^{11} s^{-1} about their molecular short axes and 10^6 s^{-1} about their molecular long axes as determined by neutron scattering studies. For polymers or dendrimers, the dynamic motions, or relaxations, are much slower.

The rapid dynamic motions in liquids and nematic liquid crystals mean that interactions of the molecules are limited and time dependent; orientational distributions and clustering of the constituents become important.

The magnitudes of the various molecular relaxation processes are also dependent on external factors, such as temperature and pressure. At higher temperatures intermolecular interactions become weaker, shorter in time, and less in number, such that for the nematic phase, as the transition to the liquid is approached, the directional ordering of the constituents becomes less and the order parameter decreases.

Consider the two molecular materials, sexiphenyl, and 2′,3′′′′-dimethylsexiphenyl, shown in Figure 2.21. The two compounds have essentially rigid, rod-like structures in which the only internal relaxations are associated with rotations between the aromatic rings. Other relaxations are molecular translation, flipping, rotating, and oscillating. Sexiphenyl melts to a liquid at a temperature of over 500°C, whereas for 2′,3′′′′-dimethylsexiphenyl the melting temperature is lower at 350°C. Therefore, for sexiphenyl, the only structural feature it has that is relevant to the formation of liquid crystals is its aspect ratio (molecular length to breadth). The importance of the aspect ratio is reflected in the comparison of the clearing point of sexiphenyl compared to 2′,3′′′′-dimethylsexiphenyl, where broadening of the molecular architecture due to the incorporation of methyl groups reduces the lateral intermolecular interactions. Thus, the relative lower clearing point of 2′,3′′′′-dimethylsexiphenyl may be related to the lower aspect ratio.

Alternatively for these two materials, the lateral electrostatic and steric interactions could be affecting the clearing temperatures. For dimethylsexiphenyl the lateral methyl groups could cause steric repulsion, which would lower the level of the electrostatic $\pi - \pi$-interactions. Conversely, the methyl groups could induce a twisted arrangement of

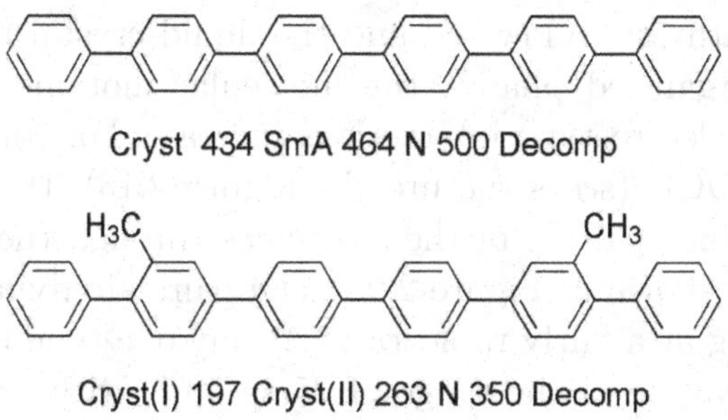

Cryst 434 SmA 464 N 500 Decomp

H₃C   CH₃

Cryst(I) 197 Cryst(II) 263 N 350 Decomp

Figure 2.21 The transition temperatures and melting points (°C) for (top) sexiphenyl, and (bottom) 2′,3′′′′-dimethylsexiphenyl.

the phenyl groups, thereby allowing a better steric fit of the molecules packing together in a lock and key arrangement, as shown in Figure 2.22. However, such steric arrangements might be the case in the solid state, but it cannot be true at temperatures over 300°C in the nematic phase, where the molecules are in thermal motion, fluctuating and gyrating wildly, so the opportunities for electrostatic and steric interactions are limited.

This is the essence of liquid in the name "liquid crystal", and to think of the molecules being organized into well-ordered arrays where electrostatic interactions are dominant is off the mark. Rather it is more likely that the methyl groups disrupt the closeness of intermolecular approach thereby increasing the free volume, meaning that the formation of the liquid state is easier. Thus for materials like 5CB, the volume change at the transition to the liquid is about 2%, whereas for the solid state it is nearer to 10%.

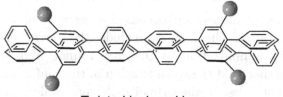

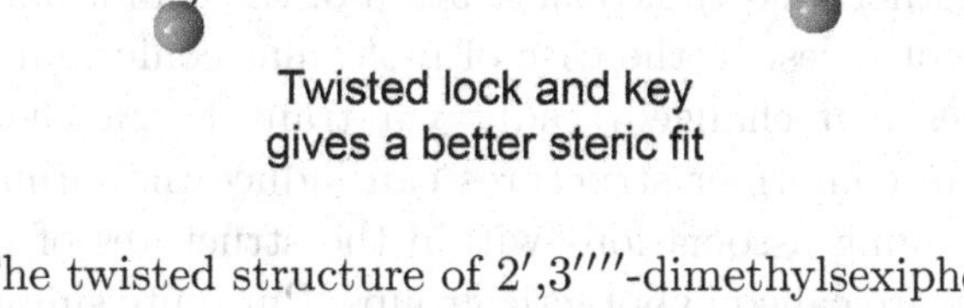

Twisted lock and key
gives a better steric fit

Figure 2.22 The twisted structure of 2′,3′′′′-dimethylsexiphenyl producing a better steric fit of the molecules when packed in a lock and key arrangement.

As temperatures are lowered and the liquid crystal transitions occur to more organized phases, the molecular motions slow and the intermolecular electrostatic interactions increase. For compounds such as 5CB and 8OCB (see structures in Figure 2.13), the $\pi - \pi$ interactions can induce pairing of the molecules through the formation of quadrupoles as shown in Figure 2.23. The pairs are dynamic, forming and dissociating at a fairly rapid rate. At any one time the concentration of pairs may be in the range of 25 to 50%, with the value being temperature dependent. This indicates that electrostatic, intermolecular interactions can affect mesophase formation and physical properties. At lower temperatures, these types of interactions increase and persist over longer periods of time.

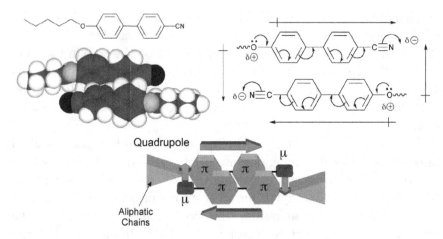

Figure 2.23 Formation of quadrupolar interactions through intermolecular polarization for cyanobiphenyls.

Other forms of interaction become increasingly important as the solid state is approached; these include the steric shapes of the molecules, the minimization of the free volume upon packing the molecules together, and the consideration of the conformational structures of the molecules. In the case of molecular conformations, molecular structures can change rapidly via trans to gauche inversions. The changes in conformer structures can induce the formation of dynamically changing associations within the structures of mesophases. These clusters are called cybotactic groups. Put quite simply, the rotational/conformational shapes of the molecules decide how the molecules can pack together, and thereby determine the relative locations of the polar and polarizable molecular groups and how they interact. These interactions will be discussed in more detail in later chapters.

EXERCISES

2.1 For aligned calamitic and discotic nematic phases, sketch the individual powder X-ray diffraction patterns, indicating the specific d-spacings and relate them to molecular structure.

2.2 Three compounds exhibit the following liquid crystal polymorphism:
    Compound **1** - Nematic, smectic A and smectic X phases
    Compound **2** - Nematic, smectic A and smectic Y phases
    Compound **3** - Nematic, smectic B and smectic Z phases
Smectic phase Z appears to be miscible with the phases X and Y, whereas X and Y are immiscible. Sketch the binary phase diagrams for each pair of compounds. Explain the relationship between the smectic X, Y, and Z phases, and give possibilities for the classifications of the phases.

2.3 Two aromatic liquid crystals have structures with terminal cyano groups. Compound **1** exhibits nematic and smectic A phases, whereas compound **2** is only nematogenic. A mixture of the two compounds in a ratio of 70:30 wt% in favor of compound **1** exhibits a phase sequence on cooling from the isotropic liquid of nematic - smectic A - nematic. In terms of the intermolecular interactions explain this phase behavior, and sketch a potential binary diagram between the two compounds given that the 50:50 wt% mixture is purely nematogenic.

2.4 Nematogen A has a melting point of 30°C and an isotropization point of 50°C, whereas nematogen B has a melting point of 40°C and an isotropization point of 60°C. The eutectic temperature is 20°C. Draw and label a phase diagram for binary mixtures between A and B, and use the Phase Rule to identify the most suitable mixture composition for applications in nematic displays.

2.5 In the table below, there are comparisons for the packing factors (PFs) of two materials, 4-undecyloxy-4′-cyanobiphenyl (11OCB) and 4-decyloxy-4′-cyanobiphenyl (12OCB).

$C_{11}H_{23}O$—⟨ ⟩—⟨ ⟩—CN   11OCB

$C_{12}H_{25}O$—⟨ ⟩—⟨ ⟩—CN   12OCB

nOCB	Solid	Change in PFs	Sm A	Change in PFs	Liquid
11OCB	0.717	0.071	0.646	0.015	0.631
12OCB	0.681	0.046	0.635	0.005	0.630

Compare and contrast the difference in the packing factors of the phases of the individual compounds and between the phases of the compounds.

CHAPTER 3

Nature's Anisotropic Fluids – It's All about Direction

3.1 ANISOTROPY

Solids either can be isotropic or anisotropic, depending on the molecules occupying lattice sites and the crystal lattice itself. For example, solids with cubic symmetry are isotropic, whereas solids with all the other possible symmetries are anisotropic. This anisotropy manifests itself in the elastic, electric, magnetic, and optical properties of a material, in that measurements of an elastic modulus, dielectric constant, magnetic susceptibility, or index of refraction give different results depending on the direction along which they are measured.

Because liquid crystals are fluids, they also show anisotropy in their flow behavior, like logs flowing down a river. This can easily be understood by imagining measuring the viscosity of a liquid crystal by placing it between two flat plates and measuring the force necessary to move one plate past the other at a certain velocity. In Figure 3.1, the plates lie in the x-y plane and are separated by a distance d. The bottom plate is fixed and the force acting on the top plate is in the x-direction, F_x. The velocity of the top plate is also in the x-direction, v_x.

Figure 3.1 shows three situations in which the director of the liquid crystal points along different directions. It is not difficult to imagine that the force necessary to move the top plate at a specified velocity is

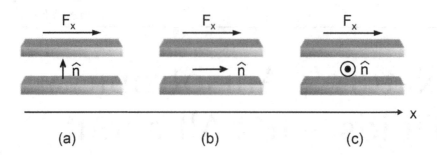

Figure 3.1 Measuring the three viscosities of a nematic liquid crystal: (a) η_1, (b) η_2, and (c) η_3.

different for each situation. If two assumptions are made, namely that the flow is laminar and that the velocity of the fluid next to each plate moves with the same velocity as the plate, then the general definition of viscosity η as the shearing stress divided by the velocity gradient becomes

$$\eta = \frac{\frac{F_x}{A}}{\frac{dv_x}{dz}} = \frac{F_x d}{A v_x}, \qquad (3.1)$$

where A is the cross-sectional area of the plates. The resulting three viscosities, sometimes called the Miesowicz coefficients, are called η_1 (director perpendicular to the flow and parallel to the velocity gradient), η_2 (director parallel to the flow), and η_3 (director perpendicular to the flow and perpendicular to the velocity gradient). A typical temperature dependence of the viscosities is given in Figure 3.2, where it is clear that η_1 is much greater than the other two.

All three viscosities decrease with increasing temperature, which is quite typical for fluids. This general temperature dependence can be eliminated by calculating the reduced viscosity, *i.e.*, the viscosity divided by the average of the three viscosities. These are graphed in Figure 3.3, where it is clear that the difference between the reduced viscosity η_1^r and the other two reduced viscosities follows the temperature dependence of the order parameter as shown in Figure 1.2. Also notice that the viscosities in a liquid crystal tend to be one to two orders of magnitude larger than that of room temperature water, which is about 0.001 Pa s.

In terms of elastic or electromagnetic properties, if two of the three directions in a material are equivalent, the material is said to be uniaxial. The nematic and smectic A phases of liquid crystals are uniaxial,

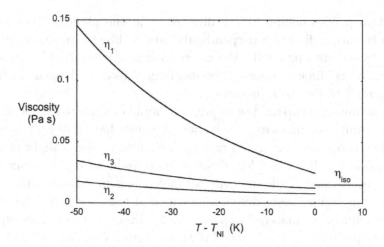

Figure 3.2 Typical variation of the three Miesowicz viscosities with temperature. T_{NI} is the temperature of the nematic-isotropic phase transition.

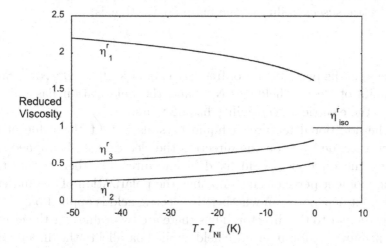

Figure 3.3 Typical variation of the three reduced viscosities with temperature. T_{NI} is the temperature of the nematic-isotropic phase transition.

since all directions perpendicular to the director are equivalent and different from the direction of preferred orientational order. Solids with hexagonal, tetragonal, and trigonal symmetry are also uniaxial. If all three directions in a material are inequivalent, then the material is biaxial. The liquid crystalline smectic C phase is biaxial because one direction perpendicular to the director is in the plane of the layers while the other direction perpendicular to the director makes an angle equal to the tilt angle with the layers. Solids of orthorhombic, mono-clinic, and triclinic symmetry are also biaxial (see the Bravais lattices in Figure 2.18 for the structures).

It is convenient to use the response of liquid crystals to electric fields as an example of anisotropic behavior. Imagine that an electric field $\vec{E}$ is applied to a nematic liquid crystal. This field causes a slight charge separation on the molecules, creating weak induced dipole moments. Since the average dipole moment per molecule is in the same direction for each molecule, these dipole moments add together and cause each unit volume of the material to possess a net dipole moment. This dipole moment per unit volume is called the polarization $\vec{P}$ of the material and points in the same direction as the average molecular dipole moment (from the negative separated charge to the positive separated charge). For small enough electric fields, $\vec{P}$ is proportional to $\vec{E}$. The constant of proportionality (in units of the permittivity of free space) is called the electric susceptibility of the material χ_e, that is,

$$\vec{P} = \epsilon_0 \chi_e \vec{E}, \tag{3.2}$$

where ϵ_0 is the permittivity of free space, 8.85×10^{-12} C^2/Nm^2. Since the units of electric field are N/C and the polarization has units of C/m^2, the electric susceptibility has no units.

The additional feature of liquid crystals is that the value of the polarization depends on the direction the electric field is applied. As-suming the director is held fixed in one direction, then the average dipole moment per molecule and thus the polarization of the material is different depending on whether the electric field is applied parallel or perpendicular to the director. Thus there are two values for the electric susceptibility, $\chi_\parallel$ for an electric field applied parallel to the director and $\chi_\perp$ for an electric field applied perpendicular to the director (the sub-script e has been dropped). If the director lies along the z-axis, then an electric field with components E_x, E_y, and E_z produces a polarization given by

$$P_x = \epsilon_0 \chi_\perp E_x \qquad P_y = \epsilon_0 \chi_\perp E_y \qquad P_z = \epsilon_0 \chi_\parallel E_z. \tag{3.3}$$

This situation is illustrated for the case where $P_x = 0$ and $E_x = 0$ in Figure 3.4, where it is clear that $\vec{P}$ and $\vec{E}$ are no longer parallel as they would be for an isotropic material.

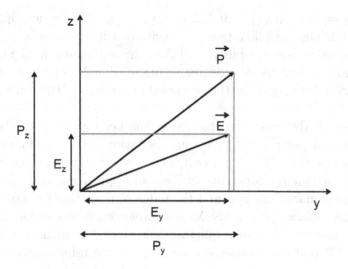

Figure 3.4 The polarization and electric field in a liquid crystal.

These three equations can be written as one equation by defining the electric susceptibility as a tensor. A tensor is an entity that operates on a vector to give a second vector. If the components of vectors are written as a column vector, then a tensor can be represented as a square matrix with the rules of matrix multiplication defining how the tensor operates on the vector. In this way, the last three equations can be written as

$$\begin{pmatrix} P_x \\ P_y \\ P_z \end{pmatrix} = \epsilon_0 \begin{pmatrix} \chi_\perp & 0 & 0 \\ 0 & \chi_\perp & 0 \\ 0 & 0 & \chi_\| \end{pmatrix} \begin{pmatrix} E_x \\ E_y \\ E_z \end{pmatrix} \quad \text{or simply} \quad \vec{P} = \epsilon_0 \overset{\leftrightarrow}{\chi_e} \vec{E}, \quad (3.4)$$

where $\overset{\leftrightarrow}{\chi_e}$ is the electric susceptibility tensor.

The matrix representing the electric susceptibility tensor is diagonal because one of the axes, the z-axis, points along the director. If one of the axes does not point along the director, then $\overset{\leftrightarrow}{\chi_e}$ is no longer diagonal. However, in order for it to have physical meaning, it must be symmetric ($\chi_{ij} = \chi_{ji}$).

In order to draw out the anisotropy in the electric susceptibility tensor, χ is usually written as an average part plus an anisotropic part

$$\overset{\leftrightarrow}{\chi} = \bar{\chi} \begin{pmatrix} 1 & 0 & 0 \\ 0 & 1 & 0 \\ 0 & 0 & 1 \end{pmatrix} + \frac{1}{3}\Delta\chi \begin{pmatrix} -1 & 0 & 0 \\ 0 & -1 & 0 \\ 0 & 0 & 2 \end{pmatrix}, \quad (3.5)$$

where $\bar{\chi} = (2\chi_\perp + \chi_\parallel)/3$ and $\Delta\chi = \chi_\parallel - \chi_\perp$. Thus a description of the electric susceptibility tensor is contained in two numbers: $\bar{\chi}$, the average electric susceptibility, and $\Delta\chi$, the anisotropy in the electric susceptibility. Because the degree of orientational order in a nematic liquid crystal decreases as the temperature increases, the same is true for $\Delta\chi$.

Although the preceding discussion has been in terms of the electric field and polarization, the same behavior is true for many other tensor properties. The best examples are the magnetic susceptibility (the proportionality constant between an applied magnetic field and the induced magnetization) and the index of refraction for light polarized along different directions. More complex examples include various nuclear magnetic resonance splittings and shifts for magnetic fields applied in different directions and Raman scattering using polarized light.

3.2 TENSOR ALGEBRA

Tensor quantities are so important in the description of both the properties and orientational order of liquid crystals that it is important to understand them fully. In fact, a proper description of the orientational order in liquid crystals must discuss the relationship between molecular tensor properties and tensor properties of the bulk material. This involves understanding how tensors transform when the coordinate system is rotated, which is the main topic of this section. The algebra involving tensors is also extremely useful when discussing the polarization optics of liquid crystals.

There is no question that the anisotropy of liquid crystals makes a mathematical description of their properties more complex. Thus the rest of this chapter introduces the mathematics necessary for this description and then uses that mathematics in order to achieve a proper understanding. The mathematics is complicated at times, but there is no other way to explore the behavior of liquid crystals. Keep in mind that often approximations are warranted, resulting in a much more simple description. For example, we will see that two order parameters are necessary to describe fully the orientational order in a nematic liquid crystal. But in many cases, considering only one of the order parameters is sufficient. Another example concerns the fact that there are three elastic constants necessary to describe the behavior of a nematic liquid crystal. But in many cases they can be approximated as equal, simplifying the analysis without introducing much error.

Figure 3.5 shows how the components of the vector $\vec{A}$ transform when the coordinate axes are rotated (counter-clockwise) by an angle θ about the z-axis. To avoid confusion, $\vec{A}$ refers to the vector assuming its components are relative to the original axes (x, y, z), and $\vec{A}'$ refers to the same vector assuming its components are relative to the new axes (x', y', z'). The equations for the new components of $\vec{A}$ $(A_{x'}, A_{y'},$ and $A_{z'})$ in terms of the old components of $\vec{A}$ $(A_x, A_y$ and $A_z)$ are

$$\begin{aligned} A_{x'} &= A_x \cos\theta + A_y \sin\theta \\ A_{y'} &= -A_x \sin\theta + A_y \cos\theta \\ A_{z'} &= A_z. \end{aligned} \tag{3.6}$$

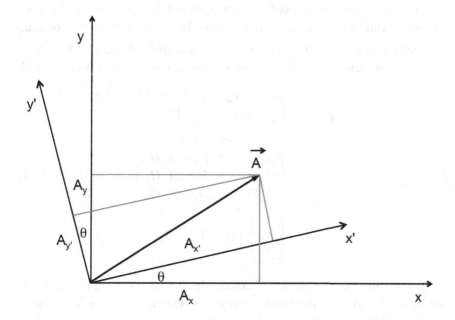

Figure 3.5 Components of a vector in two coordinates systems.

These equations can be summarized using a rotation matrix $\overset{\leftrightarrow}{R}_z(\theta)$ operating on the original components of $\vec{A}$ written as a column vector.

$$\vec{A}' = \overset{\leftrightarrow}{R}_z(\theta)\vec{A} \quad \text{or} \quad \begin{pmatrix} A_{x'} \\ A_{y'} \\ A_{z'} \end{pmatrix} = \begin{pmatrix} \cos\theta & \sin\theta & 0 \\ -\sin\theta & \cos\theta & 0 \\ 0 & 0 & 1 \end{pmatrix} \begin{pmatrix} A_x \\ A_y \\ A_z \end{pmatrix} \tag{3.7}$$

It is not difficult to come up with the matrices representing rotations about the x- and y-axes.

If a vector is written as a row vector rather than a column vector, then the rules of matrix multiplication demand that the rotation matrix operates from the right in order to produce another row vector. The proper rotation matrix to use in this case is the transpose of the one used on column vectors. The transpose of a matrix is one in which the rows and columns have been switched.

$$\vec{A}' = \vec{A} \overset{\leftrightarrow}{R^t}_z(\theta) \tag{3.8}$$

$$\begin{pmatrix} A_{x'} & A_{y'} & A_{z'} \end{pmatrix} = \begin{pmatrix} A_x & A_y & A_z \end{pmatrix} \begin{pmatrix} \cos\theta & -\sin\theta & 0 \\ \sin\theta & \cos\theta & 0 \\ 0 & 0 & 1 \end{pmatrix}$$

The next step is to realize that symmetric tensors can be constructed from two vectors. In matrix algebra, this is done by forming the outer product of the two vectors. Thus any symmetric tensor $\overset{\leftrightarrow}{T}$ can be constructed from the proper choice of the two vectors $\vec{A}$ and $\vec{B}$.

$$
\begin{aligned}
\overset{\leftrightarrow}{T} &= \begin{pmatrix} T_{xx} & T_{xy} & T_{xz} \\ T_{yx} & T_{yy} & T_{yz} \\ T_{zx} & T_{zy} & T_{zz} \end{pmatrix} \\
&= \begin{pmatrix} A_x B_x & A_x B_y & A_x B_z \\ A_y B_x & A_y B_y & A_y B_z \\ A_z B_x & A_z B_y & A_z B_z \end{pmatrix} \\
&= \begin{pmatrix} A_x \\ A_y \\ A_z \end{pmatrix} \begin{pmatrix} B_x & B_y & B_z \end{pmatrix}
\end{aligned}
\tag{3.9}
$$

The outer product, which is defined in the previous equation, must not be confused with the inner product of two vectors, which produces a scalar rather than a tensor.

$$\begin{pmatrix} A_x & A_y & A_z \end{pmatrix} \begin{pmatrix} B_x \\ B_y \\ B_z \end{pmatrix} = A_x B_x + A_y B_y + A_z B_z \tag{3.10}$$

In order to see how a tensor transforms under a rotation of the coordinate axes, all that need be done is to transform properly the two vectors used to construct the tensor.

$$\overset{\leftrightarrow}{T'} = \begin{pmatrix} A_{x'} \\ A_{y'} \\ A_{z'} \end{pmatrix} \begin{pmatrix} B_{x'} & B_{y'} & B_{z'} \end{pmatrix}$$

$$= \overset{\leftrightarrow}{R_z}(\theta) \begin{pmatrix} A_x \\ A_y \\ A_z \end{pmatrix} \begin{pmatrix} B_x & B_y & B_z \end{pmatrix} \overset{\leftrightarrow}{R^t}_z(\theta) \qquad (3.11)$$

$$= \overset{\leftrightarrow}{R_z}(\theta) \overset{\leftrightarrow}{T} \overset{\leftrightarrow}{R^t}_z(\theta)$$

Whereas only one rotation matrix must be applied to vectors, two rotation matrices must be applied to tensors. Calculating the new components of a tensor when the coordinate axes are rotated can be tedious, but in principle it is no more difficult than determining the new components of a vector.

3.3   ORDER PARAMETERS AND THEIR MEASUREMENT

The anisotropy of a macroscopic phase originates with the molecular anisotropy on a microscopic scale. For example, it is the different induced dipole moments on the molecule that cause the liquid crystal to display anisotropy in the electric susceptibility. Assuming each molecule contributes to the bulk properties independently (an assumption more true for the magnetic susceptibility than the electric susceptibility), then it is only the orientational order of the molecules that determines the relationship between microscopic and macroscopic properties.

Imagine a general molecular tensor property $\overset{\leftrightarrow}{T}$ expressed in a coordinate system fixed to the molecule that causes $\overset{\leftrightarrow}{T}$ to be diagonal.

$$\overset{\leftrightarrow}{T} = \begin{pmatrix} T_{xx} & 0 & 0 \\ 0 & T_{yy} & 0 \\ 0 & 0 & T_{zz} \end{pmatrix} \qquad (3.12)$$

Let the z-axis be the long axis of the molecule. In general, the director, which is denoted by the unit vector $\hat{n}$, is not aligned with the long axis of the molecule, so let us denote the direction of $\hat{n}$ in the molecular coordinate system by the polar angle θ and the azimuthal angle ϕ. This is shown in Figure 3.6(a), where the director is shown relative to the molecular coordinate system (x, y, z).

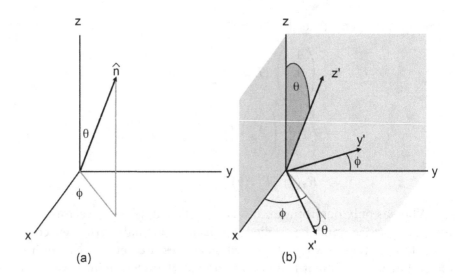

Figure 3.6 Microscopic and macroscopic coordinate systems. (a): the director $\hat{n}$ relative to the microscopic coordinate system (x, y, z) attached to the molecule. (b) The macroscopic coordinate system (x', y', z') attached to the laboratory relative to the microscopic coordinate system (x, y, z) attached to the molecule. $\hat{n}$ is parallel to the z'-axis.

In order to see how this molecule contributes to $\overset{\leftrightarrow}{T'}$, the tensor property in the laboratory coordinate system (x', y', z'), the components of the molecular tensor property in a coordinate system in which the z'-axis points along the director must be calculated. To obtain such a coordinate system, the molecular coordinate system must first be rotated by ϕ around the z-axis and then this coordinate system must be rotated by θ about the new y-axis (all counter-clockwise). This is shown in Figure 3.6(b), where the x'-, y'-, and z'-axes denote the macroscopic or laboratory coordinate system with the z'-axis along the director.

The first rotation, $\overset{\leftrightarrow}{R_z}(\phi)$, results in a tensor $\overset{\leftrightarrow}{T^i}$, in a new coordinate system given by the x^i-, y^i-, and z-axes. The symbol i indicates that this is an intermediate coordinate system and that another rotation must be performed.

$$\overset{\leftrightarrow}{T^i} = \overset{\leftrightarrow}{R_z}(\phi)\overset{\leftrightarrow}{T}\overset{\leftrightarrow}{R^t}_z(\phi) \quad \text{where} \quad \overset{\leftrightarrow}{R_z}(\phi) = \begin{pmatrix} \cos\phi & \sin\phi & 0 \\ -\sin\phi & \cos\phi & 0 \\ 0 & 0 & 1 \end{pmatrix} \quad (3.13)$$

The result of this operation is

$$\overleftrightarrow{T^i} = \begin{pmatrix} T_{xx}\cos^2\phi + T_{yy}\sin^2\phi & (T_{yy} - T_{xx})\sin\phi\cos\phi & 0 \\ (T_{yy} - T_{xx})\sin\phi\cos\phi & T_{xx}\sin^2\phi + T_{yy}\cos^2\phi & 0 \\ 0 & 0 & T_{zz} \end{pmatrix}. \quad (3.14)$$

The second rotation, $\overleftrightarrow{R}_{y^i}(\theta)$, is about the y^i-axis.

$$\overleftrightarrow{T'} = \overleftrightarrow{R}_{y^i}(\theta)\overleftrightarrow{T^i}\overleftrightarrow{R}^t{}_{y^i}(\theta) \quad \text{where} \quad \overleftrightarrow{R}_{y^i}(\theta) = \begin{pmatrix} \cos\theta & 0 & -\sin\theta \\ 0 & 1 & 0 \\ \sin\theta & 0 & \cos\theta \end{pmatrix}. \quad (3.15)$$

Instead of writing out all nine components of $\overleftrightarrow{T'}$, let us concentrate on the diagonal components only.

$$\begin{aligned} T_{x'x'} &= T_{xx}\cos^2\theta\cos^2\phi + T_{yy}\cos^2\theta\sin^2\phi + T_{zz}\sin^2\theta \\ T_{y'y'} &= T_{xx}\sin^2\phi + T_{yy}\cos^2\phi \qquad\qquad (3.16) \\ T_{z'z'} &= T_{xx}\sin^2\theta\cos^2\phi + T_{yy}\sin^2\theta\sin^2\phi + T_{zz}\cos^2\theta \end{aligned}$$

One property of these tensors is quite evident now. The trace of a tensor is simply the sum of the diagonal components. For the microscopic tensor, the trace of $\overleftrightarrow{T}$ is just $T_{xx} + T_{yy} + T_{zz}$. Adding up the diagonal components of $\overleftrightarrow{T'}$ also gives $T_{xx} + T_{yy} + T_{zz}$. Thus the trace is invariant under rotations of the coordinate system.

The anisotropy of the macroscopic phase is defined as its value along the director $T_{z'z'}$ minus the average of its values along two perpendicular directions orthogonal to the director, $T_{x'x'}$ and $T_{y'y'}$.

$$\Delta T' = T_{z'z'} - \frac{1}{2}(T_{x'x'} + T_{y'y'}) = \frac{3}{2}T_{z'z'} - \frac{1}{2}\text{trace}(T'). \quad (3.17)$$

The anisotropy can be expressed in terms of the microscopic tensor using Eq. (3.16).

$$\begin{aligned} \Delta T' &= \left(\frac{3}{2}\sin^2\theta\cos^2\phi - \frac{1}{2}\right)T_{xx} + \left(\frac{3}{2}\sin^2\theta\sin^2\phi - \frac{1}{2}\right)T_{yy} \\ &\quad + \left(\frac{3}{2}\cos^2\theta - \frac{1}{2}\right)T_{zz}. \quad (3.18) \end{aligned}$$

This expression gains even more simplicity if one realizes that

$\sin\theta\cos\phi$, $\sin\theta\sin\phi$, and $\cos\theta$ are just the cosines of the angles between $\hat{n}$ and the molecular x-, y-, and z-axes, respectively.

In a liquid crystal, the molecules are always in motion, so at any instant the director is given by a different value of θ and ϕ for each molecule. The total macroscopic anisotropy is just an average of $\Delta T'$ over all the molecules. Computation of this average is simplified since the components of $\overset{\leftrightarrow}{T}$ are the same for all molecules. Thus

$$\langle\Delta T'\rangle = \left\langle\frac{3}{2}\sin^2\theta\cos^2\phi - \frac{1}{2}\right\rangle T_{xx} + \left\langle\frac{3}{2}\sin^2\theta\sin^2\phi - \frac{1}{2}\right\rangle T_{yy}$$
$$+ \left\langle\frac{3}{2}\cos^2\theta - \frac{1}{2}\right\rangle T_{zz}, \tag{3.19}$$

where $\langle\cdots\rangle$ denotes an average over many molecules. In performing the average in this way, it has been assumed that the intramolecular motions, which play a role in determining the components of $\overset{\leftrightarrow}{T}$, are not correlated with the intermolecular motions, which determine the three averaged terms.

At this point it appears that three averages, or order parameters, are necessary to relate the molecular tensor property to the macroscopic tensor property of the bulk phase. In fact, only two are necessary, since the three terms in brackets add to zero. Knowledge of any two of the three terms allows the third to be determined. Thus either before or after performing the average, the bulk anisotropy could have been written using just two of these terms. The convention is to use the following averages for the two order parameters,

$$S = \left\langle\frac{3}{2}\cos^2\theta - \frac{1}{2}\right\rangle$$
$$D = \left\langle\frac{3}{2}\sin^2\theta\cos^2\phi - \frac{1}{2}\right\rangle - \left\langle\frac{3}{2}\sin^2\theta\sin^2\phi - \frac{1}{2}\right\rangle \tag{3.20}$$
$$= \frac{3}{2}\left\langle\sin^2\theta\cos 2\phi\right\rangle.$$

in which case the anisotropy can be written as

$$\langle\Delta T'\rangle = \left[T_{zz} - \frac{1}{2}(T_{xx} + T_{yy})\right]S + \left[\frac{1}{2}(T_{xx} - T_{yy})\right]D. \tag{3.21}$$

S is the order parameter described in Chapter 1. It describes to what degree the long axes of the molecules are aligned with the director on average. The significance of D can be found by remembering that $\sin\theta\cos\phi$ and $\sin\theta\sin\phi$ are the cosines of the angle between the director and the molecular x- and y-axes. Thus D is a measure of the difference in the tendencies of the molecular x- and y-axes to point along the director. This should not be confused with a biaxial phase, to be discussed in the next section, where the molecular x- and y-axes have preferred orientations. Even in a uniaxial phase there can be a difference in the average projections of the two transverse molecular axes on the director. Whereas S can be as large as 0.9, D is usually less than 0.1. Theoretically, both order parameters are necessary when relating bulk tensor properties to molecular tensor properties.

It is interesting to notice how this general relationship is affected by the symmetry of the molecule. A molecule that is cylindrically symmetric about its long axis is difficult to imagine, but if the molecule rotates or flips its orientation rapidly enough, its contribution to the macroscopic properties is as if it were a cylindrically symmetric molecule. In such a case, $T_{xx} = T_{yy}$ and the macroscopic anisotropy $\langle\Delta T'\rangle$ depends only on S.

The most straightforward way to measure S is to assume that the small value of D and/or the small value of $T_{xx} - T_{yy}$ cause the second term to be negligible. If the anisotropy of the molecular tensor, $T_{zz} - (T_{xx} + T_{yy})/2$, is known, then S can be determined from a measurement of $\langle\Delta T'\rangle$. Sometimes the components of the molecular tensor can be found from experiments in the crystal phase. All the relationships hold for the crystal case also, except that there is no motional averaging, only an average over a unit cell of the crystal. If the crystal structure is known, then measuring the bulk anisotropy allows the molecular anisotropy to be calculated. Of course, if this value is used in the liquid crystal phase, the assumption is being made that the conformation of the molecule is the same in the crystal and liquid crystal phases. Birefringence, diamagnetic susceptibility, and dichroic absorption measurements are typically performed.

Measurement of S and D together requires that two macroscopic anisotropies be measured and that the molecular anisotropies of both are known. This is most easily accomplished in nuclear magnetic resonance experiments, where the dipolar or quadrupolar splittings act as tensor anisotropies. All that need be known is the conformation of the molecule.

What about the off-diagonal elements of $\overset{\leftrightarrow}{T'}$? These involve various averages of angular functions. If the z-axis is along the director, then the off-diagonal elements of the third row and third column of $\overset{\leftrightarrow}{T}{}'$ are zero. If the macroscopic phase is uniaxial, then $T_{x'x'} = T_{y'y'}$ and the other off-diagonal elements are also zero.

3.4   UNIAXIAL VS. BIAXIAL ORDER

Almost all nematic phases are uniaxial, meaning that $\langle T_{x'x'} \rangle = \langle T_{y'y'} \rangle$. As a result, any tensor property must have the form given by Eq. (3.5),

$$\langle \overset{\leftrightarrow}{T}{}' \rangle = \langle \bar{T}' \rangle \begin{pmatrix} 1 & 0 & 0 \\ 0 & 1 & 0 \\ 0 & 0 & 1 \end{pmatrix} + \frac{1}{3} \langle \Delta T' \rangle \begin{pmatrix} -1 & 0 & 0 \\ 0 & -1 & 0 \\ 0 & 0 & 2 \end{pmatrix}, \qquad (3.22)$$

where

$$\begin{aligned}
\langle \bar{T}' \rangle &= \frac{1}{3} \left(2\langle T_{x'x'} \rangle + \langle T_{z'z'} \rangle \right) = \frac{1}{3} \left(2\langle T_{y'y'} \rangle + \langle T_{z'z'} \rangle \right) \\
\langle \Delta T' \rangle &= \langle T_{z'z'} \rangle - \langle T_{x'x'} \rangle = \langle T_{z'z'} \rangle - \langle T_{y'y'} \rangle.
\end{aligned} \qquad (3.23)$$

But what if $\langle T_{x'x'} \rangle \neq \langle T_{y'y'} \rangle$? Such a phase is called biaxial, because, in addition to orientational ordering of the molecular long axes along the director $\hat{n}$, there is some orientational ordering of one of the short molecular axes along a direction perpendicular to the director (often labeled $\hat{m}$). Of course, this means that there also is some orientational ordering of the other short molecular axis along the direction perpendicular to both $\hat{n}$ and $\hat{m}$. Thus in a biaxial nematic phase, there are three mutually orthogonal directions of orientational order.

To describe a tensor property of a biaxial nematic phase, besides $\langle \bar{T} \rangle$ and $\langle \Delta T' \rangle$, a third parameter is needed to specify the difference between $\langle T_{x'x'} \rangle$ and $\langle T_{y'y'} \rangle$. Let us call it $\langle \Delta T'_{\perp} \rangle$ and define it as follows,

$$\langle \Delta T'_{\perp} \rangle = \langle T_{x'x'} \rangle - \langle T_{y'y'} \rangle. \qquad (3.24)$$

Using the general definitions of $\langle \bar{T}' \rangle$ and $\langle \Delta T' \rangle$,

$$\begin{aligned}
\langle \bar{T}' \rangle &= \frac{1}{3} \left(\langle T_{x'x'} \rangle + \langle T_{y'y'} \rangle + \langle T_{z'z'} \rangle \right) \\
\langle \Delta T' \rangle &= \langle T_{z'z'} \rangle - \frac{1}{2} \left(\langle T_{x'x'} \rangle + \langle T_{y'y'} \rangle \right),
\end{aligned} \qquad (3.25)$$

and the definition of $\langle \Delta T'_{\perp} \rangle$, the tensor property can be expressed similarly to Eq. (3.5).

$$\overset{\leftrightarrow}{\langle T' \rangle} = \langle \bar{T}' \rangle \begin{pmatrix} 1 & 0 & 0 \\ 0 & 1 & 0 \\ 0 & 0 & 1 \end{pmatrix} + \frac{1}{3} \langle \Delta T' \rangle \begin{pmatrix} -1 & 0 & 0 \\ 0 & -1 & 0 \\ 0 & 0 & 2 \end{pmatrix}$$

$$+ \frac{1}{2} \langle \Delta T'_\perp \rangle \begin{pmatrix} 1 & 0 & 0 \\ 0 & -1 & 0 \\ 0 & 0 & 0 \end{pmatrix}. \tag{3.26}$$

The second term on the right describes the tensor anisotropy due to uniaxial order, whereas the third term on the right describes an additional tensor anisotropy due to biaxial order.

In order for a biaxial nematic phase to form, the molecules cannot be cylindrically symmetric about their long axes, and even if they are not cylindrically symmetric, the rotation and flipping motions of the molecule cannot be fast enough so the motion of the two short molecular axes average out to be identical. In short, the three axes of the molecules must all be different, because each of these axes possesses some orientational order relative to a macroscopic axis. A simplistic way to visualize this is to consider the molecule to be in the shape of a board with a length, breadth, and width that are all different. There are many liquid crystalline molecules in which the time-averaged length is much greater than the time-averaged breadth and time-averaged width. On the other hand, a possible molecule that forms a liquid crystal phase in which the time-averaged breadth is significantly larger than the time-averaged width, i.e., 3:1, is rare.

The same sort of analysis that led to the order parameters S and D is more complicated for a biaxial nematic. In order to specify all possible orientations of the two directions orthogonal to the director in the coordinate system of the molecule, three angles are needed. Usually a standard set of three angles called Euler angles are used for this. For example, if in addition to the rotations by ϕ and θ introduced previously, a third rotation by ψ about the z-axis (after the θ rotation) is performed, then these three rotations comprise a set of Euler angles (ϕ, θ, ψ). The mathematics gets quite complicated, but the first interesting result is that the anisotropy $\langle \Delta T' \rangle$ is still given by Eq. (3.19). Because the definition of $\langle \Delta T' \rangle$ involves an average of $\langle T_{x'x'} \rangle$ and $\langle T_{y'y'} \rangle$, the rotation by ψ plays no role. Thus $\langle \Delta T' \rangle$ depends only on the uniaxial order parameters S and D in a biaxial nematic phase.

On the other hand, the expressions for $T_{x'x'}$ and $T_{y'y'}$ are more complicated than those in Eq. (3.16), and each involves all three angles and all three components of the molecular tensor property. The difference between $\langle T_{x'x'} \rangle$ and $\langle T_{y'y'} \rangle$ therefore depends on the motional averages of various functions of the three angles. These become the additional order parameters that describe different aspects of the biaxial nature of the tensor property.

Although this discussion has concerned biaxial nematics, it must be pointed out that such a phase is extremely rare. Conventional nematic liquid crystals are usually composed of biaxial molecules ($T_{xx} \neq T_{yy} \neq T_{zz}$) but are uniaxial anisotropic fluids ($T_{x'x'} = T_{y'y'}$). On the other hand, biaxial phases are common in smectic liquid crystals. Although the smectic A phase is uniaxial, the smectic C phase is biaxial, because no two axes describing the orientational or positional order are identical.

3.5 ELECTRIC AND OPTICAL ANISOTROPY

Before going on, it is useful to explore two important examples of anisotropy in liquid crystals. We have already discussed the anisotropy of the electric susceptibility, $\overset{\leftrightarrow}{\chi_e}$ with $\Delta\chi = \chi_\parallel - \chi_\perp$. The electric permittivity is closely related to the electric susceptibility through $\overset{\leftrightarrow}{\epsilon} = \epsilon_0(\overset{\leftrightarrow}{1} + \overset{\leftrightarrow}{\chi_e})$. Thus there is an anisotropy in the electric permittivity given by $\Delta\epsilon = \epsilon_\parallel - \epsilon_\perp = \epsilon_0\Delta\chi$, where $\epsilon_\parallel$ is the electric permittivity parallel to the director and $\epsilon_\perp$ is the electric permittivity perpendicular to the director. As is discussed in Chapter 12, if $\Delta\epsilon$ is positive, the director tends to align parallel to an applied electric field, and if $\Delta\epsilon$ is negative, the director tends to align perpendicular to an applied electric field.

The anisotropy in the electric permittivity affects the propagation of light in liquid crystals. Light polarized parallel to the director travels with a velocity given by $c/n_\parallel$, where c is the velocity of light in a vacuum and $n_\parallel$ is the index of refraction for light of this polarization. Likewise, light polarized perpendicular to the director travels with a velocity given by $c/n_\perp$, where $n_\perp$ is the index of refraction for light of this polarization. The anisotropy in the index of refraction, $\Delta n = n_\parallel - n_\perp$, is also called the birefringence of the liquid crystal. The fact that liquid crystals are birefringent is the most important aspect surrounding the optical behavior of liquid crystals, and of critical importance in developing applications such as LCDs. This is discussed at length in Chapter 13.

Whether $\Delta\epsilon$ or Δn is positive or negative for a specific liquid crystal material depends on the electronic structure of the molecules and exactly how the molecules are organized. The degree and orientation of any separation of charge is important in determining the sign of $\Delta\epsilon$, whereas the orientation of any polarizable groups largely determines the sign of Δn and can be important for the sign of $\Delta\epsilon$. One of the hallmarks of the research into liquid crystals has been understanding how these anisotropies depend on molecular structure and phase organization. As a result, materials with each combination of anisotropies ($\Delta\epsilon > 0$ and $\Delta n > 0$, $\Delta\epsilon > 0$ and $\Delta n < 0$, *etc.*) have been developed.

3.6 MICROSCOPIC STRUCTURE

X-ray diffraction has been the most useful technique in investigating the microscopic structure of liquid crystals. As introduced in Chapter 2, X-ray diffraction from liquid crystals is just like Bragg scattering from crystals, in that periodicities in the structure of the phase give rise to constructive interference and therefore peaks in the scattering of X-rays.

X-ray diffraction from spatial periodicities (crystal planes or smectic layers) can be understood by looking at the phase difference from incident waves scattered off regions of the material separated by one repeat distance. Figure 3.7 shows an incident plane wave of X-rays with wavelength λ striking a structure with period d in a direction perpendicular to the X-ray beam. Imagine that some of the X-rays are scattered at angle 2θ toward a detector very far away compared to the distance d. The path length difference for the scattered waves from two neighboring regions is $d\sin(2\theta)$. If this distance is equal to a multiple of the wavelength, then constructive interference occurs for the waves scattering in this direction from all of the regions. Thus a peak is present when

$$n\lambda = d\sin(2\theta) \approx 2d\sin\theta, \tag{3.27}$$

where $n = 1, 2, 3, \ldots$ and where the last equality is true only if the angle 2θ is very small (which is almost always the case).

By convention, this condition is usually written in terms of the scattering wavevector q,

$$q = \frac{4\pi}{\lambda}\sin\theta, \tag{3.28}$$

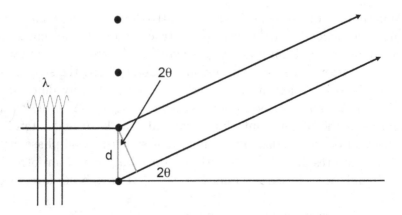

Figure 3.7 X-ray diffraction from crystal planes or smectic layers.

so the condition for a peak in the X-ray diffraction intensity is just

$$q = n\frac{2\pi}{d}, \tag{3.29}$$

where n is called the order of the peak. Notice that the peak occurs on either side of the incident beam direction in the same direction as the periodicity. Since the actual distance from the direction of the incident beam to the peak equals twice the distance from sample to detector times $\sin\theta$ (if θ is small), this distance is proportional to q. X-ray diffraction data showing the locations of the various peaks are therefore calibrated in terms of q rather than θ or the actual distance at the detector. In fact, this two-dimensional space calibrated in terms of the components of q perpendicular to the incident beam is part of a general concept called reciprocal space.

Imagine an X-ray experiment on a smectic A liquid crystal in which the beam propagates parallel to the planes of the layers. If the layer thickness equals the molecular length L and the direction in the sample normal to the layers is the z-axis, then the diffraction pattern has two peaks, one at $q_z = +2\pi/L$ and the other at $-2\pi/L$. It is instructive to calculate how large 2θ is for such an experiment. If the X-ray wavelength is 0.15 nm and the length of the molecules is 3.5 nm, then 2θ equals 2.5°. Hence these measurements involve very small scattering angles, making the approximations used quite appropriate.

What does the diffraction pattern from a smectic C liquid crystal look like? If the sample is completely aligned with the normal to the layers along the z-axis, something called a monodomain sample, then there are two peaks at values for q_z of $\pm 2\pi/d$, where d is equal to the molecular length L times the cosine of the tilt angle. In some cases the director of a smectic C liquid crystal can be aligned, but the layer normals of the many domains distribute themselves about a cone centered on the director with the tilt angle ϕ as the generating angle. If the director is chosen as the z-axis, then the component of the layer repeat distance along the z-axis is $d \cos \phi = L$. The distance over which the layers repeat themselves along the y-axis (the direction perpendicular to the z-axis and beam direction) ranges from a minimum of $L/\tan \phi$ to a maximum of ∞ in both the $+y$ and $-y$ directions. This is illustrated in Figure 3.8(a).

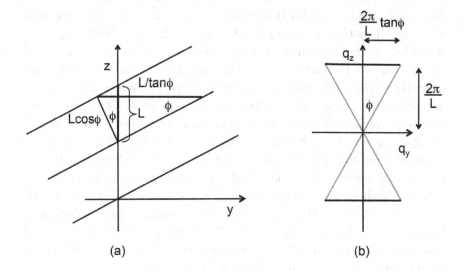

(a)   (b)

Figure 3.8   (a) Smectic C liquid crystal; (b) resulting X-ray diffraction pattern.

The resulting X-ray diffraction pattern shows a maximum with $q_z = 2\pi/L$, but with q_y ranging from $-(2\pi/L)\tan \phi$ to $+(2\pi/L)\tan \phi$. This pattern is shown in Figure 3.8(b). Notice that the angle between the line to the edge of the diffraction maximum and the z-axis is just the tilt angle (as it must be).

$$\sin^{-1}\left(\frac{(2\pi/L)\tan \phi}{2\pi/(L \cos \phi)}\right) = \sin^{-1}(\tan \phi \cos \phi) = \phi \qquad (3.30)$$

In exactly the same way, the positional order parallel to the smectic planes can be investigated. The only necessary change is to direct the incident X-rays along either the layer normal or the director so that repeating arrangements of molecular positions within the plane produce the diffraction maximum. The scattering angles are larger in this case, since the distance between molecules is determined by the molecular width (about 0.4 nm) rather than the length of the molecule. A typical scattering angle for a peak is therefore about 20°. Such experiments determine the arrangement of the molecules from the symmetry of the diffraction pattern and the various distances between the peaks in the diffraction pattern.

This discussion has not included the fact that several orders of diffraction peaks are present due to a periodic array of scattering centers. In fact, higher order peaks are very small in liquid crystals and understanding the reason for this requires a more mathematical treatment of X-ray diffraction. In such a treatment, the X-ray diffraction pattern is related to the periodicities in the electron density of the material. Every sinusoidal variation in the electron density produces scattering intensity in the corresponding region in reciprocal space. The reason a periodic array of scattering centers produces many orders is that many sinusoidal variations in the electron density are necessary to represent an electron density with sharp changes. Thus X-ray diffraction is just an example of the use of Fourier analysis. Figure 3.9 demonstrates this by showing some of the sinusoidal functions necessary to produce a function with sharp changes. Each of these sinusoidal functions produces a maximum in the X-ray diffraction pattern, with the maximum falling at a q value of 2π divided by the spatial period of the sinusoidal function. Thus higher diffraction orders are simply the result of the higher harmonics in the periodicity of the density.

The significance of the lack of higher orders in liquid crystals is now clear. The electron density variation in smectic liquid crystals is nearly sinusoidal, with only small amounts of higher harmonics. This implies that any discussion of smectic layers and positional order within smectic planes should be accompanied with a strong reminder that in fact the variation in the density is quite smooth, varying almost sinusoidally with nothing close to sharp changes.

X-ray diffraction is also useful in investigating the order in nematic liquid crystals. The size of the molecule itself implies that there should be a slight variation in the electron density, with the spatial period being roughly a molecular length along the director and roughly a molecular width perpendicular to the director. The motion of the molecules ensures that these periodicities are very weak, but they should produce

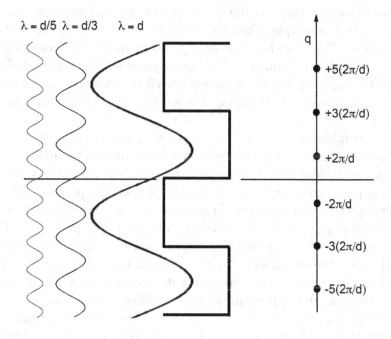

$\lambda = d/5$ $\lambda = d/3$ $\lambda = d$

q

+5(2π/d)

+3(2π/d)

+2π/d

-2π/d

-3(2π/d)

-5(2π/d)

Figure 3.9   Diffraction maxima for the sinusoidal components of a square wave electron density function.

a diffuse maximum in the diffraction pattern at $q_z = \pm 2\pi/L$ and $q_y = \pm 2\pi/a$, where L and a are the length and width of the molecule, respectively. These diffuse peaks are observable, and the width of the peaks themselves is a direct measure of the distance over which the liquid crystalline order is correlated. For example, as the temperature in lowered in the nematic phase and a smectic phase is approached, the diffuse peaks along the z-axis grow narrow and increase in magnitude as a precursor of the order that establishes itself at the nematic-smectic phase transition.

3.7   CONTINUUM THEORY

Our discussion of liquid crystals so far has assumed that the director is perfectly uniform. In fact, this is rarely the case; careful use of surface treatments and/or external fields are usually required to obtain this situation. In general, the director in a liquid crystal varies with position. This variation is extremely modest on a microscopic level, so there is no reason to suspect that any orientational or positional order parameter varies in such a sample. Thus the continuum theory of liquid crystals concerns itself only with elastic energy considerations and assumes all order parameters are constant. This assumption breaks down in regions where there is a defect in the orientational or positional order. Such regions are ignored in this discussion but are covered in the following section.

The equilibrium state of the director in non-chiral systems is a non-varying director. Deviations from this condition are analogous to changes in the length of a spring from its equilibrium length. A logical starting point, therefore, is to express the free energy per unit volume of the liquid crystal in terms of the square of the spatial derivative of the director, just as the elastic energy of a spring is proportional to the square of the deviation of its length from the equilibrium length. What makes the situation in liquid crystals a bit more complicated is that there are three components of the director and three directions in space, opening up the possibility of many different derivatives.

Fortunately, there are restrictions on the free energy that limit the number of possible terms. First, there can be no term for which $\hat{n}$ and $-\hat{n}$ give different energy values. Second, there can be no linear terms except in chiral systems. These terms change if the coordinate system is rotated about the director, if the director is reversed, or if the coordinate system is inverted. Finally, terms that can be written as a divergence and integrated over the volume of the sample can be changed to a surface integral. These contributions to the free energy must be taken into account if certain surface constraints on the director are present.

Finding all of the terms that satisfy these three criteria is a painstaking operation and there is no need to repeat it here. Thankfully, the allowed terms can be grouped together in a nice way, allowing for relatively simple calculations. The free energy per unit volume of a non-chiral nematic liquid crystal can be written as follows,

$$F_V = \frac{1}{2}K_1 \left[\nabla \cdot \hat{n}\right]^2 + \frac{1}{2}K_2 \left[\hat{n} \cdot (\nabla \times \hat{n})\right]^2 + \frac{1}{2}K_3 \left|\hat{n} \times (\nabla \times \hat{n})\right|^2, \quad (3.31)$$

where K_1, K_2, and K_3 are constants analogous to the spring constant. These three constants describe how "stiff" the liquid crystal is to distortions of the director. Obviously, the greater the order parameter, the greater these constants. If the free energy per unit volume has units of joules/meter^3, then since second derivatives of the director have units of meter^{-2}, the units for the constants must be joules/meter or newtons. Typical values for these constants in a thermotropic nematic liquid crystal are about 10^{-11} newtons (10 pN), with K_3 being two or three times larger than K_1 and K_2.

The meaning of this expression for the free energy per unit volume is clear if some simple cases are examined. Let us evaluate the derivatives in each term at a point in a liquid crystal where the director is along the z-axis. This means that the first derivative of n_z with respect to x, y, or z is zero at this point.

$$[\nabla \cdot \hat{n}]^2 = \left[\left(\frac{\partial n_x}{\partial x} \right)_{y,z} + \left(\frac{\partial n_y}{\partial y} \right)_{x,z} \right]^2$$

$$[\hat{n} \cdot (\nabla \times \hat{n})]^2 = \left[\left(\frac{\partial n_y}{\partial x} \right)_{y,z} - \left(\frac{\partial n_x}{\partial y} \right)_{x,z} \right]^2 \tag{3.32}$$

$$|\hat{n} \times (\nabla \times \hat{n})|^2 = \left(\frac{\partial n_x}{\partial z} \right)_{x,y}^2 + \left(\frac{\partial n_y}{\partial z} \right)_{x,y}^2$$

Figure 3.10 shows the director distortion for which only the first term on the right-hand side of each of the above equations is non-zero. These distortions have been named splay, twist, and bend (easily remembered as "STAB"); thus K_1, K_2, and K_3 are the splay, twist, and bend elastic constants. Values for the three elastic constants in two types of nematic liquid crystals are given in Table 3.1.

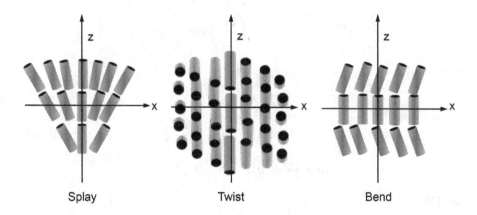

Splay   Twist   Bend

Figure 3.10 The three types of elastic distortion.

Table 3.1 Values for the three elastic constants at room temperature for a thermotropic and a chromonic nematic liquid crystal.

Type	Compound	K_1 (pN)	K_2 (pN)	K_3 (pN)
thermotropic	5CB	6.6	3.0	10
chromonic	Sunset Yellow FCF	8.4	0.8	8.1

The director configuration of a chiral nematic liquid crystal is a good example of how the free energy expression can be used. The components of the director of the chiral nematic in Figure 1.10 assuming (1) the x-axis is vertical with its origin at the bottom of the figure, (2) the horizontal direction is the z-axis, and (3) the helix is right-handed are

$$
\begin{aligned}
n_x &= 0 \\
n_y &= -\sin\left(\frac{2\pi x}{P}\right) \\
n_z &= \cos\left(\frac{2\pi x}{P}\right),
\end{aligned}
$$
(3.33)

where P is the pitch. Notice that evaluation of the divergence and curl of $\hat{n}$ yields

$$
\begin{aligned}
\nabla \cdot \hat{n} &= 0 \\
(\nabla \times \hat{n})_y &= \left(\frac{2\pi}{P}\right)\sin\left(\frac{2\pi x}{P}\right) \\
(\nabla \times \hat{n})_z &= -\left(\frac{2\pi}{P}\right)\cos\left(\frac{2\pi x}{P}\right),
\end{aligned}
$$
(3.34)

and that

$$
\begin{aligned}
[\nabla \cdot \hat{n}]^2 &= 0 \\
[\hat{n} \cdot (\nabla \times \hat{n})]^2 &= \left(\frac{2\pi}{P}\right)^2 \\
|\hat{n} \times (\nabla \times \hat{n})|^2 &= 0
\end{aligned}
$$
(3.35)

so that

$$
F_V = \frac{1}{2}K_2\left(\frac{2\pi}{P}\right)^2 = \frac{1}{2}K_2 q_0^2,
$$
(3.36)

where $q_0 = 2\pi/P$ is the wavevector representing the twist of the chiral nematic. As expected, twist is the only distortion present. In a nematic liquid crystal, this free energy per unit volume is higher than the free energy per unit volume of the undistorted case ($F_V = 0$), implying that a twisted nematic liquid crystal spontaneously relaxes to an untwisted state if at all possible.

If the liquid crystal is composed of chiral molecules, then one linear term is allowed in the free energy. This is the term that changes sign when the axes are inverted, which is allowed in a chiral (non-centrosymmetric) phase. Thus for chiral liquid crystals, the proper free energy expression is

$$F_V = k_2[\hat{n} \cdot (\nabla \times \hat{n})] + \frac{1}{2}K_1[\nabla \cdot \hat{n}]^2 + \frac{1}{2}K_2[\hat{n} \cdot (\nabla \times \hat{n})]^2$$
$$+ \frac{1}{2}K_3|\hat{n} \times (\nabla \times \hat{n})|^2, \tag{3.37}$$

where k_2 is a new elastic constant with dimensions of newton/meter. Using this expression, the free energy per unit volume of the chiral nematic is

$$F_V = -k_2\left(\frac{2\pi}{P}\right) + \frac{1}{2}K_2\left(\frac{2\pi}{P}\right)^2 = -k_2 q_0^2 + \frac{1}{2}K_2 q_0^2. \tag{3.38}$$

Setting the first derivative of F_V with respect to P or q_0 equal to zero, the values of P and q_0 that minimize the free energy are

$$P = 2\pi K_2/k_2 \qquad\qquad q_0 = k_2/K_2. \tag{3.39}$$

Thus the pitch of a chiral nematic liquid crystal is determined by the ratio of these two elastic constants. The value of the free energy per unit volume with this value for the pitch is

$$F_V = -\frac{1}{2}\frac{k_2^2}{K_2}; \tag{3.40}$$

for any other value of the pitch (or for any other distortion for that matter), the free energy per unit volume is higher. This is illustrated in Figure 3.11, where it can be seen that this value for q_0 represents the minimum of the free energy per unit volume. For comparison, the free energy per unit volume as a function of q_0 for a nematic liquid crystal is also shown in Figure 3.11.

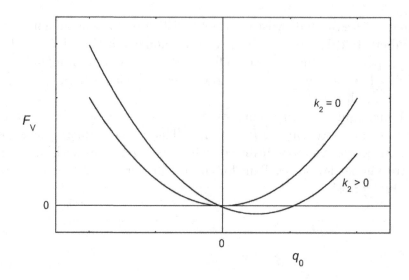

Figure 3.11 Free energy densities for a non-chiral $(k_2 = 0)$ and a chiral $(k_2 > 0)$ liquid crystal.

3.8 DISCLINATIONS AND DEFECTS

In addition to variations in the director, a typical liquid crystal sample usually contains many points where the director is undefined. Theoretically, these defects could be points, lines, or sheets where the direction of orientational order discontinuously changes in passing through one of these defects. Point defects tend to occur in restricted geometries and at surfaces. Sheet defects tend to spread out such that the change in orientational order is continuous over a slab containing the sheet rather than discontinuous at a set of two-dimensional points defining a sheet. Such structures are called walls. Line defects are the most ubiquitous in liquid crystals, so the discussion starts with them. They have been given the name disclinations.

Let us restrict our attention to a plane perpendicular to the disclination. In Figure 3.12, this plane contains the x- and y-axes and the disclination cuts through the plane at the origin. At every point in the plane, the director has a specific orientation, given by the angle θ measured counter-clockwise from the x-axis. In general, θ is a function of x and y, *i.e.*, $\theta = \theta(x, y)$. The components of the director are

$$
\begin{array}{rcl}
n_x & = & \cos[\theta(x, y)] \\
n_y & = & \sin[\theta(x, y)] \\
n_z & = & 0.
\end{array} \tag{3.41}
$$

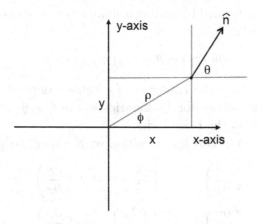

Figure 3.12 Using Cartesian and cylindrical coordinates to specify the configuration of the director.

The divergence and curl of the director are

$$
\begin{array}{rcl}
\nabla \cdot \hat{n} & = & -\sin\theta \left(\dfrac{\partial\theta}{\partial x}\right)_y + \cos\theta \left(\dfrac{\partial\theta}{\partial y}\right)_x \\[2ex]
(\nabla \times \hat{n})_z & = & \cos\theta \left(\dfrac{\partial\theta}{\partial x}\right)_y + \sin\theta \left(\dfrac{\partial\theta}{\partial y}\right)_x.
\end{array} \tag{3.42}
$$

If for simplicity the three elastic constants are set equal to K (called the one constant approximation), then the free energy per unit volume is just

$$
F_V = \frac{1}{2} K \left[\left(\frac{\partial\theta}{\partial x}\right)_y^2 + \left(\frac{\partial\theta}{\partial y}\right)_x^2 \right]. \tag{3.43}
$$

Director configurations around a disclination must minimize the free energy. This implies minimizing the free energy per unit volume integrated over the x-y plane, thus yielding a free energy per unit length F_L for the disclination,

$$
F_L = \int\int \frac{1}{2} K \left[\left(\frac{\partial\theta}{\partial x}\right)_y^2 + \left(\frac{\partial\theta}{\partial y}\right)_x^2 \right] \partial x \partial y. \tag{3.44}
$$

Finding the function $\theta(x, y)$ that minimizes F_L is done using the calculus of variations. This technique is utilized many times in studying the behavior of liquid crystals, so it is important to summarize it here.

Let us denote the function that minimizes F_L as $\theta_m(x, y)$. In order for it to be a minimum, small variations from $\theta_m(x, y)$ cannot change F_L to first order in the parameter producing the change. In other words, if a new $\theta(x, y)$ is formed by adding a small amount of another function, $\eta(x, y)$ to it, *i.e.*, if

$$\theta(x, y) = \theta_m(x, y) + \alpha\eta(x, y), \tag{3.45}$$

where $\alpha << 1$, then the change in F_L cannot depend linearly on α. Since there must always be boundaries where $\theta(x, y)$ is fixed, $\eta(x, y)$ must equal zero on the boundaries.

Differentiation of this last equation with respect to x and y gives

$$\left(\frac{\partial\theta}{\partial x}\right)_y = \left(\frac{\partial\theta_m}{\partial x}\right)_y + \alpha\left(\frac{\partial\eta}{\partial x}\right)_y$$

$$\left(\frac{\partial\theta}{\partial y}\right)_x = \left(\frac{\partial\theta_m}{\partial y}\right)_x + \alpha\left(\frac{\partial\eta}{\partial y}\right)_x. \tag{3.46}$$

Using these when differentiating F_L with respect to α and setting the result equal to zero yields the following expression in the limit of α going to zero.

$$F_L = \int\int\frac{1}{2}K\left[2\left(\frac{\partial\theta_m}{\partial x}\right)_y\left(\frac{\partial\eta}{\partial x}\right)_y + 2\left(\frac{\partial\theta_m}{\partial y}\right)_x\left(\frac{\partial\eta}{\partial y}\right)_x\right]\partial x\partial y$$

$$= 0 \tag{3.47}$$

The left-hand side can now be integrated by parts

$$\left[\left(\frac{\partial\theta_m}{\partial x}\right)_y + \left(\frac{\partial\theta_m}{\partial y}\right)_x\right]\eta - \int\int\left[\left(\frac{\partial^2\theta_m}{\partial x^2}\right)_y + \left(\frac{\partial^2\theta_m}{\partial y^2}\right)_x\right]\eta\partial x\partial y$$

$$= 0. \tag{3.48}$$

Since $\eta(x, y) = 0$ at the boundary, the first term is zero when evaluated at the limits. Since $\eta(x, y)$ is arbitrary, the only way for the second term to be zero is if the expression in brackets is zero. Thus the function $\theta_m(x, y)$ that minimizes F_L must satisfy the equation

$$\left[\left(\frac{\partial^2\theta_m}{\partial x^2}\right)_y + \left(\frac{\partial^2\theta_m}{\partial y^2}\right)_x\right] = 0, \tag{3.49}$$

which is just Laplace's equation in two dimensions. It is illustrative to write this in polar coordinates (ρ, ϕ) as defined in Figure 3.12

$$\left(\frac{\partial^2 \theta_m}{\partial \rho^2}\right)_\phi + \frac{1}{\rho}\left(\frac{\partial \theta_m}{\partial \rho}\right)_\phi + \frac{1}{\rho^2}\left(\frac{\partial^2 \theta_m}{\partial \phi^2}\right)_\rho = 0, \qquad (3.50)$$

where

$$x = \rho\cos\phi \qquad y = \rho\sin\phi$$
$$\rho = \sqrt{x^2 + y^2} \qquad \phi = \tan^{-1}(y/x). \qquad (3.51)$$

One solution to this equation, called an axial disclination, depends linearly on ϕ and does not depend on ρ; that is, $\theta_m(\phi) = s\phi + \theta_0$, where s and θ_0 are constants. One restriction on the function $\theta_m(\phi)$ is that it must be single-valued. Since the director in a liquid crystal can point in either of two directions, this means that $\theta_m(\phi)$ must change by some multiple of π when ϕ is increased by 2π. Obviously, the constant s must equal a multiple of $\pm 1/2$, giving the final form of the equation as

$$\theta_m(\phi) = s\phi + \theta_0, \quad \text{where} \quad s = \pm\frac{1}{2}, \pm 1, \pm\frac{3}{2}, \cdots. \qquad (3.52)$$

The free energy per unit volume of an axial disclination can be found by writing F_V in cylindrical coordinates. Using

$$\left(\frac{\partial \rho}{\partial x}\right)_y = \cos\phi \qquad \left(\frac{\partial \rho}{\partial y}\right)_x = \sin\phi$$

$$\left(\frac{\partial \phi}{\partial x}\right)_y = -\frac{1}{\rho}\sin\phi \qquad \left(\frac{\partial \phi}{\partial y}\right)_x = \frac{1}{\rho}\cos\phi, \qquad (3.53)$$

the derivatives of θ_m with respect to ρ and ϕ can be found.

$$\left(\frac{\partial \theta_m}{\partial x}\right)_y = \left(\frac{\partial \theta_m}{\partial \rho}\right)_\phi \left(\frac{\partial \rho}{\partial x}\right)_y + \left(\frac{\partial \theta_m}{\partial \phi}\right)_\rho \left(\frac{\partial \phi}{\partial x}\right)_y$$

$$= \cos\phi \left(\frac{\partial \theta_m}{\partial \rho}\right)_\phi - \frac{1}{\rho}\sin\phi \left(\frac{\partial \theta_m}{\partial \phi}\right)_\rho$$

$$\left(\frac{\partial \theta_m}{\partial y}\right)_x = \left(\frac{\partial \theta_m}{\partial \rho}\right)_\phi \left(\frac{\partial \rho}{\partial y}\right)_x + \left(\frac{\partial \theta_m}{\partial \phi}\right)_\rho \left(\frac{\partial \phi}{\partial y}\right)_x \qquad (3.54)$$

$$= \sin\phi \left(\frac{\partial \theta_m}{\partial \rho}\right)_\phi + \frac{1}{\rho}\cos\phi \left(\frac{\partial \theta_m}{\partial \phi}\right)_\rho$$

Using these expressions, the free energy becomes

$$F_V = \frac{1}{2}K \left[\left(\frac{\partial \theta_m}{\partial \rho}\right)^2_\phi + \frac{1}{\rho^2} \left(\frac{\partial \theta_m}{\partial \phi}\right)^2_\rho \right], \tag{3.55}$$

which for an axial disclination is

$$F_V = \frac{1}{2}K \frac{s^2}{\rho^2}, \tag{3.56}$$

showing that the free energy per unit volume is proportional to s^2 and diverges at the disclination itself ($\rho \to 0$). Integration over an annular region centered on the disclination with smaller radius ρ_1 and larger radius ρ_2 yields an expression for the free energy per unit length.

$$F_V = \pi K s^2 \ln \frac{\rho_2}{\rho_1} \tag{3.57}$$

The fact that the free energy per unit volume diverges at the center of the disclination has interesting ramifications. Clearly the structure of the disclination must be different from what is given in Eq. (3.51) at the center of the disclination. One hypothesis is that a core of isotropic liquid forms at the center with a radius that minimizes the free energy of the disclination. Another hypothesis is that the order parameter, which we have assumed is constant everywhere, decreases to zero as the center of the disclination is approached. This means that both K and ρ in Eq. (3.55) go to zero in a way that minimizes the free energy of the disclination.

It is not difficult to visualize the director configuration for some of these axial disclinations. If s, which is called the strength of the disclination, is positive, then the director rotates counter-clockwise in traversing a counter-clockwise path around the disclination. If s is negative, the director rotates clockwise in traversing this same counter-clockwise path. If $s = \pm 1/2$, then the director rotates by π for one full loop around the disclination. If $s = \pm 1$, then the director rotates by 2π. The director configurations for several of these disclinations are illustrated in Figure 3.13.

What do these axial disclinations look like in a liquid crystal? The best way to observe them is with the sample between crossed polarizers. As will be discussed at length later, liquid crystals are birefringent, so only light that is polarized parallel or perpendicular to the director is transmitted unchanged. This means that regions of the liquid

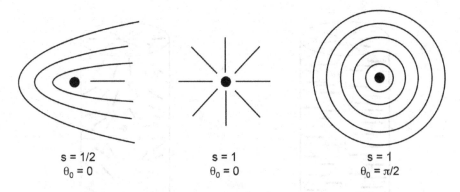

$$s = 1/2 \qquad\qquad s = 1 \qquad\qquad s = 1$$
$$\theta_0 = 0 \qquad\qquad \theta_0 = 0 \qquad\qquad \theta_0 = \pi/2$$

Figure 3.13 Three director configurations around an axial disclination. θ_0 is the angle between the director just to the right of the disclination and the horizontal axis.

crystal where the director is parallel or perpendicular to one of the crossed polarizer axes are dark (as is true for the isotropic phase). The rest of the liquid crystal is bright. This results in dark bands emanating from a disclination where the director points in either of two perpendicular directions. Thus there will be two dark bands emanating from opposite sides of the disclination for $s = \pm 1/2$, and four bands emanating from $s = \pm 1$ disclinations. In addition, if the two polarizers are rotated counter-clockwise together, then the dark bands rotate counter-clockwise for disclinations of positive strength and clockwise for disclinations of negative strength. Such features are commonly seen in nematic liquid crystals.

Just because the boundary conditions on a liquid crystal favor the creation of a certain disclination, there is no guarantee that another director configuration may not form. A good example is a nematic liquid crystal confined to a cylindrical capillary tube, the inside surfaces of which have been treated so the director is normal to the glass. The expectation is that a $s = +1$ axial disclination forms along the axis of the capillary tube, as shown in Figure 3.14(a). In most cases, the director in the center region of the tube develops a component along the tube axis and "escapes to the third dimension". Since both directions along the tube axis are equivalent, point defects can develop along the axis of the tube as shown in Figure 3.14(b).

There are other disclinations besides axial disclinations that form in nematic liquid crystals. In axial disclinations, the rotation axis of the director in traversing a loop around the disclination is parallel to the

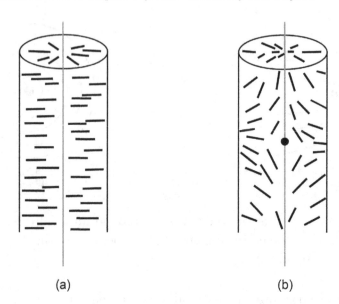

(a)	(b)

Figure 3.14 Vertical cross-sectional view of an axial disclination (a) and an "escaped" axial disclination (b) in a capillary tube. The dot in (b) is a point defect where the director is undefined, *i.e.*, there is no orientational order as in the isotropic phase.

disclination. In a twist disclination, the rotation axis is perpendicular to the disclination. Figure 3.15 shows $+1/2$ and $+1$ strength twist disclinations in which the rotation axis for the director as it circles the disclination is around the vertical axis and the disclination lies in the middle plane pointing in and out of the page.

Due to the fact that the director is not uniform, an entirely new class of disclinations forms in chiral nematic liquid crystals. Likewise, the spatial periodicity of both chiral nematic and smectic liquid crystals allows for defects in the periodic structure in addition to defects in the director configuration. These additional defects are quite different and resemble dislocations in solids.

When liquid crystals are confined, especially if the size of the confining region is hundreds of micrometers or less, often the liquid crystal cannot adopt a director configuration with no distortion. Instead, the director varies throughout the region in whatever configuration minimizes the sum of the volume elastic free energy and the free energies associated with disclinations, point defects, and surface interactions. Figure 3.14 already demonstrated this for a liquid crystal confined to

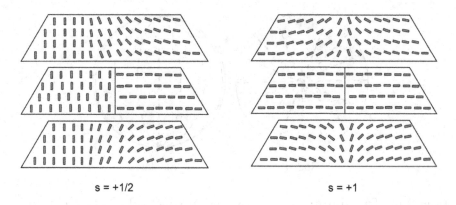

s = +1/2   s = +1

Figure 3.15 Two twist disclinations of different strengths. The disclination is shown in the middle planes coming out of the page. As one circles the disclination in a plane perpendicular to it, the director rotates about the vertical axis, which is perpendicular to the disclination.

a cylindrical region. As another example, if a nematic liquid crystal is confined inside a spherical surface, which could be an interface with a solid or an immiscible liquid, with surface interactions that force the director of the liquid crystal to be perpendicular to the spherical surface, then a radial director configuration sometimes results as is shown in Figure 3.16. There are no disclinations running through the liquid crystal, but there is a point defect, sometimes called a hedgehog defect, at the center. On the other hand, if the surface interactions force the director to be parallel to the spherical surface, a bipolar director configuration sometimes results as shown in Figure 3.16. Again, there are no disclinations, but two point defects form at opposite points on the spherical surface. These surface defects are often called boojums. In both of these spherical cases, there are other director configurations that may be more stable than the radial or bipolar configurations shown. Which configuration is stable in any given situation depends on the values of the splay, twist, and bend elastic constants together with the detailed free energies associated with disclinations, defects, and surface interactions.

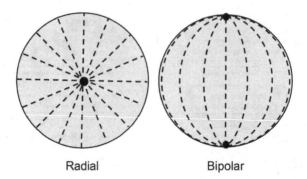

Radial   Bipolar

Figure 3.16 Two director configurations in spherical confinement with point defects. The radial and biolar configurations sometimes result from surface interactions that force the director to be perpendicular and parallel to the surface, respectively.

EXERCISES

3.1 The nematic liquid crystal 5CB has χ_e components $\chi_\| = 20.3$ and $\chi_\perp = 6.4$ at room temperature. (a) What is the value of the electric susceptibility anisotropy $\Delta\chi$? Assuming the director (which is aligned with the z-axis) is not affected by an external electric field, what is the electric polarization P if an electric field E of 5 N/C is applied (b) parallel to the director, and (c) perpendicular to the director?

3.2 Let an external electric field be applied in the y-z plane as shown in Figure 3.4 to 5CB at room temperature. Again, assuming the director is not affected by the electric field, what is the angle the polarization P makes with the y-axis if the electric field E is applied at 45° to the y-axis?

3.3 Let the vector $\vec{A}$ in Figure 3.5 be the director of a nematic liquid crystal with components in the x-y reference frame $n_x = 0.643$ and $n_y = 0.766$. What are the components of $\hat{n}$ in the x'-y' reference frame if $\theta = 30°$?

3.4 Eq. (3.14) contains the components of the tensor given in Eq. (3.12) but in a reference frame that has been rotated about the director by an angle ϕ. Such a rotation should leave the tensor components unchanged if the original tensor is uniaxial. Show that if the tensor given in Eq. (3.12) is uniaxial, its components in the new reference frame, as given by Eq. (3.14), are unchanged.

3.5 Imagine a unit-less molecular tensor quantity with $T_{xx} = 2$, $T_{yy} = 1$, and $T_{zz} = 6$. (a) What is the macroscopic anisotropy of the tensor quantity if $S = 0.55$ and $D = 0.1$? (b) What is the macroscopic anisotropy of the tensor quantity if the contribution from D is ignored?

3.6 Let the components of the director be given as follows.

$$n_x = \frac{x}{\sqrt{x^2 + z^2}}$$
$$n_y = 0$$
$$n_z = \frac{z}{\sqrt{x^2 + z^2}}$$

(a) Find $\nabla \cdot \hat{n}$ and $\nabla \times \hat{n}$. (b) Which of the distortions (splay, twist, and bend in Figure 3.10) are present?

3.7 Which of the distortions shown in Figure 3.10 is represented by the following director configuration?

$$n_x = \frac{-z}{\sqrt{x^2 + z^2}}$$
$$n_y = 0$$
$$n_z = = \frac{x}{\sqrt{x^2 + z^2}}$$

3.8 Typical values for the two twist elastic constants in a chiral nematic liquid crystal are $k_2 = 3.0 \times 10^{-6}$ N/m and $K_2 = 4.5 \times 10^{-12}$ N. (a) What is the pitch of this liquid crystal? (b) What is the free energy per unit volume of this liquid crystal with this value of the pitch? (c) What is the free energy per unit volume of this liquid crystal if it is forced not to twist?

3.9 The radial director configuration (see Figure 3.16) is not always the most stable one (lowest free energy) when the surface interactions force the director to be perpendicular to a spherical surface. Another possibility is what is known as a Saturn ring configuration in which half a +1 disclination runs around the surface in a great circle (like the equator of Earth). (a) Sketch the director for such a configuration. (b) What type of deformation dominates in a radial director configuration? (c) What type of deformation dominates in a Saturn ring configuration? (d) Why might one configuration be more stable than the other?

CHAPTER 4

Theoretical Insights – So Many Interesting Possibilities

4.1 LANDAU-DE GENNES THEORY

The Landau-de Gennes theory is a phenomenological model for the nematic-isotropic phase transition. It is based on Landau's general description of phase transitions and was first developed by de Gennes. It is a theory that attempts to describe the nematic-isotropic phase transition rather than calculate the behavior of a system starting from molecular interactions. The strengths of the Landau-de Gennes theory are its simplicity and its ability to capture the most important elements of the nematic-isotropic transition. In addition, the theory has been both extended to many other transitions and embellished in many ways.

We should first discuss phase transitions in general before concentrating on the nematic-isotropic transition. In a phase transition driven by a change in temperature, there is a less-ordered higher temperature phase and a more-ordered lower temperature phase. One way of describing the change that takes place at the phase transition is to define an order parameter that is zero in the higher temperature phase and non-zero in the lower temperature phase. For example, the orientational order parameter S (defined in Chapter 1) is an appropriate choice for the nematic-isotropic phase transition since it is zero in the isotropic phase and non-zero in the nematic phase. Likewise, the positional order parameter ψ (also defined in Chapter 1) is zero in the nematic phase and non-zero in the smectic A phase, so it can be used to describe the transition between these two phases. If the order parameter changes discontinuously at the phase transition, it is called a discontinuous or first-order phase transition. The nematic-isotropic transition is an example of a discontinuous phase transition. If the order parameter grows from zero continuously as the temperature is decreased through the phase transition, then it is called a continuous or second-order phase transition. The smectic A-nematic transition sometimes is continuous.

In most cases there is a change in symmetry at a phase transition. The symmetry of a phase is described by the operations (translations, rotations, reflections, *etc.*) that can be applied to the phase and leave the structure unchanged. Typically there are one or more such operations which leave the higher temperature phase unchanged but cause a change in the lower temperature phase. Thus a symmetry is usually spontaneously broken at a phase transition. For example, rotations about any axis leave the isotropic phase unchanged. However, only rotations about the director leave the nematic phase unchanged. Another example is the fact that a translation in any direction and by any amount leaves the nematic phase unchanged, whereas only translations perpendicular to the director by any amount and along the director by an amount equal the layer spacing leave the smectic A phase unchanged.

There does not have to be a change of symmetry at a phase transition. The liquid-gas transition is a good example of a phase transition in which both the higher and lower temperature phases have the same symmetry (isotropic). One choice for an order parameter is the difference between the density of the phase relative to the density of the gas phase. This order parameter is obviously zero in the gas phase and non-zero in the liquid phase. Since there is no change in symmetry, the only way a transition can exist is if there is a discontinuous change in the order parameter. In many substances, the discontinuity in the density difference between the liquid and gas phases decreases as the transition temperature increases under increasing pressure. At the point where this difference becomes zero, the transition ceases to exist, since there is no way to tell one phase from the other. This is called a critical point, and is illustrated in Figure 4.1. However, if the phase transition involves a change in symmetry, then it is possible to tell one phase from the other even if the order parameter is not discontinuous at the phase transition, since a change of symmetry can always be detected. In this case the order parameter discontinuity might decrease as temperature and pressure increase, with the transition simply changing from discontinuous to continuous at the point where the discontinuity becomes zero (called a tricritical point). In this case the phase transition continues to exist as shown in Figure 4.1.

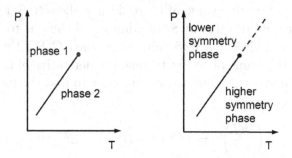

Figure 4.1 Phase transition ending in a critical point (left). Phase transition with a tricritical point (right).

The Landau-de Gennes theory for the nematic-isotropic transition starts by assuming that the order parameter S is small in the nematic phase in the vicinity of the transition and thus the difference between the free energy per unit volume of the two phases can be expanded in powers of S. Of course a term linear in the order parameter is not allowed since this would mean that the free energy per unit volume of the nematic phase would be less than the free energy per unit volume of the isotropic phase at all temperatures. If we are considering a system at constant temperature and constant pressure, then the Gibbs free energy per unit volume is the appropriate one,

$$G(S,T) = G_{iso} + \frac{1}{2}A(T)S^2 + \frac{1}{3}BS^3 + \frac{1}{4}CS^4, \qquad (4.1)$$

where G_{iso} is the free energy per unit volume of the isotropic phase. The parameter $A(T)$ is the most important in determining when $G(S,T)$ is greater or less than G_{iso}, so it is given the most simple temperature dependence possible,

$$A(T) = A_0(T - T^*), \qquad (4.2)$$

where A_0 and T^* are constants. Since the phase transition takes place in the vicinity of where $A(T)$ changes sign, T^* locates this temperature region. Again for simplicity reasons, the parameters B and C are assumed to be constant.

What does this free energy per unit volume look like as a function of S and T? We could just graph $G(S,T)$ for various values of S and T, but it is more instructive to do some analysis first. For example, clearly $C > 0$, since a negative value of C would not allow $G(S,T)$ to have a minimum for a finite value of S. In addition, if there is to be a stable nematic phase with a non-zero value of S, then $G(S,T)$ must have a minimum at this value of S. Let us take the derivative of $G(S,T)$ with respect to S and set it equal to zero in order to find the local maxima and minima,

$$\left(\frac{\partial G}{\partial S}\right)_T = A(T)S + BS^2 + CS^3 = 0 \tag{4.3}$$

or

$$S = 0 \quad \text{or} \quad S = \frac{-B \pm \sqrt{B^2 - 4A(T)C}}{2C}. \tag{4.4}$$

The local extremum at $S = 0$ represents the isotropic phase. For stability this must be a minimum. The other two values of S represent a local maximum and a local minimum for non-zero values of S. If $S = 0$ is a local minimum, then the solution using the negative sign must be either a local maximum or local minimum with the solution using the positive sign being the other. Let us look at the temperature for which the free energies per unit volume of the nematic and isotropic phases are equal. For this value of the temperature T_C, both the $S = 0$ solution and the proper $S > 0$ solution must be local minima with the same value of the free energy per unit volume, G_{iso}. Setting the two free energies per unit volume equal to each other results in the following relationships.

$$S = 0 \quad \text{or} \quad \frac{1}{2}A(T) + \frac{1}{3}BS + \frac{1}{4}CS^2 = 0. \tag{4.5}$$

Combining this requirement together with the requirement that the derivative be zero yields

$$S = 0 \quad \text{or} \quad S_C = -\frac{2B}{3C}. \tag{4.6}$$

Thus the two minima when the free energies per unit volume are equal occur at $S_C = 0$ (isotropic phase) and $S_C = -2B/(3C)$ (nematic phase). Clearly B must be less than zero if S_C is to be greater than zero. Thus there is a discontinuous jump in the order parameter at the transition, indicating that the transition is discontinuous. If we substitute the non-zero value for S_C into the free energy, we find that the two free energies per unit volume are equal when

$$A(T_C) = A_0(T_C - T^*) = \frac{2B^2}{9C} \quad \text{or} \quad T_C = T^* + \frac{2B^2}{9A_0C}. \quad (4.7)$$

Thus the transition takes place at a temperature slightly higher than T^*.

We can check to see if these two values for S at the transition correspond to local minima by evaluating the second derivative. At $T = T_C$,

$$\left(\frac{\partial^2 G}{\partial S^2}\right)_{T=T_C} = A(T_C) + 2BS + 3CS^2 = \frac{2B^2}{9C} \quad (4.8)$$

at both $S_C = 0$ and $S_C = -2B/(3C)$. Since $2B^2/(9C) > 0$, the second derivative is greater than zero and these points represent local minima as expected. The expression for the second derivative also allows us to see the significance of the temperature T^*. At this temperature the second derivative is zero at $S = 0$, indicating that $S = 0$ is no longer a local minimum for temperatures below T^*. Thus T^* represents a temperature slightly below the transition temperature at which the isotropic phase becomes thermodynamically unstable. Of course, the system is in the nematic phase at this temperature since the free energy of the nematic phase is lower, but T^* can be interpreted as the lower limit for supercooling of the isotropic phase.

Is there an upper limit for superheating of the nematic phase? This temperature would correspond to the second derivative being zero at the non-zero value for S. Setting both the first and second derivatives equal to zero yields

$$S = -\frac{2A(T^{**})}{B} \quad \text{and} \quad A(T^{**}) = A_0(T^{**} - T^*) = \frac{B^2}{4C}, \quad (4.9)$$

where we have defined T^{**} to be this stability limit for the nematic phase. Solving for T^{**} gives

$$T^{**} = T^* + \frac{B^2}{4A_0C} = T_C + \frac{B^2}{36A_0C}. \quad (4.10)$$

Thus T^* differs from T_C by 8 times as much as T^{**} differs from T_C.

We are now ready to appreciate graphs of the free energy per unit volume of the nematic phase as a function of S for various values of T. Figure 4.2 shows $G(S,T) - G_{iso}$ for five values of the temperature, T^*, T_C, T^{**}, and values both below T^* and above T^{**}.

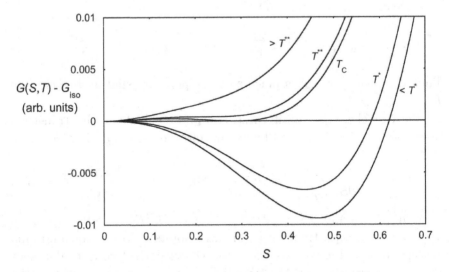

Figure 4.2 Free energy per unit volume curves as a function of the order parameter for five different temperatures.

We can now return to the two non-zero expressions for S and determine which is the local minimum and which is the local maximum. Substituting $A(T_C)$ into the expression for S yields $-B/(3C)$ for the plus sign and $-2B/(3C)$ for the negative sign. Clearly the negative sign corresponds to the local minimum so we can write down the temperature variation for the order parameter in the isotropic phase,

$$S = \frac{-B - \sqrt{B^2 - 4A_0(T - T^*)C}}{2C}$$

$$= \frac{3}{4}S_C\left[1 + \sqrt{1 - \frac{8(T - T^*)}{9(T_C - T^*)}}\right], \qquad (4.11)$$

where the second equation comes from substituting for A_0, B, and C using Eqs. (4.6) and (4.7). Eq. (4.11) is plotted in Figure 4.3, where the discontinuity at the transition is evident, along with the gradual increase in S as the temperature decreases. Notice that the parameters can be chosen so that S more or less follows the data for a typical liquid crystal.[1] This is not surprising considering that there are three independent parameters (T_C, S_C, and T^*).

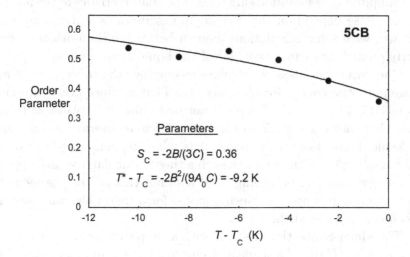

Figure 4.3   Order parameter as a function of temperature for a specific set of parameters in the Landau-de Gennes theory along with experimental data. T_C is the nematic-isotropic transition temperature.

Finally, since the entropy per unit volume Σ is the negative derivative of the Gibb's free energy per unit volume with respect to temperature, the difference in entropy per unit volume at the transition is just the difference in the derivatives evaluated in each phase at T_C. Thus the latent heat of the transition is

$$L = -T_C(\Sigma - \Sigma_{iso}) = T_C \left[\frac{\partial (G - G_{iso})}{\partial T} \right]_{S=S_C} = \frac{2A_0 B^2 T_C}{9C^2}. \quad (4.12)$$

Measurement of the latent heat together with the order parameter allows the parameters A_0, B, and C to be given in units of energy per volume per degree (for A_0) and energy per volume (for B and C).

4.2   MAIER-SAUPE THEORY

A molecular theory is one that starts with the behavior of single molecules and attempts to determine the characteristics of the macroscopic phase. Since a macroscopic sample of liquid crystal contains a

---

[1] A. Sanchez-Castillo, M. A. Osipov, and F. Giesselmann, *Phys. Rev. E* **81**, 021707 (2010).

huge number of molecules, it is extremely difficult to perform a calculation keeping track of all these molecules. As a result, approximations and simplifying assumptions are used. The characteristics of the macroscopic phase depend on the specific approximations and assumptions utilized, with some calculations doing a better job than others in correctly predicting certain behavior of the liquid crystal phase.

One important class of calculations employs the concept of an average potential energy for all molecules. That is, since every molecule is embedded in a "sea" of many other molecules, it is plausible to assume that each one experiences the same forces on average as any other molecule. Instead of trying to calculate what happens to a huge number of molecules simultaneously, such a theory calculates what happens to a single molecule on average and assumes this is what happens to every molecule on average. Such a model for a thermodynamic system is called a mean-field theory.

The Maier-Saupe theory starts with a simple expression for the potential energy $U_i(\theta_i)$ of a single molecule in the "sea" of other molecules,

$$U_i(\theta_i) = -\frac{A}{V^2} S \left(\frac{3}{2} \cos^2 \theta_i - \frac{1}{2} \right), \qquad (4.13)$$

where θ_i is the angle between the long axis of the molecule and the director, V is the volume of the sample, and A is a constant independent of temperature. Such a potential energy is not selected at random, but results from a few assumptions about the properties of the molecules and the interactions between them. First, the molecules are assumed not to possess permanent dipole moments. This means that the dominant force between the molecules is an interaction between induced dipoles. A momentary dipole moment on one molecule induces a momentary dipole moment on a neighboring molecule, resulting in an attractive dispersion force. Such a force varies with distance to the minus sixth power, which is why the mean field potential energy $U_i(\theta_i)$ varies as the inverse square of the volume. Second, it is assumed that the molecules are cylindrically symmetric about their long axes. Thus the potential energy between two molecules can only depend on the angle between their long axes with an angular dependence proportional to the second Legendre polynomial of this angle. Third, it is assumed that the degree of orientational order of the molecules enters into the mean-field potential energy in a linear way. That is why S, the orientational order parameter, appears in the expression for $U_i(\theta_i)$ as it does.

Since the potential energy of each molecule is known, the fact that the system must be in thermodynamic equilibrium demands that the probability of a molecule being oriented at an angle θ_i from the director is given by the Boltzmann factor,

$$P_i(\theta_i) = \frac{1}{Z} e^{-\frac{U_i(\theta_i)}{k_B T}}, \tag{4.14}$$

where k_B is the Boltzmann constant, T is the absolute temperature, and Z is given by

$$Z = \int_0^\pi e^{-\frac{U_i(\theta_i)}{k_B T}} \sin\theta_i d\theta_i \int_0^{2\pi} d\phi_i, \tag{4.15}$$

where ϕ_i is the azimuthal angle. Knowing the probability distribution, the average of a function of θ_i or ϕ_i can be found. One special function is the second Legendre polynomial, the average of which is just S,

$$
\begin{aligned}
S &= \left\langle \frac{3}{2}\cos^2\theta_i - \frac{1}{2} \right\rangle = \int_0^\pi \left(\frac{3}{2}\cos^2\theta_i - \frac{1}{2} \right) P_i(\theta_i)\sin\theta_i d\theta_i \int_0^{2\pi} d\phi_i \\
&= \frac{1}{Z} \int_0^\pi \left(\frac{3}{2}\cos^2\theta_i - \frac{1}{2} \right) e^{-\frac{U_i(\theta_i)}{k_B T}} \sin\theta_i d\theta_i \int_0^{2\pi} d\phi_i. \tag{4.16}
\end{aligned}
$$

Thus we have a self-consistent equation involving S, T, and V. For example, using this equation to see how S depends on T with V being constant gives us a prediction of the temperature dependence of the order parameter.

To calculate this dependence explicitly requires a bit of work. The first step is to perform the ϕ_i integral and then separate the integral over the second Legendre polynomial into two integrals. The result is

$$S = \frac{3}{2} \frac{\int_0^\pi \cos^2\theta_i \sin\theta_i e^{-\frac{U_i(\theta_i)}{k_B T}} d\theta_i}{\int_0^\pi e^{-\frac{U_i(\theta_i)}{k_B T}} \sin\theta_i d\theta_i} - \frac{1}{2}. \tag{4.17}$$

Changing the integration variable to $x = \cos\theta_i$ yields

$$S = \frac{3}{2} \frac{\int_{-1}^{+1} x^2 e^{\frac{3ASx^2}{2k_B TV^2}} e^{-\frac{AS}{k_B TV^2}} dx}{\int_{-1}^{+1} e^{\frac{3ASx^2}{2k_B TV^2}} e^{-\frac{AS}{k_B TV^2}} dx} - \frac{1}{2}. \tag{4.18}$$

Since the integrals are even functions in x, the integral from -1 to $+1$ is just twice the integral from 0 to $+1$. In addition, the second exponential function in each integral is not dependent on the integration variable, so it can be factored out of both integrals. With the substitution

$$m = \frac{3AS}{2k_BTV^2},$$

(4.19)

the equation can be written

$$S = \frac{3}{2} \frac{\int_0^1 x^2 e^{mx^2}}{\int_0^1 e^{mx^2}} - \frac{1}{2}.$$

(4.20)

But the numerator can be integrated by parts,

$$\int_0^1 x^2 e^{mx^2} dx = \frac{e^m}{2m} - \frac{1}{2m} \int_0^1 e^{mx^2} dx,$$

(4.21)

so the expression now is

$$S = \frac{3}{4m} \left[\frac{e^m}{\int_0^1 e^{mx^2} dx} - 1 \right] - \frac{1}{2}.$$

(4.22)

A convenient way to see how S depends on T is to choose a value for m, and use computer software to find S from Eq. (4.22). Since $S/m = (2k_BTV^2)/(3A)$, which is a quantity proportional to T, S/m can be used as the temperature axis. The greatest value of S/m is 0.14856. If we call this T_{max}, then a graph of S versus T/T_{max} can be drawn by making the horizontal axis $(S/m)/T_{max}$. This has been done in Figure 4.4.

As is clear from Figure 4.4, the highest value of T occurs when $S = 0.32$. Below T_{max} there are sometimes two values for S. We must look at the physical state corresponding to these solutions to make some sense of the curve in Figure 4.4. For example, if the free energy of the nematic phase is greater than the free energy of the isotropic phase, then the nematic phase is unstable even if a self-consistent solution with non-zero S exists.

The Helmholtz free energy is given by

$$F = U - T\Sigma,$$

(4.23)

where Σ is the entropy of the phase. The average energy of a molecule is

$$< U_i >= \left\langle -\frac{AS}{V^2} \left(\frac{3}{2} \cos^2 \theta_i - \frac{1}{2} \right) \right\rangle = -\frac{AS}{V^2} S = -\frac{AS^2}{V^2}.$$

(4.24)

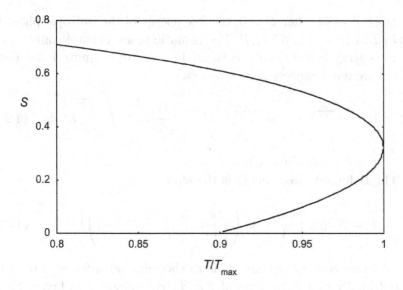

Figure 4.4 The order parameter as a function of temperature for the Maier-Saupe theory. T_{max} represents the highest value of S/m, which is proportional to the temperature.

where $< \cdots >$ denotes a thermal average using the probability $P_i(\theta_i)$ given before. Since this energy stems from an interaction between two molecules, to find the total energy for N molecules, we cannot simply multiply the average energy per molecule by N, since we would be counting the interaction energy of each pair of molecules twice. Therefore, the total energy for N molecules is

$$U = -\frac{1}{2}N\frac{AS^2}{V^2}. \tag{4.25}$$

The entropy of a single molecule is just

$$\Sigma_i = -k_B \ln P_i(\theta_i) = \frac{U_i}{T} + k_B \ln Z, \tag{4.26}$$

so the total entropy for N molecules is

$$\Sigma = N < \Sigma_i >= \frac{N < U_i >}{T} + Nk_B \ln Z = -N\frac{AS^2}{TV^2} + Nk_B \ln Z. \tag{4.27}$$

Thus the Helmholtz free energy for N molecules is

$$F = N\left[\frac{1}{2}\frac{AS^2}{V^2} - k_B T \ln Z\right]. \tag{4.28}$$

Since $Z = 4\pi$ when $S = 0$, the free energy of the isotropic phase is $-Nk_BT \ln 4\pi = -2.53Nk_BT$. The nematic phase is stable only when its free energy is less than this value. In order to compute F for non-zero S, we must express Z in terms m,

$$Z = 2\pi e^{-\frac{AS}{2k_BTV^2}} \int_{-1}^{+1} e^{\frac{3}{2}\frac{AS^2}{k_BTV^2}x^2} dx = \frac{2\pi}{\sqrt{m}}e^{-\frac{m}{3}} \int_{-\sqrt{m}}^{+\sqrt{m}} e^{y^2} dy, \quad (4.29)$$

where $x = \cos\theta$ and $y = \sqrt{m}x$.

The Helmholtz free energy is therefore

$$F = Nk_BT \left[\frac{mS}{3} - \ln\left(\frac{2\pi}{\sqrt{m}}e^{-\frac{m}{3}} \int_{-\sqrt{m}}^{+\sqrt{m}} e^{y^2} dy \right) \right], \quad (4.30)$$

and the best way to evaluate F is to choose a value for m, calculate S, and finally find F in units of Nk_BT. For values of m higher than 2.93 (S values higher than 0.43), F of the liquid crystal phase is lower than $Nk_BT \ln 4\pi$, which is the free energy of the isotropic phase. For values of S below 0.43, the isotropic phase has a lower free energy and is therefore the stable phase. Thus the Maier-Saupe theory predicts that the order parameter should decrease discontinuously from 0.43 to 0 at the transition to the isotropic phase. This behavior is shown in Figure 4.5.

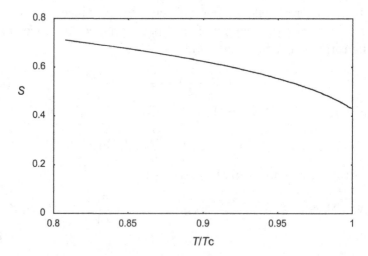

Figure 4.5 Order parameter versus temperature for the Maier-Saupe theory. T_C is the nematic-isotropic transition temperature.

Notice that this prediction is a universal one. Unlike the Landau-de Gennes theory, there are no parameters that can be adjusted to allow for different order parameter behavior. Since different substances have slightly different temperature dependences, the Maier-Saupe theory describes only a small portion of them well.

4.3   ONSAGER THEORY

The starting point for the Maier-Saupe theory is the assumption that there is an attractive interaction between the molecules. A very different starting point is used in Onsager theory, namely that there is only a repulsive interaction between the molecules, with the repulsion being infinite whenever parts of two molecules occupy the same volume in space and zero otherwise. Thus the molecules are considered to be hard cylinders or hard rods that cannot penetrate each other. Notice that with no interaction energy, all configurations of a system of hard rods result in the same system energy (assuming no rods interpenetrate). Thus for such a system, entropy considerations determine the behavior, with temperature playing no role.

For hard rods, both translational and orientational contributions to the entropy have to be taken into account. If the concentration of rods is low, the translational contribution to the entropy can be obtained from the ideal gas result, as long as the rods are assumed to occupy some volume. This contribution increases as the orientational order of the rods increases, since there is more volume available to the rods if they are ordered. On the other hand, the orientational contribution to the entropy decreases as the orientational order of the rods increases, since the presence of orientational order decreases the freedom of the rods to orient in all directions. Thus the translational and orientational contributions compete with each other. A large length to diameter ratio and a large density of rods favors the translational entropy resulting in orientationally ordered rods (the nematic phase). A small length to diameter ratio and a small density of rods favors the orientational entropy resulting in randomly orientated rods (the isotropic phase).

Let us first look at the translational contribution to the entropy in more detail. The Sacker-Tetrode equation for the translational entropy per molecule of an ideal gas can be written

$$\frac{\Sigma_{trans}}{k_B N} = \ln\left(\frac{V}{N\Lambda^3}\right) + \frac{5}{2}, \qquad (4.31)$$

where Σ is the system entropy, T is the absolute temperature, V is the system volume, N is the number of molecules, and Λ is the thermal wavelength. In much the same way as is done for the van der Waals gas, if we assume the volume not available to other molecules due to each molecule is v, then the relationship is slightly different,

$$\frac{\Sigma_{trans}}{k_B N} = \ln\left(\frac{V - Nv}{N\Lambda^3}\right) + \frac{5}{2}. \tag{4.32}$$

The factor Nv is the volume that the molecules cannot occupy, so such a term is often called the excluded volume of the system. If it is assumed that the concentration of rods is dilute, $i.e.$, $Nv/V \ll 1$, then the lowering of the transitional entropy due to the excluded volume is easier to see,

$$\frac{\Sigma_{trans}}{k_B N} = \left[\ln\left(\frac{V}{N\Lambda^3}\right) + \frac{5}{2}\right] - \frac{Nv}{V}. \tag{4.33}$$

Finally, if the rods have a length of L and a diameter of D, then the volume fraction of the rods relative to the system is given by

$$\phi = \frac{N\pi L D^2}{4V}, \tag{4.34}$$

and the translational entropy can be written as

$$\frac{\Sigma_{trans}}{k_B N} = \ln\left(\frac{\pi L^2 D}{4\Lambda^3}\right) + \frac{5}{2} - \ln\left(\frac{L\phi}{D}\right) - \frac{4\phi}{\pi L D^2}v. \tag{4.35}$$

The orientational contribution to the entropy can be examined by starting with a form of the Boltzmann equation and adapting it to rods at polar angle θ and azimuthal angle ϕ relative to the director,

$$\frac{\Sigma_{orient}}{k_B N} = -\sum_i p_i \ln p_i = -\int f(\theta) \ln f(\theta) d\Omega, \tag{4.36}$$

where the sum is the general relationship involving microstates (labeled by i) and the integral is the classical relationship in terms of the orientational distribution function $f(\theta)$. Because the integral of $f(\theta)$ over all solid angles is one, this is usually written in a slightly different form,

$$\frac{\Sigma_{orient}}{k_B N} = \ln 4\pi - \int f(\theta) \ln\left[4\pi f(\theta)\right] d\Omega. \tag{4.37}$$

The next step is to find the excluded volume v for two rods making an angle γ with each other. As can be seen from Figure 4.6, two rods at angle γ exclude a volume given by area $L^2|\sin\gamma|$ times depth $2D$. This volume, $2L^2D|\sin\gamma|$, must then be integrated over all possible angles for two rods and divided by two in order to arrive at the excluded volume per rod,

$$v = L^2D < |\sin\gamma| >= L^2D \int\int f(\theta)f(\theta')|\sin\gamma|d\Omega d\Omega'. \qquad (4.38)$$

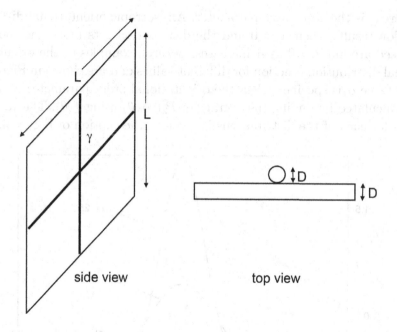

Figure 4.6 Excluded volume for two rods of length L and diameter D at an angle of γ. The area shown in the side view equals $L^2|\sin\gamma|$, and the depth shown in the top view equals $2D$.

Putting both contributions together in terms of the Helmholtz free energy, $F = U - T\Sigma$ yields

$$\frac{F}{Nk_BT} = \frac{-T(\Sigma_{trans} + \Sigma_{orient})}{Nk_BT} = -\ln\left(\frac{\pi LD^2}{4\Lambda^3}\right) - \frac{5}{2} - \ln 4\pi \qquad (4.39)$$

$$+ \ln\left(\frac{L\phi}{D}\right) + \frac{4}{\pi}\frac{L\phi}{D}\langle|\sin\gamma|\rangle + \int f(\theta)\ln\left[4\pi f(\theta)\right]d\Omega.$$

For any value of L and D, it is only the second line that needs to be minimized to find $f(\theta)$, and the solution can only depend on the combination of variables $L\phi/D$.

The easiest way to proceed is to use an expression for $f(\theta)$ that approximates the actual solution, but contains an adjustable parameter that can be chosen to minimize the sum of the last three terms in Eq. (4.39). One such function is

$$f(\theta) = \frac{\alpha}{4\pi} \frac{\cosh(\alpha \cos\theta)}{\sinh\alpha}, \qquad (4.40)$$

where α is the adjustable parameter. An isotropic orientational distribution results when $\alpha = 0$ and the distribution gets more and more peaked around $\theta = 0$ as α increases. Several examples of the orientational distribution function for different values of α are shown in Figure 4.7. Once α is specified, then the orientational order parameter S can be calculated by finding the average of $P_2(\cos\theta)$ using $f(\theta)$. The value of S for each of the distributions is given in the caption of Figure 4.7.

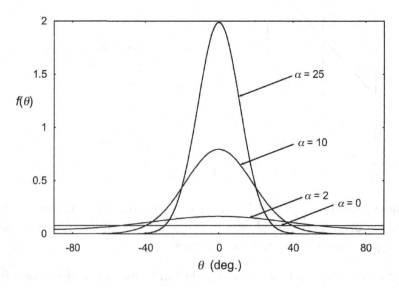

Figure 4.7 Distribution of rod orientations for the Onsager theory with different values of α in Eq. (4.40). For $\alpha = 0$, 2, 10, and 25, $S = 0$, 0.19, 0.73, and 0.89, respectively.

Since assigning a value to α determines both $f(\theta)$ and S, the integral in Eq. (4.38), which is just $< |\sin\gamma| >$, can be calculated. The dependence of $< |\sin\gamma| >$ on S is displayed in Figure 4.8, where it is clear that the angles the rods make with each other decreases as the order parameter increases.

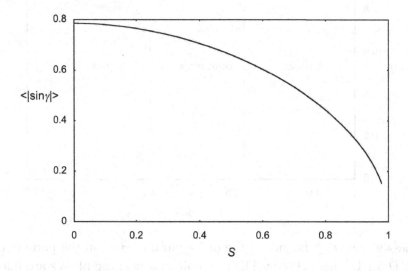

Figure 4.8 Dependence of the average of the absolute value of the sine of the angle between rods according to Onsager theory. As the order parameter increases, the average angle between rods decreases and therefore the average sine of this angle decreases.

In much the same manner, the free energy can be calculated as a function of S using the last line of Eq. (4.39) if a value for $L\phi/D$ is assigned. One finds that for values of $L\phi/D$ less than 3.3, the free energy increases as S increases. In this region, the minimum free energy is at $S = 0$ and the isotropic phase is stable. On the other hand, if $L\phi/D$ is greater than 4.5, the minimum free energy occurs at values of S greater than 0.84. By keeping track of the value of S that minimizes the free energy, the dependence of the order parameter on the parameter $L\phi/D$ can be examined, and this is displayed in Figure 4.9.

Finally, knowing that the boundaries of the coexistence regions occur at $L\phi/D$ equal to 3.3 and 4.5, it is quite easy to make a phase diagram showing what phases are stable in terms of the aspect ratio L/D and the volume fraction ϕ. On a graph of ϕ versus L/D, these boundaries will be hyperbolas, as shown in Figure 4.10. It is clear from the plot that the higher the aspect ratio, the lower the volume fraction at which the nematic phase appears.

The Landau-de Gennes, Maier-Saupe, and Onsager theories for the

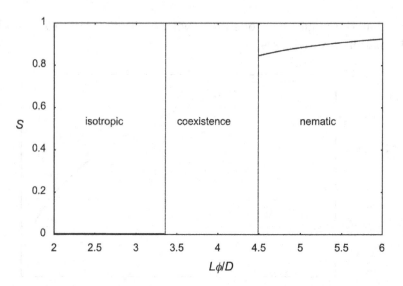

Figure 4.9 The dependence of the order parameter S on the parameter $L\phi/D$ for Onsager theory. The isotropic and nematic phases are both present in the coexistence region, with $S = 0$ for the isotropic regions and $S = 0.84$ for the nematic regions.

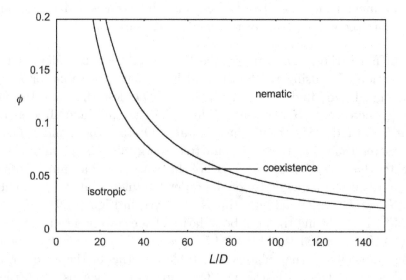

Figure 4.10 Phase diagram according to Onsager theory.

nematic phase represent three general models, namely phenomenological, mean-field, and hard rod, respectively. All three are simple idealizations that generally do a poor job in describing the behavior of real systems. But each of them provides an extremely useful starting point for more sophisticated theories. For example, both attractive interactions of the Maier-Saupe type and repulsive interactions as described by Onsager theory seem to be important in understanding the nematic phase of thermotropic liquid crystals. Theories that contain both types of interactions can therefore describe these systems well.

4.4 EXTENSIONS TO THE SMECTIC PHASE

The Landau-de Gennes theory for the nematic isotropic transition can be extended to the smectic A-nematic transition. The order parameter for this transition is $|\psi|$, the amplitude of the density wave describing the formation of layers in the smectic A phase. Since the difference between a value of $|\psi|$ and $-|\psi|$ only amounts to a shift of one half layer spacing in the location of all the layers (and therefore no change in the free energy per unit volume), the expansion in terms of powers of $|\psi|$ can only contain even powers. Hence the free energy per unit volume in the smectic A phase can be written as

$$G(|\psi|, T) = G_{nem} + \frac{1}{2}\alpha(T)|\psi|^2 + \frac{1}{4}\beta|\psi|^4 + \frac{1}{6}\gamma|\psi|^6, \qquad (4.41)$$

where the coefficient in front of the quadratic term has the same temperature dependence as before

$$\alpha(T) = \alpha_0(T - T^*), \qquad (4.42)$$

and β and γ are greater than zero. G_{nem} is the free energy per unit volume of the nematic phase just above the transition to the smectic A phase. Since there is no cubic term as in the description of the nematic-isotropic transition, this transition is continuous involving no discontinuity in $|\psi|$. This is easily seen from the expression for the value of the order parameter at the nematic-isotropic transition. If the coefficient in front of the cubic term is zero, then the order parameter equals zero at the transition and increases continuously as the temperature decreases.

In order to include the fact that there is also orientational order in the smectic phase, a term is added to the free energy expressing that a change in S at the smectic A transition must change the free energy per unit volume. Let $\delta S = S - S_C$, where S_C is the value of S in the nematic phase just above the transition to the smectic A phase.

Two terms involving δS are usually added, one quadratic in δS and the other a cross term in $|\psi|^2 \delta S$. The free energy per unit volume is then

$$
\begin{aligned}
G(|\psi|, T) &= G_{nem} + \frac{1}{2}\alpha(T)|\psi|^2 + \frac{1}{4}\beta|\psi|^4 + \frac{1}{6}\gamma|\psi|^6, \\
&\quad + \frac{1}{2}\frac{(\delta S)^2}{\chi} - \mu(\delta S)|\psi|^2,
\end{aligned} \tag{4.43}
$$

where χ and μ are positive constants. χ depends on the width of the nematic phase. If the nematic phase is wide, S is large at the smectic A-nematic transition and χ is small. If the width of the nematic phase is small, then χ is large. Since $G(|\psi|, T)$ must be a minimum with respect to δS, the partial derivative of $G(|\psi|, T)$ with respect to δS must be zero. This allows us to express δS in terms of $|\psi|^2$,

$$
\delta S = \chi \mu |\psi|^2. \tag{4.44}
$$

Substituting this into the free energy expression yields

$$
G(|\psi|, T) = G_{nem} + \frac{1}{2}\alpha(T)|\psi|^2 + \frac{1}{4}\beta'|\psi|^4 + \frac{1}{6}\gamma|\psi|^6, \tag{4.45}
$$

where

$$
\beta' = \beta - 2\chi\mu^2. \tag{4.46}
$$

Clearly the sign of β' can change depending on whether χ is large or small (whether the nematic range is narrow or wide). If the nematic range is narrow and χ is large, then β' is negative and the transition is discontinuous. This is evident in Figure 4.11, where different free energy per unit volume curves have been drawn with a negative value for β'. Notice that $|\psi|$ drops from a positive value to zero at the transition.

If the nematic range is wide and χ is small, then β' is positive and the transition is continuous. Free energy per unit volume curves for this case are shown in Figure 4.12. The continuous nature of the transition is evident by how the minimum moves to $|\psi| = 0$ continuously. Thus the point where $\beta' = 0$ is where the transition changes from discontinuous to continuous (a tricritical point).

The Maier-Saupe theory can also be extended to describe the smectic A-nematic transition in what is called the McMillan model. Two order parameters are introduced into the mean-field potential energy function, the usual orientational order parameter S and an order parameter σ related to the amplitude of the density wave describing the

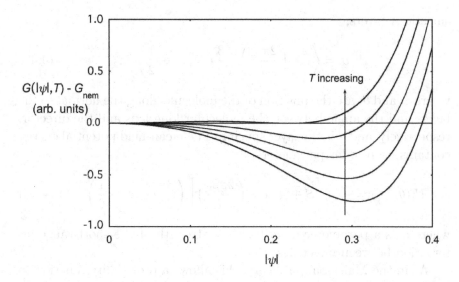

Figure 4.11 Free energies per unit volume for five temperatures. The highest temperature represents the nematic - smectic transition. β' is negative so the transition is discontinuous.

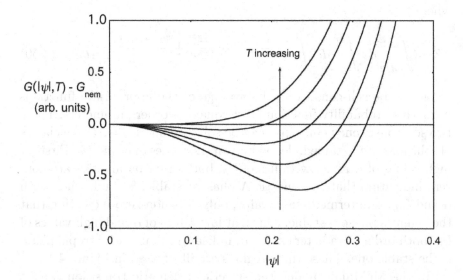

Figure 4.12 Free energies per unit volume for five temperatures. The highest temperature represents the nematic - smectic transition. β' is positive so the transition is continuous.

smectic A layers,

$$\sigma = \left\langle \cos\left(\frac{2\pi z_i}{d}\right)\left(\frac{3}{2}\cos^2\theta_i - \frac{1}{2}\right)\right\rangle, \qquad (4.47)$$

where z_i and θ_i are the position of the molecule along the normal to the layers and the angle between the molecular long axis and the director, respectively, and d is the layer spacing. The mean-field potential energy contains both order parameters,

$$U_i(\theta_i, z_i) = -U_0\left[S + \alpha\sigma\cos\left(\frac{2\pi z_i}{d}\right)\right]\left(\frac{3}{2}\cos^2\theta_i - \frac{1}{2}\right), \qquad (4.48)$$

where α is a parameter describing the strength of the short-range interaction between molecules.

As in the Maier-Saupe theory, this allows a probability function to be defined

$$P_i(\theta_i, z_i) = \frac{1}{Z}e^{-\frac{U_0\left[S + \alpha\sigma\cos\left(\frac{2\pi z_i}{d}\right)\right]\left(\frac{3}{2}\cos^2\theta_i - \frac{1}{2}\right)}{k_B T}}, \qquad (4.49)$$

where

$$Z = \int_{-d/2}^{d/2} dz \int_0^{2\pi} d\phi \int_0^\pi e^{-\frac{U_0\left[S + \alpha\sigma\cos\left(\frac{2\pi z_i}{d}\right)\right]\left(\frac{3}{2}\cos^2\theta_i - \frac{1}{2}\right)}{k_B T}} \sin\theta d\theta. \qquad (4.50)$$

Again, to be self-consistent, the average of the appropriate functions using this probability function must equal the order parameters. These two self-consistency equations can be solved numerically. Three classes of solutions are obtained, depending on the values of α and U_0. First, at high values of α and low values of U_0, both order parameters are non-zero indicating that the smectic A phase is stable. Second, when both α and U_0 are intermediate in value, only S is non-zero indicating that the nematic phase is stable. Third, at low values of α and high values of U_0, both order parameters are zero indicating that the isotropic phase is the stable one. These three regions are illustrated in Figure 4.13.

In the McMillan model, the smectic A-nematic transition can be continuous or discontinuous. If α is less than 0.7, then σ decreases to zero continuously and S is continuous at the smectic A-nematic transition. If α is between 0.7 and 0.98, then σ jumps to zero discontinuously and S has a small discontinuity at the smectic A-nematic transition. When α is greater than 0.98, the smectic phase transforms directly

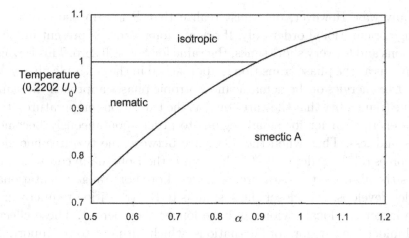

Figure 4.13 Phase diagram for McMillan's model.

into the isotropic phase with discontinuities in both order parameters. So just as in the extended Landau-de Gennes theory for the smectic A phase, a tricritical point is predicted at $\alpha = 0.7$, which corresponds to a ratio in the smectic A-nematic transition temperature to the nematic-isotropic transition temperature of 0.87. A great deal of experimental work has been done on the smectic A-nematic transition, and the results seem to indicate that the tricritical point occurs when the ratio of the two transition temperatures is significantly larger than 0.87.

4.5   PRETRANSITIONAL FLUCTUATIONS

In our discussions so far, the transition from one phase to another occurs at a specific temperature and the order parameter increases in the lower temperature phase starting either from zero or a non-zero value. What we have not discussed is what occurs above the phase transition in the higher temperature phase. Since the order parameter decreases more and more rapidly as the phase transition is approached from below, it is clear that the proximity of the phase transition is evident. The question is whether there is any evidence that the phase transition is near when approaching the transition temperature in the higher temperature phase. As we will see, there are "pretransitional effects" in the higher temperature phase also.

Let us use the nematic to isotropic transition as an example. In the isotropic phase the order parameter S is zero, so it cannot vary in any way. However, one must realize that S is a measure of long-range orientational order only. If orientational order is present in local regions and for very short times, the value for S is still zero. This is what happens as the phase transition is approached in the isotropic phase. As the free energies of the nematic and isotropic phases approach the same value (remember that they are equal at the transition temperature), the free energy barrier for small regions to order spontaneously becomes less and less. Thus when the difference between the two free energies becomes on the order of $k_B T$, where k_B is the Boltzmann constant and T is the absolute temperature, some local or short-range orientational order develops. The closer the system is to the transition temperature, the larger the local regions and the longer they persist. These effects fall under the category of fluctuations, which turn out to be important in the study of phase transitions in general.

In order to investigate the nature of these fluctuations, we must consider the case in which S is not constant in space. Allowing S to vary in the most general way possible calls for some complex mathematics. All of the important physical behavior, however, can be seen by simply looking at the case in which S varies in only one direction, say the x direction. Now an expression for the free energy per unit volume must include an additional term in the derivative of S, since it is energetically unfavorable for S to vary spatially. If we keep only the first term in the Landau expansion and add this additional term, the most important aspects of the pretransitional behavior can be examined. Thus the free energy per unit area perpendicular to the x-axis in the isotropic phase can be written

$$G_{iso}(T) = G_0 + \frac{1}{2} \int_{-\infty}^{+\infty} \left(A(T)[S(x)]^2 + D\left[\frac{dS(x)}{dx}\right]^2 \right) dx, \quad (4.51)$$

where D is a constant, and $A(T)$ has the same temperature dependence as before, $A(T) = A_0(T - T^*)$. Notice that the form of the additional term is similar to many of the other distortion type free energy terms, namely one-half times a coefficient times a distortion term squared. G_0 now represents the free energy in the absence of fluctuations.

The next step is to express the order parameter through its Fourier transform. This simply means that whatever the function $S(x)$ is, it can be represented by a sum of sinusoidal functions of x, provided the sum contains an infinite number of sine waves of varying wavelengths (making it an integral rather than a sum). These sinusoidal functions must have exactly the correct amplitudes so they add up to the function

$S(x)$. These amplitudes are themselves represented by a function $S(q)$, where q is the wavevector or spatial frequency, $q = 2\pi/\lambda$. In fact, $S(x)$ and $S(q)$ can both be expressed as integrals over sinusoidally varying functions,

$$S(x) = \sqrt{\frac{1}{2\pi}} \int_{-\infty}^{+\infty} S(q) e^{iqx} dq,$$

$$S(q) = \sqrt{\frac{1}{2\pi}} \int_{-\infty}^{+\infty} S(x) e^{-iqx} dx. \qquad (4.52)$$

The Fourier transform of $S(x)$, namely $S(q)$, can be thought of as the amplitude of a sinusoidal variation of $S(x)$ with a wavevector equal to q and therefore a wavelength equal to $2\pi/q$.

In order to write the free energy in terms of $S(q)$, we must evaluate $[S(x)]^2$ and $[dS(x)/dx]^2$. These are

$$[S(x)]^2 = \frac{1}{2\pi} \int \int S(q) S(q') e^{i(q+q')x} dq dq',$$

$$\left[\frac{dS(x)}{dx}\right]^2 = -\frac{1}{2\pi} \int \int qq' S(q) S(q') e^{i(q+q')x} dq dq', \qquad (4.53)$$

where the limits of the integration are assumed to be over all q and the order of integration and differentiation has been switched in the evaluation of the second expression. Thus the free energy per unit area can be written

$$G_{iso}(T) - G_0 = \qquad (4.54)$$

$$\frac{1}{2} \int \int (A(T) - Dqq') S(q) S(q') \left[\frac{1}{2} \int e^{i(q+q')x} dx\right] dq dq'.$$

The expression in brackets is simply one way of writing the Dirac delta function $\delta(q + q')$, that is,

$$\delta(q + q') = \frac{1}{2\pi} \int e^{i(q+q')x} dx. \qquad (4.55)$$

The Dirac delta function has the property that it equals zero everywhere except at zero where it is infinite. It also has the property that integrals containing the Dirac delta function are equal to the integrand evaluated at the value of the integration variable where the argument of the Dirac delta function is equal to zero. In the free energy expression, the argument of the Dirac delta function is zero when $q' = -q$, so

$$G_{iso}(T) - G_0 = \frac{1}{2} \int \int \left(A(T) - Dqq' \right) S(q)S(q')\delta(q+q')dqdq' \quad (4.56)$$

and

$$G_{iso}(T) - G_0 = \frac{1}{2} \int \int \left(A(T) + Dq^2 \right) S(q)S(-q)dq, \quad (4.57)$$

where the integral over q' has been performed in this last expression. From the definition of $S(q)$, it is clear that $S(-q)$ is simply the complex conjugate of $S(q)$,

$$S(-q) = \sqrt{\frac{1}{2\pi}} \int_{-\infty}^{+\infty} S(x)e^{+iqx}dx$$

$$= \left[\sqrt{\frac{1}{2\pi}} \int_{-\infty}^{-\infty} S(x)e^{-iqx}dx \right]^* = [S(q)]^* \quad (4.58)$$

Therefore the free energy per unit area can be written as an integral over q rather than an integral over x

$$G_{iso}(T) - G_0 = \int \frac{1}{2} \left(A(T) + Dq^2 \right) |S(q)|^2 dq. \quad (4.59)$$

Realizing that the free energy can be written as an integral over sinusoidal variations of $S(x)$ with different values of q is useful because the equipartition theorem can be invoked. Since these sinusoidal variations represent different ways or degrees of freedom for the variation of $S(x)$, then on average the free energy must be partitioned between them equally. Thus

$$\left\langle \frac{1}{2} \left(A(T) + Dq^2 \right) |S(q)|^2 \right\rangle = \frac{1}{2} \left(A(T) + Dq^2 \right) \left\langle |S(q)|^2 \right\rangle$$

$$\propto \frac{1}{2} k_B T, \quad (4.60)$$

where the brackets denote a thermal average. Thus, although the average value for $S(q)$ in the isotropic phase is zero, the mean-square value of $S(q)$ is non-zero and dependent on both the temperature and the wavevector of the variation of $S(q)$,

$$\left\langle |S(q)|^2 \right\rangle \propto \frac{k_B T}{A(T) + Dq^2} = \frac{k_B T}{A_0(T - T^*) + Dq^2}. \quad (4.61)$$

Another way of writing this last expression is to define a new T^* temperature, dependent on q, for which the denominator is zero. Hence

$$T^*(q) = T^* - \frac{D}{A_0}q^2, \qquad (4.62)$$

and

$$\langle |S(q)|^2 \rangle \propto \frac{k_B T}{A_0(T - T^*(q))}. \qquad (4.63)$$

Thus, as the transition temperature is approached in the isotropic phase, the mean-square value of $S(q)$ increases as if approaching infinity at a temperature below the transition temperature. This temperature at which the mean-square value of $S(q)$ would diverge if the transition to the nematic phase did not take place depends on the wavevector of the spatial variation in $S(q)$. Figure 4.14 shows a typical variation of the mean-square value of $S(q)$ in the isotropic phase. The amount of light scattered by these fluctuations is directly proportional to the mean-square value of $S(q)$ and this variation has been confirmed through light scattering experiments. Such experiments show that the difference between the transition temperature and T^* is roughly one kelvin.

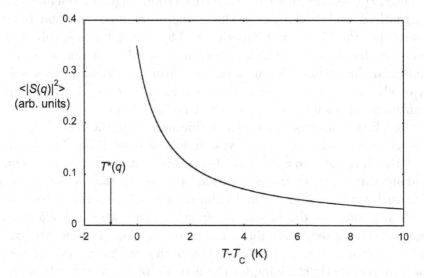

Figure 4.14 Mean square value of the order parameter in the isotropic phase. T_C is the nematic-isotropic phase transition temperature and $T^*(q)$ is the temperature at which the mean square value of the order parameter would diverge based on its behavior in the isotropic phase.

The combination of parameters $D/A(T)$ has units of length squared. Therefore the square root of $D/A(T)$ represents a length, $\xi(T)$,

$$\xi(T) = \sqrt{\frac{D}{A(T)}} = \sqrt{\frac{D}{A_0(T - T^*)}} = \xi_0 \sqrt{\frac{T^*}{T - T^*}}, \qquad (4.64)$$

where

$$\xi_0 = \sqrt{\frac{D}{A_0 T^*}}. \qquad (4.65)$$

If the free energy is written using $\xi(T)$,

$$G_{iso}(T) - G_0 = \frac{1}{2} \int_{-\infty}^{+\infty} A(T) \left([S(x)]^2 + [\xi(T)]^2 \left[\frac{dS(x)}{dx} \right]^2 \right) dx, \qquad (4.66)$$

then it is clear that $\xi(T)$ is related to the length over which the orientational order is correlated, since the larger this length, the larger the free energy contribution due to spatial variations in the order parameter. $\xi(T)$ is therefore called the correlation length for orientational fluctuations, and it increases as the temperature in the isotropic phase approaches the transition temperature. The parameter ξ_0 is called the bare correlation length and it should have a value on the order of a molecular dimension. Measurements confirm this, yielding values for ξ_0 of about 1 nm. With typical values for $T - T^*$ of 1 K near the transition, values for $\xi(T)$ can get as large as 20 nm.

Recall that we represented these fluctuations of the order parameters as a sum of sinusoidally varying functions of position. This means that the index of refraction can also be thought of as a sum of sinusoidally varying functions of position. Just as we saw in Section 3.6 that such variations on the molecular scale result in x-ray diffraction, these variations in the index of refraction cause a similar effect for light. There is a significant difference however. The variations are fixed at the molecular level (e.g., the location of the smectic layers), whereas they are constantly changing for the nematic phase. The result is that the light is not diffracted in specific directions, but scattered in all directions and varies with time.

When this scattering of light is examined in an experiment, a significant simplification occurs. In order for light to change its direction (i.e., be scattered), it must change its wavevector, so there must be something in the sample capable of doing that. For a sample with fluctuating orientational order, it is the sinusoidally varying index of refraction that can do this. But if light with a wavevector $\vec{k}$ is scattered by an angle θ as shown in Figure 4.15, the wavevector of the light must change by $|\vec{q}| = 2|\vec{k}|\sin(\theta/2)$. Thus a light scattering experiment picks out fluctuations in the sample with a specfic value of the wavevector $\vec{q}$, making the analysis much simpler. In fact, by changing the scattering angle θ, the fluctuations for different $\vec{q}$ can be probed.

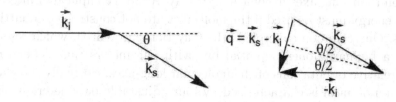

Figure 4.15 Wavevectors in a light scattering experiment. $\vec{k}_i$ is the wavevector of the incident light, $\vec{k}_s$ is the wavevector of the scattered light, and $\vec{q}$ is the change in the wavevector of the light that must be supplied by orientational order fluctuations in the sample.

4.6 SIMULATION TECHNIQUES

All of the theoretical models discussed so far have been analytical, in that various properties are computed using different starting assumptions and different techniques of calculation. There is another important way in which theoretical information on a large system of molecules can be gathered, and that involves simulating a physical system. Since large numbers of molecules are involved, such simulations routinely involve computer calculations. With the recent advances in computational power available at reasonable cost, more and more simulations of liquid crystal systems are being performed, yielding important information on the phases formed, the transitions between them, and the properties of the various phases. It must be pointed out right at the beginning that computer simulations of liquid crystals are inherently difficult due to the fact that the range of both the length scales and timescales involved span orders of magnitude. A bond excitation can take place in picoseconds and extend over tens of picometers. On the other hand, rearrangements of the director can take microseconds and extend over micrometers. Creating computational techniques that are accurate across such a range of time and space while taking a reasonable amount of computation time is a huge challenge.

The first step in performing a simulation is to define the system under study and the interactions between the entities that comprise the system. For example, one might choose a system of rod-like molecules, each of which is constrained to a point in a cubic lattice. The orientation of each molecule is allowed to vary and the interaction energy between neighboring molecules is assumed to depend on the angle between their long axes in some simple way. A more complicated interaction energy must be used if the molecules are not constrained to lattice sites. One common example is the Gay-Berne potential, which resembles a Lennard-Jones potential but with parameters that depend on the relative orientations of the molecular long-axes. Normally the Gay-Berne potential is characterized by four adjustable parameters, which can be tuned to represent the interaction of specific molecules. Once the system and interactions are defined, the problem then becomes one of simulating the thermodynamics of such a system. There are several ways to do this. Here we discuss only the molecular dynamics and Monte Carlo techniques.

One can consider a molecular dynamics simulation as a "brute force" method to calculate the motion of the molecules (or the simplistic entities representing molecules) according to Newton's laws. That is, starting from a specific location, orientation, translational velocity, and rotational velocity for each molecule, Newton's laws are integrated to give these four quantities for each molecule some short time later. This is quite a numerical calculation, since the force and torque on each molecule due to all of the others must be calculated before the change in the four quantities can be determined for that molecule. Evaluation of the forces and torques, for which all pairs of molecules must be considered, takes much more time than the determination of the new quantities, for which each molecule must be considered. Many methods have been devised to speed up such calculations, since it is important that the number of molecules in the system be as large as possible. In many cases larger time steps are employed than at first would seem reasonable, because it can be shown that the quantities of interest are not strongly affected by this large time step. After allowing the system to come to an equilibrium state, average values for macroscopic quantities can be evaluated by averaging these quantities for many time steps. In addition, time-dependent quantities can be determined by observing how certain parameters of the system change with time.

The number of molecules in the system, the complexity of the interactions between the molecules, the translational and orientational constraints on the molecules, the desired time resolution, the method of calculation employed, and the computer being used all determine the time necessary to perform a molecular dynamics simulation. Simulations involving many thousand molecules and performed on the fastest computers available are now being done quite frequently.

The Monte Carlo technique is similar to a molecular dynamics simulation except that the steps are not time steps. Again, a starting configuration for the system is defined in which the four quantities mentioned above are specified for each molecule. Then an algorithm is used to change this configuration in a random way. The free energy of this new configuration is then calculated and a decision is made on whether to include this new configuration in the average or not. If the free energy of the new configuration is less than the starting configuration, then this configuration is included in the average. If the energy of the new configuration is higher than the starting configuration by some amount, ΔU, then another algorithm is employed to make sure its chance of being used in the average is proportional to the Boltzmann factor $e^{-\Delta U/kT}$. With the decision made as to whether this new configuration is used in the average or not, the process repeats itself again, starting with the random generation of another new configuration. The energy of each newly generated configuration is compared to the energy of the last configuration accepted for use in the average. This continues until many configurations representing the full range of variation within the system have been averaged to produce macroscopic quantities. The drawback of the Monte Carlo technique is that time-dependent quantities cannot be calculated since each change in the system is a random one. On the other hand, Monte Carlo simulations are much faster than molecular dynamics calculations, so more sophisticated systems involving a larger number of molecules and more complex interactions can be examined.

One additional aspect of numerical simulations must be mentioned. Every simulation must possess molecules near the boundaries of the system, and these molecules experience forces quite different from molecules not near the boundaries. In principle, such a simulation should include an infinite number of molecules so the proportion near the boundaries would have a negligible influence on the average quantities. Since this is not possible, an alternative method employing periodic boundary conditions and a finite size system is usually used. The method of periodic boundary conditions works in the following way. When calculating the interactions for molecules close to a boundary, say at $x = +d$ for a system extending from $x = -d$ to $x = +d$, it is assumed that the region with $x > +d$ is filled with molecules identical to those in the system but displaced in the $+x$ direction by an amount equal to $2d$. This technique does simulate a system of molecules surrounded by a "sea" of similar molecules, but in certain cases the actual shape of the system has an undesired effect on the results of the simulation. Thus care in the use of periodic boundary conditions is warranted.

Since the speed of computers has increased so much over the last decades, simulations now can start at the atomic level, allowing for the simulations of systems composed of specific molecules. The interactions here are between atoms, whether the atoms are on the same molecule or different molecules, and describe the behavior in a way consistent with quantum mechanics. These simulations result not only with a prediction of the phases that are stable, but the temperatures at which phase transitions take place. They also are capable of predicting material properties such as elastic constants and viscosities.

Computation time still restricts atomistic simulations considerably, to the point where calculations involving high molecular weight liquid crystals or liquid crystal organization over large length scales are impractical. There are therefore efforts to produce coarse-grained models that are capable of addressing these situations. In these models, the molecules are represented not by atoms but by a number of "sites" and the interactions between the "sites" are chosen to approximate the sum of the atomistic interactions as closely as possible. Depending on the degree of coarse-graining, the system size can be made much larger and the timescales made much longer. In this way, simulations of self-assembly and polymer liquid crystals have been developed.

4.7 DEFECT PHASES

As we discussed in Chapter 3, the introduction of disclinations into a liquid crystal phase increases the free energy of the system. Therefore, disclinations are energetically unfavorable and various techniques such as annealing and the use of applied fields are capable of reducing the number of disclinations in a sample. There are, however, a few cases where disclinations are present in a phase because they are energetically favorable. In fact, the structure of the phase depends on the existence of an ordered array of disclinations. The reason such disclinations are energetically favorable was not appreciated for many years, but understanding these exotic phases of matter illustrates perfectly how the liquid crystal state reflects a fine balance between competing factors.

The more simple example concerns the smectic A phase formed by chiral molecules. Because the molecules are chiral, it is energetically favorable for the director to form a helix. The layered structure of the smectic A phase does not permit twist or bend distortion, so in some senses the smectic A phase is a "frustrated" phase if the molecules are chiral. What happens if the chirality of the system is increased? The answer is that the temperature range of the smectic A phase sometimes decreases, with increases perhaps in the temperature range of the lower temperature chiral smectic C phase or the higher temperature chiral nematic phase. What is more interesting is that in some highly chiral smectic liquid crystals a new phase forms in place of the smectic A phase. This phase is composed of regions of smectic A liquid crystal separated from each other by an array of equally spaced grain boundaries. The director rotates by a small angle in crossing each grain boundary, so a twist in the director much like that in a chiral nematic liquid crystal develops. This phase is stable because the increase in free energy due to the introduction of grain boundaries is less than the decrease in free energy due to the introduction of twist. A representation of this twist grain boundary phase is shown in Figure 4.16.

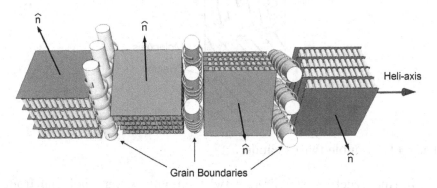

Figure 4.16 Structure of the twist grain boundary phase.

This twist grain boundary phase affects light in much the same way as the chiral nematic phase does. If the distance between grain boundaries is about 25 nm and the rotation of the director in crossing a grain boundary is about 20°, then the pitch of the structure is $(360°/20°)(25$ nm) or 450 nm. Such a structure should show selective reflection for visible light and this in fact is observed. Twist grain boundary phases based on the smectic C structure instead of the smectic A structure also exist.

The second example of a phase that spontaneously incorporates defects requires slightly more thought. In Chapter 3 we showed that the chiral nematic (or single twist) structure minimized the free energy for a chiral material. This is certainly true for a homogeneous structure. Yet might it be possible that another twisted structure has a lower free energy at least locally? Such a possibility is the "double twist" structure shown in Figure 4.17.

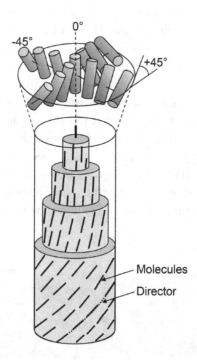

Figure 4.17 Double twist cylinder.

In this structure, the director twists in moving perpendicular from the z-axis along any radius. Thus at the center the director points along the z-axis, but has rotated by 45° at all points in the x-y plane located a distance equal to one-eighth the pitch away from the origin. If the structure does not vary in the z-direction, then this defines a double twist cylinder.

Using cylindrical coordinates (ρ, ϕ, z), the components of the director for a right-handed structure can be written as

$$n_\rho = 0 \qquad n_\phi = -\sin(q_0\rho) \qquad n_z = \cos(q_0\rho), \qquad (4.67)$$

where $q_0 = 2\pi/P$. The free energy expression involves terms containing both the divergence and curl of the director

$$
\begin{aligned}
\nabla \cdot \hat{n} &= 0 \\
\hat{n} \cdot (\nabla \times \hat{n}) &= -\left[q_0 + \frac{\sin(2q_0\rho)}{2\rho} \right] \\
|\hat{n} \times (\nabla \times \hat{n})| &= \frac{\sin^2(q_0\rho)}{\rho}.
\end{aligned}
\qquad (4.68)
$$

Thus this structure contains both twist and bend distortion, but notice that some of the terms become very negative when ρ is small. In addition, there are terms originally discarded from the distortion free energy discussion in Chapter 3 because they involved the divergence of some quantity. Since the volume integral of a divergence can be converted to an integral over the surface of the volume, and since the surface area increases as the square of a linear dimension and the volume increases as the cube of a linear dimension, such terms have negligible contribution for bulk samples. These terms cannot be dropped from the free energy expression here, however, since the radius of the double twist cylinder is not infinite. In fact, when the double twist terms are substituted into the free energy expression containing these divergence terms, the free energy per unit volume for the double twist cylinder is lower than for the single twist structure if the radius of the cylinder is small enough. Therefore, a structure made of double twist cylinders with disclinations running through it is stable if the decrease in free energy due to the formation of many double twist cylinders is larger than the increase in free energy due to the creation of disclinations. Such phases do occur, with the double twist cylinders packing together to form a cubic structure, but only when the chirality is high and the temperature is just a degree or so below the transition to the isotropic phase. Where perpendicular double twist cylinders touch in this cubic structure, the director is continuous due to the 45° rotation in each cylinder. However, disclinations run through the structure, forming a cubic lattice. The spacing between disclinations can be on the order of the wavelength of visible light, so these structures diffract light just as cubic crystals diffract X-rays. The first observed phases of this type strongly diffracted blue light, so they were named blue phases. Since these first observations, blue phases that strongly diffract light in other parts of the visible spectrum and in the ultraviolet have been discovered. These blue phases are truly remarkable. Single crystals of a blue phase can be "grown" from another phase. The shape of these crystals shows the cubic symmetry and facets can sometimes be seen. When observing these single crystals, it is difficult to keep in mind that the molecules inside and outside the single crystal are diffusing much like they do in a liquid, passing in and out of the single crystal quite randomly. Yet the observations demonstrate that when a molecule is inside the single crystal, it maintains the orientational order of the structure composed of double twist cylinders (disclinations and all). When it is outside the structure, it maintains the orientational order of the surrounding phase.

4.8   SELF-ASSEMBLY THEORY

Lyotropic liquid crystals were discussed in Chapter 1. These occur when molecules in a solvent spontaneously self-assemble to form structures with orientational and sometimes positional order. Exactly which structures form and how ordered they are depend on the concentration more than any other parameter. It turns out that it is not difficult to formulate a simple theory to describe this process.

Let us consider the case where a specific number of molecules N is necessary to form a structure. An example is the formation of spherical micelles when certain soap molecules are mixed with water, since it takes a certain number of molecules to "fill up" the micelle with only soap molecules. There are two chemical species present, a molecule A_1 and an assembly of N molecules A_N. Let $C_1 = [A_1]$ be the concentration of single molecules and let $C_N = [A_N]$ be the concentration of assemblies. If the reaction is governed by an equilibrium constant K_N, then the system can be summarized as follows:

$$NA_1 \;\rightleftharpoons\; A_N \qquad K_N = \frac{[A_N]}{[A_1]^N} = \frac{C_N}{C_1^N}$$

$$C_T = C_1 + NC_N, \tag{4.69}$$

where C_T is the total concentration of molecules, both single and in assemblies. Combining the two equations results in an Nth order equation that must be solved for C_1,

$$NK_N C_1^N + C_1 - C_T = 0. \tag{4.70}$$

The units of K_N are pretty crazy, so we can simplify the situation by defining another equilibrium constant K by $K_N = K^{N-1}$. K has units of inverse concentration and this helps to make connections to other self-assembly processes. With this definition the equation to be solved becomes

$$NK^{N-1}C_1^N + C_1 - C_T = 0. \qquad (4.71)$$

Figure 4.18 shows plots of the concentration of molecules as single molecules C_1 and the concentration of molecules in assemblies $C_T - C_1$ as a function of C_T for $K = 100$ M^{-1} and $N = 6$, 60, and 600. Notice that there is a threshold concentration for the formation of assemblies, and that the threshold gets sharper as N increases. This threshold concentration for the formation of assemblies is called the critical micelle concentration or CMC. The values of $K = 100$ M^{-1} and $N = 60$ are about right for sodium dodecyl sulfate (SDS). For large N above the CMC, the concentration of single molecules is constant.

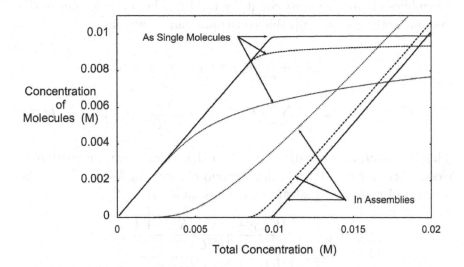

Figure 4.18 Concentration of molecules as single molecules and in assemblies according to Eq. (4.71) with $K = 100$ M^{-1} and N equal to 6 (dotted lines), 60 (dashed lines), and 600 (solid lines).

Chromonic liquid crystals, in which molecules spontaneously form linear assemblies, were described in Chapter 1. In this case, there can be any number of molecules in an assembly, and each assembly size must be considered a different chemical species. Experiments demonstrate that the process of assembly formation is close to being isodesmic, meaning the change in the free energy for the addition of a molecule to or the subtraction of a molecule from an assembly is independent of the size of the assembly. This process can therefore be described by a series of chemical reactions, all with the same equilibrium constant K,

$$A_1 + A_1 \rightleftharpoons A_2 \qquad K = \frac{[A_2]}{[A_1]^2}$$

$$A_2 + A_1 \rightleftharpoons A_3 \qquad K = \frac{[A_3]}{[A_2][A_1]}$$

$$\cdot$$
$$\cdot$$
$$\cdot$$

$$A_{i-1} + A_1 \rightleftharpoons A_i \qquad K = \frac{[A_i]}{[A_{i-1}][A_1]}. \tag{4.72}$$

Just as in the previous calculation, the system is constrained because the number of single molecules plus the number of molecules in assemblies of any size must equal the total number of molecules in the system. Letting $C_i = [A_i]$, this constraint can be written

$$
\begin{aligned}
C_T &= C_1 + \sum_{i=2}^{\infty} iC_i = C_1 + \sum_{i=2}^{\infty} iK^{i-1}C_1^i \\
&= \frac{1}{K} \sum_{i=1}^{\infty} i(KC_1)^i = \frac{C_1}{(1 - KC_1)^2}.
\end{aligned} \tag{4.73}
$$

This is a quadratic equation in C_1, so the solution can be written in closed form for either the concentration of single molecules C_1 or the fraction of molecules that are not in assemblies $\alpha_1 = C_1/C_T$,

$$
\begin{aligned}
C_1 &= \frac{2KC_T + 1 - \sqrt{4KC_T + 1}}{2K^2C_T} \\
\alpha_1 &= \frac{2KC_T + 1 - \sqrt{4KC_T + 1}}{2(KC_T)^2}.
\end{aligned} \tag{4.74}
$$

Once C_1 or α_1 is known, the concentration of the assemblies of different sizes and the fraction of molecules in assemblies of different sizes can be calculated from the equilibrium constant relations given in Eq. (4.72),

$$C_i = K^{i-1}C_1^i \quad \text{and} \quad \alpha_i = \frac{iC_i}{C_T} = \frac{i(KC_1)^i}{KC_T} = i(KC_T)^{i-1}\alpha_1^i. \tag{4.75}$$

Figure 4.19 shows the fraction of molecules in assemblies of different sizes as a function of the total concentration. Unlike the previous example in which a threshold concentration is necessary to form assemblies, the isodesmic assembly process is continuous, with assemblies present at every concentration. A small number of small assemblies form at low concentrations, and the number and average size of assemblies increases as the concentration increases. The inset shows the fraction of molecules in assemblies of different sizes for two total concentrations, $C_T = 0.005$ M^{-1} and 0.045 M^{-1}.

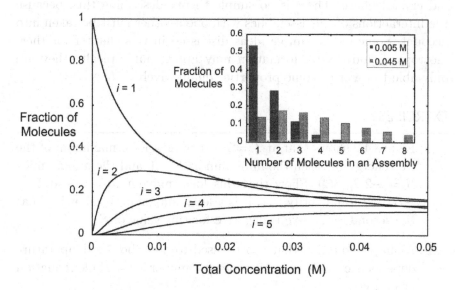

Figure 4.19 Fraction of molecules in assemblies of size i as a function of the total concentration C_T for $K = 100$ M^{-1}. The inset shows the fraction of molecules in each assembly size for $C_T = 0.005$ M^{-1} and 0.045 M^{-1}.

The average length of an assembly $\langle i \rangle$ is a function of K and C_T and is given by

$$\langle i \rangle = \sum_{i=1}^{\infty} iC_i \Big/ \sum_{i=1}^{\infty} C_i = \frac{1}{1 - KC_1} = \sqrt{C_T/C_1} = \sqrt{1/\alpha_1}, \quad (4.76)$$

where C_1 and α_1 as a function of K an C_T are given by Eq. (4.74).

According to Eq. (4.75), C_i, the concentration of assemblies of size i, decreases by a factor of KC_1 every time i increases by 1. This means that C_i is an exponentially decreasing function of i with a characteristic size of $-1/\ln(KC_1)$. On the other hand and as shown in Figure 4.19, the fraction of molecules in assemblies of size i is generally a peaked function of i, with the peak occurring at around the average assembly size.

It is important to keep in mind that this discussion is only concerned with the formation of assemblies and does not address under what conditions a liquid crystal phase forms. In general, at some point when the concentration of assemblies is great enough and the shape of the assemblies is anisotropic enough, the assemblies order into a liquid crystal phase. There is no simple theory describing this, because the interactions of the assemblies with one another must be taken into account. Some of the simple ideas discussed in this chapter for thermotropic liquid crystal formation may apply, but typically they are only able to describe some properties qualitatively.

EXERCISES

4.1 Figure 4.2 shows that at $T = T_C$ there is a local maximum of the free energy between the minimum at $S = 0$ and the minimum at $S = -2B/(3C)$. Show that this local maximum occurs at $S = -B/(3C)$, and verify that it is a local maximum by showing that the second derivative is negative for this value of S.

4.2 A simple equation that can be used to describe the temperature dependence of the nematic order parameter is the Haller equation given by

$$S(T) = \left(1 - \frac{T}{T_H}\right)^{\beta} = \left(1 - \frac{T_C + [T - T_C]}{T_C + [T_H - T_C]}\right)^{\beta},$$

where β is the exponent governing the basic temperature dependence and T_H is the temperature at which S would equal zero if there were no phase transition. As before, T_C is the nematic-isotropic transition temperature, so $T_H > T_C$. Show that the Haller equation is consistent with the data for 5CB shown in Figure 4.3 if β is about 0.18 and $T_H - T_C$ is around 0.7 K. The nematic-isotropic transition temperature for 5CB is 308 K.

4.3 Let us see under what circumstances the order parameter behavior predicted by Landau-de Gennes theory resembles the prediction of Maier-Saupe theory. Rewrite Eq. (4.11) so the temperatures appear as T/T_C and T^*/T_C. Since in Maier-Saupe theory the order parameter equals 0.43 at the transition, substitute 0.43 for S_C in Eq. (4.11). For what value of T^*/T_C does this equation give behavior similar to what is shown in Figure 4.5 for the Maier-Saupe theory?

4.4 Show that for Onsager theory the volume fraction change across the coexistence region equals $1.2/(L/D)$, where L/D is the aspect ratio of the rods. Is Figure 4.10 consistent with this result?

4.5 The presence of a cubic term in Eq. (4.1) ensures that the nematic-isotropic transition is discontinuous. Therefore, to model a continuous transition, such as exists in a ferromagnet for which the order parameter is the magnetization M, the cubic term must be dropped, yielding

$$G(M,T) = G_{para} + \frac{1}{2}A_0(T - T^*)M^2 + \frac{1}{4}CM^4,$$

where $G(M,T)$ is the free energy of the ferromagnetic phase ($M \neq 0$), and G_{para} is the free energy of the paramagnetic phase ($M = 0$) that occurs at a higher temperature than the ferromagnetic phase. As with the Landau-de Gennes theory, $A_0 > 0$ and $C > 0$. (a) Show that for $T > T^*$ the minimum free energy occurs for $M = 0$. (b) Show that for $T < T^*$, the minimum free energy occurs for a non-zero value of M given by

$$M = \pm\sqrt{A_0(T^* - T)/C}.$$

Notice that there is no discontinuity in M at the transition.

4.6 Light scattering experiments are extremely useful when investigating fluctuations, because a specific value of the wavevector q is probed, depending on the wavelength of light, the scattering angle detected, and the index of refraction of the sample. Imagine that a light scattering experiment measures $\langle |S(q)|^2 \rangle$ (see Eq. (4.62)) as a function of T. (a) Describe what the graph should look like if $k_B T/\langle |S(q)|^2 \rangle$ is plotted vs. T? (b) How can the value of $T^*(q)$ be obtained from this plot? (c) How does the graph change if another value of q is used?

4.7 (a) Start with the definition of the average assembly size $\langle i \rangle$ as the quotient of the two summations in Eq. (4.76) and derive the rest of Eq. (4.76). (b) Put numbers into the expression to verify that the average size assembly is in the vicinity of the peak in the fraction of molecules histogram (inset of Figure 4.19) for $C_T = 0.045$ M^{-1}.

CHAPTER 5

Calamitic Liquid Crystals – Rods, Kinks, and Molecular Design

5.1 THE MOLECULAR BUILDING BLOCKS OF CALAMITIC LIQUID CRYSTALS

An often asked question is, "How can you predict from the molecular structure of a material the type of condensed phases it forms, and what properties would you expect those phases to exhibit?" For liquid crystals, knowing how to design materials for particular applications also requires precise nanoscale molecular engineering. In the following we examine how molecular topologies and interactions influence phase formation and thereby report back on material design. But first we need to understand some basic principles and associated definitions.

The type of liquid crystal phase that is formed by a mesomorphic material is essentially dependent on the molecular architecture; this includes the chemical structure, polarity, polarizability, and topology. A primary factor in mesophase formation is the overall, or gross, molecular shape of the material, and for anisotropic structures their aspect ratios (length to breadth). Three separate species that underpin mesophase formation are readily identified for systems where the molecules have the following rotational 3D volumes, spheroid, ellipsoid, and discoid. Spheroidal mesomorphic materials, *e.g.*, adamantane in Figure 2.2, generally give rise to plastic crystals where the molecules have long-range positional order and yet are undergoing rapid re-orientational motion about their lattice sites. Ellipsoidal or rod-like molecules give rise to calamitic liquid crystals, which include nematic (N for *nematos*, Greek for thread-like), and the smectic liquid crystals (S for *smectos*, Greek for soap-like) and anisotropic plastic crystals, as described in Chapter 2. Discoid materials produce nematic-discotic and columnar discotic liquid crystals (D for disk-like), which are discussed in the next chapter.

Materials with molecules possessing combinations of these shapes can also be mesomorphic. For instance, materials that possess aspects of both disk- and rod-like shapes can exhibit both calamitic and discotic phases. These materials are often polycatenar, and their phases are called phasmidic (after the shapes of stick insects). Similarly, molecular structures that combine features of both disks and spheres can have bowl-like shapes and can produce bowlic or pyramidal mesophases. Materials with bent architectures have been found to exhibit novel classes of mesophases, called bent-core or banana phases. Ionic materials and commonly dyes, with board-like or sanidic architectures, are often found to exhibit chromonic phases. A number of molecular templates for the formation of certain liquid-crystalline modifications are depicted in Figure 5.1, along with typical low molar mass liquid crystal materials for comparison.

Nearly all of these topologies have the potential to be chiral (handed) or form chiral mesophases. Asymmetry in a structure can be established via a stereogenic center either in the central core of the structure or in the peripheral regions. Alternatively, the molecular topology may be asymmetric without the need for a stereogenic center, for example, the case of a helical or twisted shape, as described in Chapter 7.

5.2 SELF-ORGANIZATION VERSUS SELF-ASSEMBLY

Self-organization is a process whereby individual molecules become organized into a condensed phase so that the molecules have an orientational arrangement, but are still free to move about within the phase. This is rather like the structure found in a membrane, where the lipids are organized to be parallel to one another but are able to move within the bilayer such that neighboring molecules have dynamic interactions.

Self-assembly on the other hand is where there is a strong association between two or more molecular entities such that a larger, stable supramolecular structure is formed. However, this suprastructure need

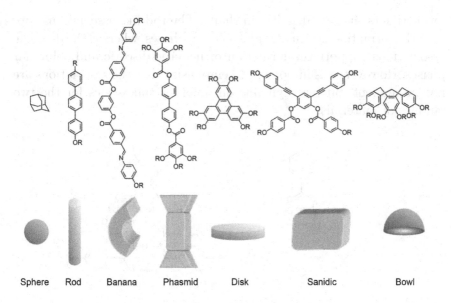

Sphere   Rod   Banana   Phasmid   Disk   Sanidic   Bowl

Figure 5.1 Topologies of materials that are conducive to exhibiting mesomorphic behavior, and examples of typical low molar mass liquid crystals. R denotes an aliphatic chain in the chemical structures.

not be a separate, identifiable, and long-lasting association. Instead the self-assembled structure might be dynamic, with associations forming and breaking apart. Thus, over relatively short-time scales, associations dominate the structure of a condensed phase. The associations may be in the form of dimeric or trimeric, *etc.*, which are formed via hydrogen-bonding, halogen-bonding, and van der Waals, dipolar, and quadrupolar interactions.

Probably the best examples of, and possibly the most numerous, material systems that self-assemble and then self-organize to form liquid crystals are those based on hydrogen-bonding networks. Classical examples include the 4-alkoxybenzoic acids, and the 4,5-*bis*(alkyloxy)phthalhydrazides, as shown in Figure 5.2. The two families of materials have individual molecules that on their own, due to topology, do not support the formation of liquid crystal phases. However, once the molecules hydrogen bond, the dimeric and trimeric associations form rod- or disk-like entities that support liquid crystallinity. Thus, the molecules of 4-alkoxybenzoic acids hydrogen bond with one another to form dimers that have large aspect ratios and are rod-like in shape, whereas the molecules of 4,5-*bis*(alkyloxy)phthalhydrazides form trimers that are disk-like in shape. The rod-like associations support the formation of nematic and smectic phases, whereas the disk-like associations support the formation of nematic-discotic and columnar phases. However, it is important to emphasize that the associations are not permanent, but are dynamic on different time-scales for the two families of materials.

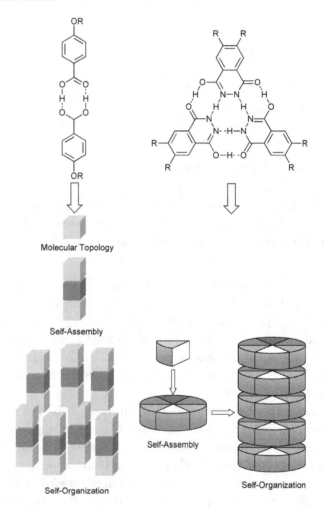

Figure 5.2 The self-assembly and self-organization of the 4-alkoxybenzoic acids, R = alkyl C_6 to C_{12} (left), to give rod-like associations that form nematic phases, in comparison to the self-assembly followed by self-organization for the 4,5-*bis*(alkyloxy)phthalhydrazides, R = alkyl C_6, C_{12}, C_{16} (right), that form columnar associations.

For liquid crystals based on materials that possess rod-like building blocks, there are a number of possible ways that the building blocks may be derived. First, the building block may be naturally rod-like; second, a block may undergo distortion by its environment so that it changes shape to become rod-like (usually found for dendrimers and substituted nanoparticles), or third, the blocks may self-assemble to form a macro-rod-like structure.

Distortion of shape is depicted in Figure 5.3(a), and illustrated by the substituted dendrimer/nanoparticle in Figure 5.3(b). In the left-hand figure a cubic building block is distorted to give a rod-like block or a board-like unit (sanidic). In the right-hand picture a spherical particle, which is symmetrically substituted, is distorted so that the substituents point up or down to give a rod.

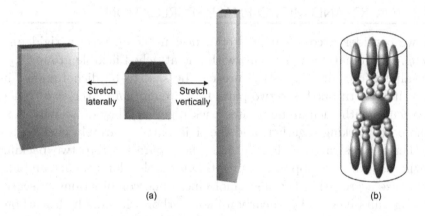

(a)   (b)

Figure 5.3 Symmetrical objects when distorted form rod-like blocks. Distortion of a cubic shape by stretching (a), and as an example, a substituted dendrimer/nanoparticle (b).

A second set of rod-like objects can be derived from symmetrical building blocks by self-assembly, as illustrated in the examples in Figure 5.4. In the arrangements shown, rod-like structures result from end-to-end associations of cubes, or by end-to-end and side-to-side associations to give boards. These are analogous to the distortions described above. There are also many other arrangements that approximate the structure of rods, such as helical twists, bends, and zig-zags.

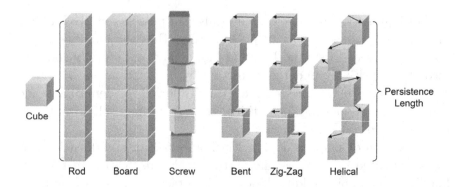

Figure 5.4 Rod-like self-assembled structures resulting from the associations of cubes.

5.3 NANO- AND MICRO-PHASE SEGREGATION

If we consider a common "material design" comprising a rigid unit (*e.g.*, an aromatic ring system) with one attached flexible group (*e.g.*, an aliphatic chain), the structure can be said to be dichotomous, in that it is partitioned into two parts that are mutually exclusive. When molecules with such architectures pack together, they do so with flexible parts packing together, and the rigid parts separately packing together as illustrated in Figure 5.5. The separation into two packing regimes is called nanophase segregation for molecular systems, and microphase segregation for macromolecules. The combinations of segregating pairs include hydrocarbon-fluorocarbon, siloxane-hydrocarbon, aromatic-aliphatic, polar-non-polar, rigid-flexible, *etc.*

Thus, nanophase segregation in dichotomous systems is somewhat similar to the hydrophobic effect in amphiphiles, except the scale regime is smaller, and the interactions weaker and subtler. Trichotomous to polychotomous materials are also possible, particularly for macromolecular materials.

An important feature of nano-segregative materials is that the points of partition allow the molecules to adopt various nonlinear shapes in comparison to the rod-like molecules in Figure 5.5. Another consequence of this structuring is that the various segments have the ability to rotate or oscillate independently of the other segments (for comparison, see the discussion on dynamics and fluctuations in Chapter 2). This has the effect of aiding the segregation process, and for typical molecules of liquid crystal phases that are trichotomous, hav-

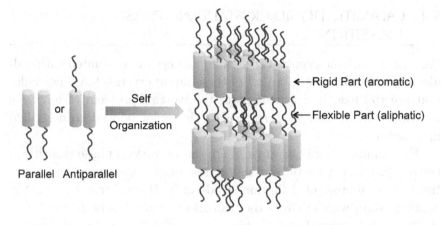

Figure 5.5 Nanophase segregation for dichotomous materials.

ing two flexible units (chains) attached to a rigid unit (core), zig-zag shaped architectures result due to the stereochemistry, as illustrated in Figure 5.6. Such systems can form layered structures where the segregation results in layering of flexible blocks sandwiched between layers of rigid units.

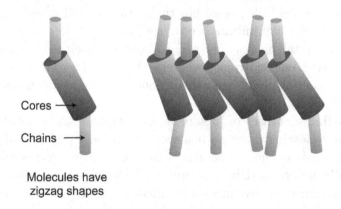

Figure 5.6 Trichotomous molecular architectures and the layering that is formed to produce nanosegregation.

The segregation process therefore has the tendency to support the formation of layered mesophases, and suppress the formation of the solid state. The consequence is that the entropy is relatively higher due to the disorder in and between the layers.

5.4 CALAMITIC LIQUID CRYSTALS AND PHASE TRANSITIONS

We now embark on a general class of liquid crystals associated with rod-like molecules. However, not all of these liquid crystals have molecules that are rod-like. In an attempt to resolve this problem of having a descriptive definition, the term "calamitic" is introduced, as described in Chapter 1.

For calamitic liquid crystals, there can be various phase transitions between nematic, smectic, and soft crystal phases, and as a consequence there are a number of definitions to describe them. First, there is the melting point, which defines the transition from the solid to the liquid crystal or soft crystal state. This transition defines where the lattice structure of the solid breaks down, with the heat of transition being the largest in the sequence of phase transitions for a material. The second transition to consider is the clearing or isotropization point. This is where a liquid crystal phase melts to a liquid. The melting and clearing points are thermodynamic constants for a material. Conversely, the recrystallization temperature is the point at which a liquid crystal phase converts back into the solid on cooling. This is not a constant for a material because recrystallization can be subject to supercooling as the transition has a degree of kinetic character, and often nucleation on dust particles can initiate crystallization.

Liquid crystal phases that occur above the melting point are defined as being enantiotropic, whereas those formed via a transition below the melting point are called monotropic, and are usually seen via supercooling.

Reporting identities of mesophases, mesophase transition temperatures, and clearing points is achieved by thermal polarized-light microscopy, whereas the temperatures for melting and recrystallization are usually determined by calorimetry. This is because observations made by microscopy give phase identification from the defect textures, whereas melting and clearing points have problems being identified by microscopy because paramorphosis of the textures makes it difficult to define exactly when such transitions occur (discussed further in Chapter 11).

Table 5.1 shows, for example, a tabulated form for the various phase changes of five imaginary materials, compounds **1** to **5**. The clearing points have been set arbitrarily to 100°C for all five, whereas the melting points have been set to 50°C for the first four, and to 110°C for compound **5**. All of the compounds exhibit a nematic phase and two smectic phases labeled smectic 1 and smectic 2. Compound **1** has two enantiotropic smectic phases and a nematic phase, with all three occurring above the melting point on heating. Compound **2** has a monotropic smectic 2 phase, an enantiotropic smectic 1 phase, and a nematic phase. Compound **3** possesses only one enantiotropic smectic 1 phase, and a nematic phase. Compound **4** has two enantiotropic smectic phases, with the smectic 2 to smectic 1 phase transition being second order. In addition it also exhibits a nematic phase. Compound **5** melts directly from the solid to the liquid, with all of the liquid crystal phases being revealed upon supercooling to recrystallization. Thus all of the liquid crystal phases are monotropic. The associated heats of transition are shown in italics under the transition temperatures. Typical values in KJ/mol are given for first-order phase transitions, with the value for the second-order phase transition being almost negligible. Dots appear in the table when phases are present, and dashes when they are not. Bracketed values for the transition temperatures indicate that they are monotropic.

Table 5.1 Reporting scheme for transition temperatures (°C), and heats of transition (KJ/mol), for 5 calamitic liquid crystals, exemplified by nematic and two unidentified smectic phases, smectic 1 and smectic 2.

Compd	Cryst		Sm2		Sm1		N		Liquid
1	·		50	·	60	·	70	·	100	·
ΔH			*30*		*3*		*1*		*0.5*	
2	·		50	(· | 40)	·	70	·	100	·
ΔH			*30*		*3*		*1*		*0.5*	
3	·		50	–	–	·	70	·	100	·
ΔH			*30*				*1*		*0.5*	
4	·		50	·	60	·	70	·	100	·
ΔH			*30*		–		*1*		*0.5*	
5	·		110	(·	60	·	70	·	100)	·
ΔH			*30*		–		*1*		*0.5*	

Table 5.2 gives an example of a homologous family of liquid crystals

called the 4-alkyl-4'-cyanobiphenyls, where the homologs have different alkyl chain lengths. It can be seen from the table that the first four homologs (methyl to butyl) possess four monotropic nematic phases, with the next three, **5 to 7** (pentyl to heptyl) exhibiting enantiotropic nematic phases. Conversely, homologs **10 to 12** (decyl to dodecyl) exhibit enantiotropic smectic A phases but no nematic phase. Only the octyl and nonyl compounds (**8** and **9**) exhibit both nematic and smectic A phases, with the mesophases being enantiotropic.

Table 5.2 The transition temperatures (°C) for the 4-alkyl-4'-cyanobiphenyls with n indicating the number of carbons in the end chain (see the top-right inset of Figure 5.7). Monotropic transitions are in brackets. Transition temperatures determined by extrapolation are denoted with an asterisk.

Compd (n)	Cryst		Smectic A		Nematic		Liquid
1	·	109	-	-	(·	45)*	·
2	·	75	-	-	(·	22)*	·
3	·	66	-	-	(·	25.5)	·
4	·	48	-	-	(·	16.5)	·
5	·	24	-	-	·	35	·
6	·	14.5	-	-	·	29	·
7	·	30	-	-	·	43	·
8	·	21.5	·	33.5	·	40.5	·
9	·	42	·	48	·	49.5	·
10	·	44	·	50.5	-	-	·
11	·	53	·	57.5	-	-	·
12	·	48	·	58.5	-	-	·

There are some interesting observations that can be made from this table. For short aliphatic chain lengths nematic phases are observed, but for longer chain lengths there is a cross over to the smectic A phase. The melting points start off at high values, which fall to a minimum value half way across the series before increasing again. Similarly, the clearing points for the nematic to liquid transition fall to a minimum level before increasing again. Conversely, the clearing points for the smectic A phase rise and level off as the chain length is increased. Thus, it appears that there is a change in properties that are associated with the balance between the proportions of the aliphatic to aromatic segments of the molecular structures related to the degree of nanophase segregation.

One of the important experimental studies that can be done is to plot the data for the transition temperatures as a function of one variable in molecular structure, which is usually the aliphatic chain length. For the 4-alkyl-4'-cyanobiphenyls the data are shown plotted graphically in Figure 5.7.[1] This illustration of the data can be used to provide important information on issues such as chemical purity, identification of phase types, and property-structure correlations. The graph in the figure shows that the melting points behave erratically as the homologous series of compounds is traversed. Phase transitions occurring above this line, however, are enantiotropic, whereas those below the line are said to be monotropic.

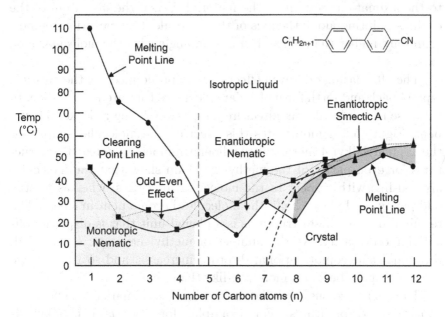

Figure 5.7 The transition temperatures plotted as a function of the aliphatic chain length for the 4-alkyl-4'-cyanobiphenyls.

Initial observations of the nematic to liquid transition temperatures might be said also to behave randomly, although less so. However, if the transition temperatures for the homologs of even number of carbon atoms in the aliphatic chain are joined up, a smooth curve re-

[1]K. J. Toyne, *Thermotropic Liquid Crystals, Critical Reports on Applied Chemistry*, G. W. Gray, ed., Society of Chemical Industry, **22**, 35 (1987).

sults. A similar observation is possible for the homologs of odd parity. This behavior is common for transitions involving nematic and smectic phases. Non-adherence of transitions to such curves usually indicates that a material is impure, even for chemical purities higher than 95%. Looking carefully at the curves for odd and even homologs indicates that there is an odd-even alternation in transition temperatures. For the 4-alkyl-4′-cyanobiphenyls the curve for the homologs of odd parity in carbon chain length appears to be higher in temperature than even parity. When there is an alkoxy chain, an even parity in the carbon chain length (plus the oxygen atom) gives clearing points of higher temperature. This relationship also applies to any divalent atom located between the rigid core and the carbon chain. This relationship is simply connected to the number of atoms in the substituent attached to the aromatic core unit of the material. When the last atom of the chain is pointing along the axis of the molecule, the transition temperatures are higher than when it makes an angle with the molecular long axis.

The alternating odd-even effect is more pronounced as the parity is flipped back and forth for molecular structures that are less rod-like. In this case the molecular architecture jumps from being rod-like to being bent. Figure 5.8 demonstrates this effect on the molecular shapes for the ω-phenylalkyl 4-methoxybenzylideneaminocinnamates. The structures for even parity of the methylene chain show that the molecules are rod-like with respect to the molecular long axes, whereas for the odd parity methylene chain, the molecules become bent in shape. The terminal unit in these structures is a phenyl unit that is displaced off-axis for odd parity. As the number of methylene units is increased, rotational and conformational disorder increases, and so the average structure again becomes more rod-like than bent.

The bent versus rod-like shape has a remarkable effect on transition temperatures. For example, for the ω-phenylalkyl 4-methoxybenzylideneaminocinnamates the differences in the liquid to nematic transition temperatures can be over 200°C as demonstrated in Figure 5.9.[2] For this family of materials only the liquid to nematic and the melting point transitions are plotted in the figure. The liquid to nematic transitions fall dramatically as the number of methylene units of even parity is increased, whereas those for odd parity remain fairly stable. This is because the molecular shape does not change too

---

[2]K.J. Harrison, *Ph.D. Thesis*, University of Hull (1972).

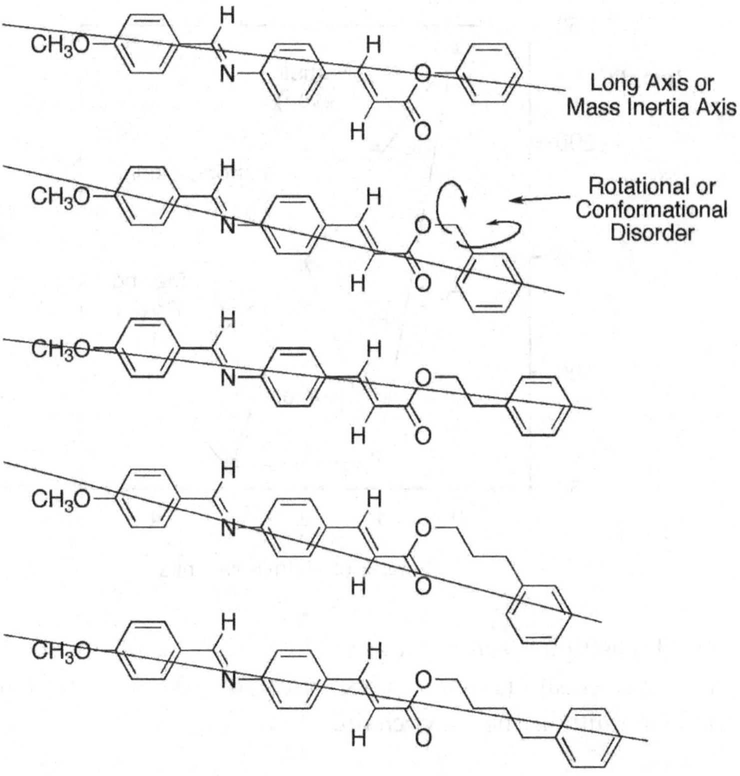

Figure 5.8 Molecular structures of the ω-phenylalkyl 4-methoxybenzylideneaminocinnamates as the methylene chain is increased in length and the parity is alternated.

much for the odd homologs, whereas for the even members the increase in the rotational and conformational disorder induces the molecules to be less rod-like.

The odd-even phenomenon also occurs for other phase transitions in calamitic systems, but the alternations are not necessarily in the same sense. For example, for the hexatic and crystal B to smectic A transitions, the alternation is inverted with the even parity being higher than the odd.

Overall, the property-structure correlations are particularly useful in the selection of homologs to use in the design of mixture formulations destined for applications of liquid crystals in displays, for which the widest temperature range of operation is required.

Returning to the graph in Figure 5.7, it shows that the homologous

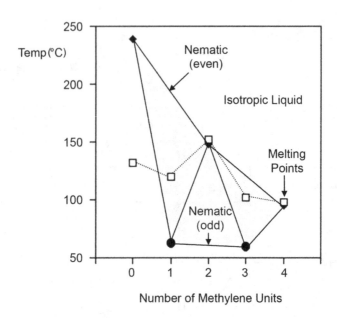

Figure 5.9 Transition temperatures for the ω-phenylalkyl 4-methoxybenzylideneaminocinnamates plotted as a function of the methylene units in the alkyl chain.

series for the cyanobiphenyls possesses two liquid crystal modifications, nematic and smectic A. The smectic A phase appears to occur at lower temperatures for homologs that exhibit the two phases. Moreover, the first of the lower homologs to exhibit the smectic A phase also determines the last homologs not to exhibit the A phase. This means that the curve for the nematic to smectic A transitions must start at a temperature lower than the recrystallization temperature of the last homolog not to exhibit the smectic A phase. This may seem obvious but the injection of phases into the homologous series determines the order of the phase stabilities for the calamitic state. For example, the order of mesophase stability falls into the following sequence upon cooling:

N → SmA → SmC → Hexatic B → Hexatic I → Crystal B → Hexatic F → Crystal J → Crystal G → Crystal E → Crystal H → Crystal K.

This means that in the graphical representation in Figure 5.7, the smectic A phase cannot appear at temperatures above the nematic phase, and therefore the sequence from the liquid should always be liquid to nematic to smectic A. This readily apparent observation can become quite complicated when multiple phases converge upon one another.

One anomaly can arise with the above analyses when mesophase reentrancy is possible. Reentrancy occurs when there is a transformation from disorder to order and back to disorder upon cooling or heating. For example, in special circumstances the nematic phase can form a smectic A phase on cooling, and then it is possible to return to the nematic phase on further cooling. This behavior is related to dynamic pairing of the molecules (with the aliphatic chains extending in opposite directions) changing as a function of temperature. At higher temperatures the molecules are proportionally unpaired in both the nematic and smectic A phases. As the temperature is lowered further, however, the increased pairing tends to create more dimers that are more bulky in the middle than near the ends. This destabilizes the layer organization and results in a return to the nematic phase.

5.5 CHEMICAL MOIETY SELECTION

For organic and organometallic materials with rod-like molecular shapes, the prototypical molecular design to achieve mesomorphism involves the incorporation of a central aromatic, heterocyclic, or alicyclic core unit to which are attached terminal aliphatic chains, thereby engendering dichotomous structures with rigid or semi-rigid sections surrounded or segregated by flexible fatty chains (see structures **1**, **2**, and **5** in Figure 5.10). When molecules with this type of architecture self-organize, they generally do so with their rigid, aromatic parts tending to pack together and their flexible/dynamic aliphatic chains orienting together, *i.e.*, "like packs with like", but it is important to realize that the relative packing interactions may be quite different in strength. For example, aliphatic chains pack via weak van der Waals interactions, whereas aromatic regions are more polarizable and the stronger interactions may be electrostatic and/or dipolar in nature. Thus the nano-segregation can exist as a dichotomy for a range of combinations, such as rigid/flexible, polarizable(π)/nonpolarizable(σ), polar/non-polar, H-bonding/non-H-bonding, hydrocarbon/fluorocarbon, ionic/non-ionic, *etc.* In fact, many liquid crystals have molecular architectures that are trichotomous (**2** and **4** in Figure 5.10), and some are even polychotomous.

Polar groups such as nitrile (cyano) or fluoro (or halogeno) (see structures **3**, **4**, and **6** in Figure 5.10) can also be important with re-

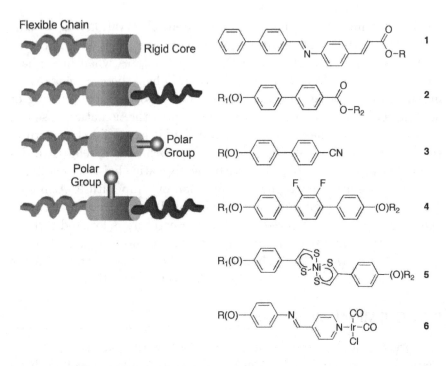

Figure 5.10 Simple design features for organic and organometallic calamitic liquid crystal materials.

spect to lateral and longitudinal electrostatic interactions. The overall system becomes "locally microphase (or nanophase) segregated" with respect to allowable polar interactions and also steric packing. Consequently, over many years, the main target of material design has been, by default, the variation in the structure of the central core region of the molecule in the belief (although not strictly true), that the core is more important in influencing mesophase incidence, mesophase temperature range, isotropization point, melting point, mesophase sequence, dielectric and optical anisotropy, elastic coefficients, and the reorientational viscosity associated with the mesophase. Thus for the synthetic chemist, rational design for fundamental science and commercial applications can become a problem of nanoscale structural engineering, with the prospect of making materials that are exceptionally pure. (For electronic devices, purity is measured in resistivity rather than by traditional chemical/analytical methods.)

The common types of "molecular block" used in the construction of calamitic mesogens are shown schematically in Figure 5.11. The blocks can be joined together in various forms to give general molecular architectures, described by the pseudonym RAZAZAR.

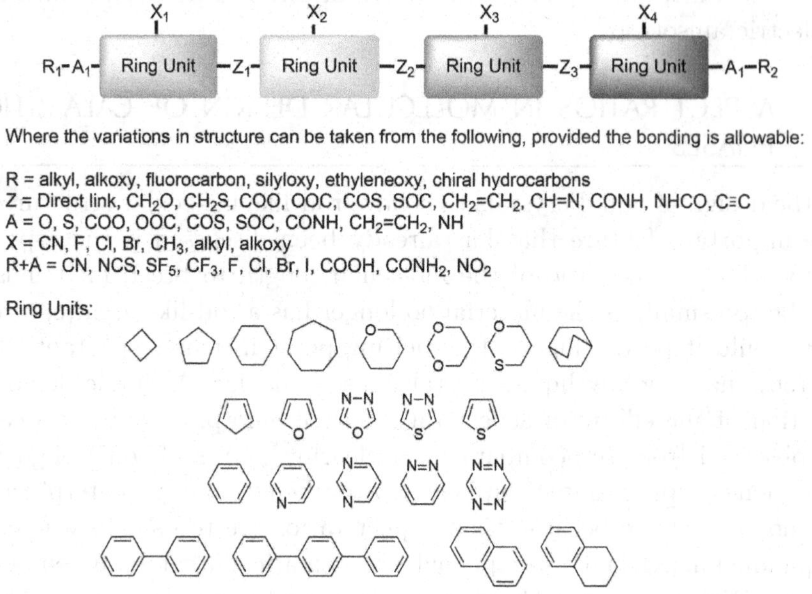
X1
X2
X3
X4
R1–A1– Ring Unit –Z1– Ring Unit –Z2– Ring Unit –Z3– Ring Unit –A1–R2
Where the variations in structure can be taken from the following, provided the bonding is allowable:
R = alkyl, alkoxy, fluorocarbon, silyloxy, ethyleneoxy, chiral hydrocarbons
Z = Direct link, CH2O, CH2S, COO, OOC, COS, SOC, CH2=CH2, CH=N, CONH, NHCO, C≡C
A = O, S, COO, OOC, COS, SOC, CONH, CH2=CH2, NH
X = CN, F, Cl, Br, CH3, alkyl, alkoxy
R+A = CN, NCS, SF5, CF3, F, Cl, Br, I, COOH, CONH2, NO2
Ring Units:

Figure 5.11 Molecular building blocks for the creation of calamitic liquid crystal materials.

The structure in the figure shows four systems, associated with rigid or quasi-rigid aromatic, heterocycle, or cyclohexanyl rings (there may be more or less), but typically most calamitics have two to four rings in their central units. The exterior chains (R) are usually aliphatic, but can have incorporated non-carbon atoms such as oxygen. Aliphatic chains can be linear or branched or even chiral via the incorporation of one or more stereogenic centers. Linking groups (Z) between the rigid core units can take a number of forms, however, they must maintain the overall linearity of the molecular structure. The linking group can also be used to provide conjugation through the π-system between the core units, thereby allowing for polarizability, or alternatively they can be used to break up the conjugation, and in some cases extend the σ-bonded system. The connecting groups between the aliphatic chains and the rigid core units (A) can be used as donor and acceptor groups, which can be employed to increase or decrease the overall molecular polarizability. Lateral groups (X) and terminating groups (R+A) can be polar or apolar, although typically they are used in their polar forms to control mesophase formation and physical properties such as dielectric anisotropy.

5.6 ASPECT RATIOS IN MOLECULAR DESIGN OF CALAMITIC PHASES

In the design of liquid crystals that have rod-like molecular structures, one important feature that has already been discussed is the aspect ratio, which is the ratio of the molecular length to breadth. It must not be too small, as the material no longer has a rod-like architecture, nor should it be too big, as the melting point increases substantially thereby masking any liquid crystal phase behavior. A simple demonstration of the effects of aspect ratio on mesomorphic properties can be observed from the comparison of clearing points of some oligo-*p*-phenylenes. For example, biphenyl, *p*-terphenyl, and *p*-quaterphenyl are not mesogenic because their aspect ratios are too small, whereas *p*-quinquephenyl and *p*-sexiphenyl are nematic but at high temperatures. Higher polyphenyls tend to decompose near their transitions to the liquid state, although there have been some reports that septiphenyl exhibits liquid crystallinity. Thus, the larger aspect ratios for the higher polyphenyls are too large for mesomorphic behavior to be observed.

Let us focus on *p*-sexiphenyl and *p*-quinquephenyl, shown together in Figure 5.12. Both are rod-like, with adjacent rings twisted with respect to one another because of the steric interactions of the *ortho*-hydrogens, as shown by the space-filling structure in the figure for *p*-sexiphenyl. This twisting is random and can be to the left or to the right, and therefore the molecules can have numerous conformational structures, which interchange between one another quite rapidly through relative rotations of the rings. Cumulatively, therefore, the molecules do not have chiral structures. *p*-Sexiphenyl is found to have a clearing point that is 85°C higher than that of *p*-quinquephenyl. In addition, *p*-sexiphenyl also exhibits a smectic A phase, which has a lamellar structure. Thus, because of the greater aspect ratio, *p*-sexiphenyl has more stable mesomorphic properties than *p*-quinquephenyl.

Through the introduction of lateral groups into sexiphenyl, we can

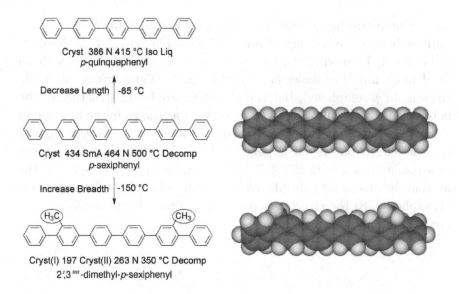

Cryst 386 N 415 °C Iso Liq
p-quinquephenyl

Decrease Length | -85 °C

Cryst 434 SmA 464 N 500 °C Decomp
p-sexiphenyl

Increase Breadth | -150 °C

Cryst(I) 197 Cryst(II) 263 N 350 °C Decomp
2′,3″‴-dimethyl-*p*-sexiphenyl

Figure 5.12 The transition temperatures and melting points for *p*-quinquephenyl, *p*-sexiphenyl, and 2′,3″‴-dimethyl-*p*-sexiphenyl, and the space filling images for the minimized structures of *p*-sexiphenyl and 2′,3″‴-dimethyl-*p*-sexiphenyl.

effectively reduce the aspect ratio simply by increasing the molecular breadth. Consider compound 2′,3″‴-dimethylsexiphenyl, which also has a rigid structure because it has two methyl groups directly attached to the polyphenyl unit. In the space-filling structure, with both methyl groups positioned on the same side of the polyphenyl unit, it can be seen that the aspect ratio is reduced because the breadth of the molecule is larger. This time the clearing point is lowered relative to *p*-sexiphenyl by 150°C.

Conversely, it has been suggested that a twisted arrangement for 2′,3″‴-dimethylsexiphenyl gives better steric (lock and key) fit while minimizing π-electron repulsion between the rings. However, given the temperature range for the nematic phase from 263 to 350°C, it is highly likely that the molecules are rotating and gyrating rapidly, and a lock and key fit, although possible, is unlikely. Instead it is more likely that the rotational volume coupled with the minimization of the free volume determines the formation and structure of the mesophase observed.

Similar comparisons can be made between polyphenylenes and polyphenyls where one or two of the phenyl rings are replaced with a heterocyclic system. For example, Figure 5.13 shows a comparison between p-sexiphenyl and 4,4'-bis(5-phenyl-1,3,4-oxadiazol-2-yl)biphenyl, **7**, in which two of the phenyl rings are replaced with an oxadiazole unit. The molecular structure of **7** still retains six aromatic rings as in p-sexiphenyl, but the five-membered ring structure of the heterocycle means that the structure is bent as shown in the figure, and the aspect ratio is again reduced in comparison, due to the slightly shorter length but much larger breadth. The clearing point for the oxadiazole analog is thus 188°C lower than for p-sexiphenyl. For the oxadiazole analog of septiphenyl (1,1'-bis(5-phenyl-1,3,4-oxadiazol-2-yl)terphenyl, **8**) the clearing point is 444°C, for which the parent may not even be liquid crystalline.

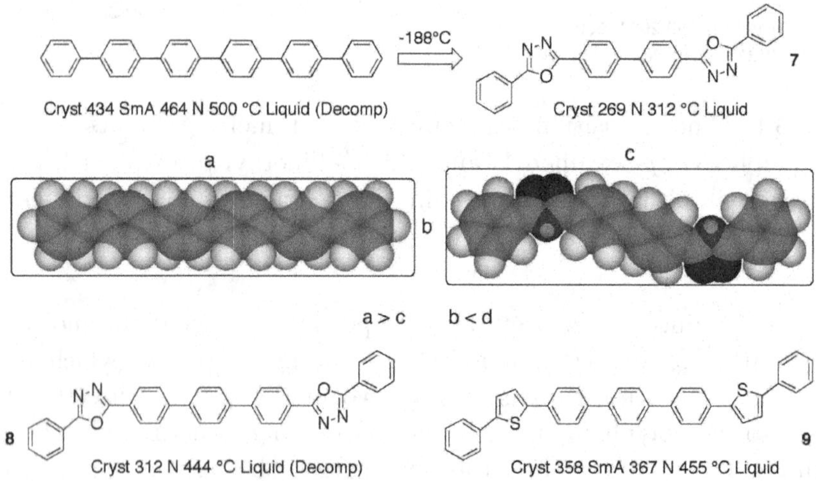

Figure 5.13 The transition temperatures and melting points for p-sexiphenyl, 4,4'-bis(5-phenyl-1,3,4-oxadiazol-2-yl)biphenyl, **7**, 1,1''-bis(5-phenyl-1,3,4-oxadiazol-2-yl)terphenyl, **8**, 1,1''-bis(5-phenyl-thiophen-2-yl)terphenyl, **9**, and the space-filling images for the minimized structures of p-sexiphenyl and compound **8**.

Similarly, the analogous material, compound **9** (1,1''-bis(5-phenyl-thiophen-2-yl)terphenyl), has the oxadiazole rings replaced by thiophenyl units, resulting in different electrostatic and polarizable interactions, and has an even lower clearing point of 367°C. This indicates that not only does the change in the aspect ratio affect the liquid crystal properties, but the electrostatic interactions, which are primarily dipolar, also contribute.

5.7 CONJUGATION, POLARIZABILITY, AND SUPRAMOLECULAR ASPECT RATIOS

Three of the more common building blocks used in the development of liquid crystals, particularly for applications, are the nitrile (cyano), phenyl, and cyclohexyl units. They are nanosegregating units because they have the respective structural properties: polar(π), rigid-polarizable(π), and semirigid-non-polarizable(σ). Coupled with flexible aliphatic chains, the resultant materials form the basis of liquid crystal mixtures found in many electronic devices from watches to phones and televisions. In combinations of {σ-bonded}-{π-bonded}-{polar} systems, these materials possess differences in aspect ratios, which are used to provide mixture formulations that allow devices to operate over wide temperature ranges, including temperatures found in Siberia to those in Death Valley.

The most famous materials derived from this design combination are the alkyl and alkoxy cyanobiphenyls, as exemplified in Figure 5.14 for 4-pentyloxy-4'-cyanobiphenyl, commonly known as 5OCB or M15. For 5OCB the core biphenyl has delocalized π-electrons, and at one end there is an electron donor unit in oxygen, and at the other end there is an electron acceptor in the form of a nitrile group. Thus, the unit has a donor-acceptor, "push-me, pull-you" core. The delocalization of a pair of electrons through the π-system is demonstrated by the structure in the top right of the figure (using curly arrows to give resonance structures and polarizability). The antiparallel arrangement of the molecules, with the donor and acceptor units of adjacent molecules overlaying one another, stabilizes the structure via induced quadrupoles as shown. In doing so the length of the paired system is effectively 1.4 times longer than the individual molecule. Thus the effective aspect ratio is higher for the paired in comparison to the unpaired system. This of course does not mean that all of the molecules in the nematic phase of 5OCB are paired; it simply means that in the dynamical nematic phase, at any one time there is a reasonable number of molecules that are paired.

For the alkyl-4'-cyanobiphenyl family, see Figure 5.15, if we replace one of the phenyl rings in the biphenyl component with a cyclohexyl ring, we could do this on either the aliphatic side or the nitrile side of the unit. As the cyclohexyl ring is σ-bonded, if the replacement is performed adjacent to the nitrile the conjugation of the π-system is broken, thereby increasing the number of nanosegregative entities and

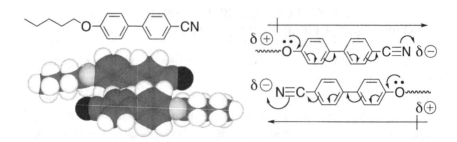

Figure 5.14 The chemical structure 4-pentyloxy-4'-cyanobiphenyl, and its polarizable behavior shown as the movement of a pair of electrons through a π-conjugated system. The antiparallel arrangement of the molecules is stabilized by quadrupolar interactions.

the complexity of the structure. Replacement adjacent to the aliphatic chain reduces the complexity and the number of segregative components. If the aliphatic chain is five methylene units long, then the material created is 4-(*trans*-4-pentylcyclohexyl)benzonitrile, commonly known as PCH5, and one of the most important materials employed in early LCDs. Its clearing point for the nematic to liquid transition is 55°C, which is 20°C higher than the biphenyl equivalent 4-pentyl-4'-cyanobiphenyl, 5CB.

Replacement of the last phenyl ring produces the material *trans*-4-(*trans*-4-pentylcyclohexyl)cyclohexylcarbonitrile, which is called CCH5. This material has almost an entire σ-bonded structure without polarizability, and a clearing point of 85°C. Why therefore should a material with potentially weaker intermolecular interactions than PCH5 or 5CB have a higher clearing point than either of the other two materials? Examination of the possible pairing of the molecules in PCH5 shows that the overlap is via one phenyl ring, as shown in Figure 5.15. Thus the effective aspect ratio increases relative to 5CB. For CCH5 the electrostatic interactions overlap the nitrile groups, and again the aspect ratio is higher than for either PCH5 or 5CB. Therefore, one would predict that the clearing points would have the following sequence: 5CB < PCH5 < CCH5, which is observed experimentally.

The structural complexity for these materials is relatively small, and there is nanophase segregation over reasonable length scales with respect to the molecules. However, by ordering the individual π-, σ-, and dipolar (μ) components in a more random way, the complexity in-

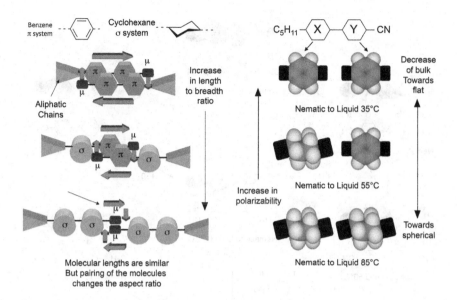

Figure 5.15 Three related materials, from top to bottom, 5CB, PCH5, and CCH5. The left-hand structures show the relative distribution of π- and σ- units, and the quadrupoles that are formed when the molecules pair. The right-hand column shows the shapes and chemical details of the makeup of the core units. In all of the figures, the wedge-shaped component is a pentyl chain.

creases. In Figure 5.16, compounds similar to 5CB, PCH5, and CCH5 are shown, along with their relative clearing points. It can be seen that the more complex the electronic structure is relative to polychotomy, the lower the clearing point. Again σ-bonded systems tend to have higher clearing points, and where there is a good fit in the steric packing, the clearing points are often higher still. For example, when there is a cyclohexyl unit adjacent to a bicyclooctanyl component, the clearing point is higher than when two cyclohexanyl segments are neighboring. The bicyclooctane unit, as shown in the bottom chemical structure of the figure, is almost spherical, whereas the cyclohexyl part is flatter. Juxtaposed packing of the two between molecules minimizes the free volume, and stabilizes the overall structure of the mesophase. Thus we can see that the complexity of the structure of a material in terms of nanosegregation can be detrimental to the formation of calamitic mesophases.

The breaking of the π-conjugation can affect mesophase formation

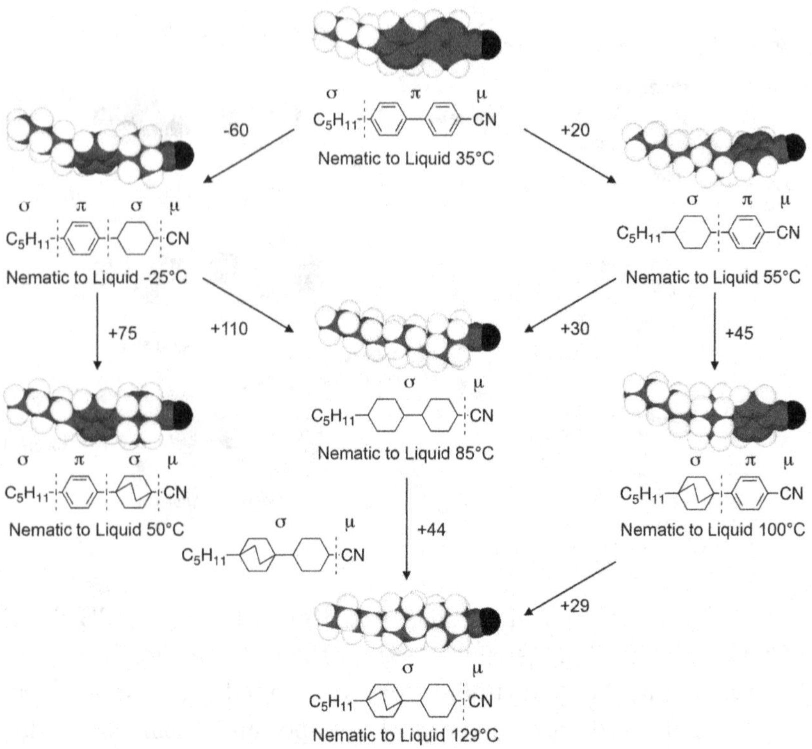

Figure 5.16 Comparison of nematogens that are composed of nitrile, cyclohexyl, bicyclohexyl, and phenyl components. The materials also possess a flexible pentyl aliphatic chain. The space filling minimized structures are given for each material.

and stability, suppressing some mesophases in favor of others. For example, following the descriptions above, consider now separating the polar group from the rigid core system. Separation breaks the conjugation, and suppresses any form of donor/acceptor activity, and flexibility introduces rotational freedom to the polar group. For the preceding materials this essentially means separating the polar nitrile group from the rigid core system. Figure 5.17 shows a comparison between two pairs of materials for which the separation induces the formation of smectic phases relative to the nematic phase, but with inconsistent changes in transition temperatures.

Inserting an ethylene spacer between the nitrile group and a dicyclohexyl core for the upper pair in the figure shows that a smectic phase is formed with a clearing point 24°C higher than that of CCH5. For the equivalent PCH5 comparison, the ethylene inserted system again has a smectic phase, but this time the clearing point is 27°C lower than the parent CCH5. A way of rationalizing this result is as follows: for extended σ-bonded systems, as with the other materials discussed previously, higher clearing points are favored, whereas for the more complex structure of $\sigma - \pi - \sigma - \mu$ units, nematic phases are favored over smectic.

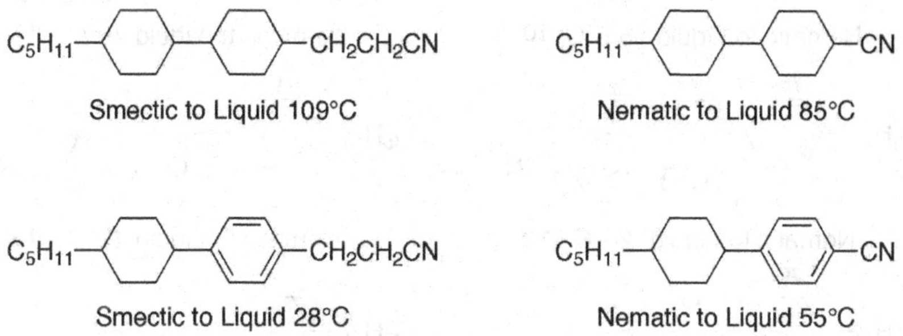

C$_5$H$_{11}$—⟨ ⟩—⟨ ⟩—CH$_2$CH$_2$CN

Smectic to Liquid 109°C

C$_5$H$_{11}$—⟨ ⟩—⟨ ⟩—CN

Nematic to Liquid 85°C

C$_5$H$_{11}$—⟨ ⟩—⟨ ⟩—CH$_2$CH$_2$CN

Smectic to Liquid 28°C

C$_5$H$_{11}$—⟨ ⟩—⟨ ⟩—CN

Nematic to Liquid 55°C

Figure 5.17 Comparison of materials for which a polar group is directly attached to the central core with materials for which the polar group is separated by an ethylene linking group from the core.

A similar pattern in transition temperatures, and hence thermal stability of the nematic phase, can be found with the groups that connect the π-systems in the central core units of mesogens. Figure 5.18 demonstrates this effect with respect to ethylene (CH$_2$CH$_2$), methoxy (OCH$_2$), and ester (COO) linking groups. The ethylene group is σ-bonded and therefore does not conjugate π-electrons between adjacent aromatic rings, the methoxy group donates electrons through the π-system to the aromatic adjacent to the oxygen atom, the ester group is conjugated through the aromatic ring to the carbonyl moiety of the ester, and the oxygen of the ester is conjugated to the other aromatic ring, so this linking group has a higher degree of conjugation with respect to the other two. The figure shows that the materials with the best distribution of microphase segregating groups, e.g., compound 11, have the highest clearing points. Next comes the material with the most extensive conjugation, compound 10. The material with the worst conjugation and distribution, compound 14, comes last, following closely compound 12, which has a slightly larger conjugated region with a donor/acceptor arrangement. Compound 13 and 15 have very similar clearing points, but this time the donor/acceptor affect is slightly detrimental to the clearing point as shown for compound **13**. This effect could also be attributed to the relative sizes of the segregating units, and to the strong polarity of oxygen.

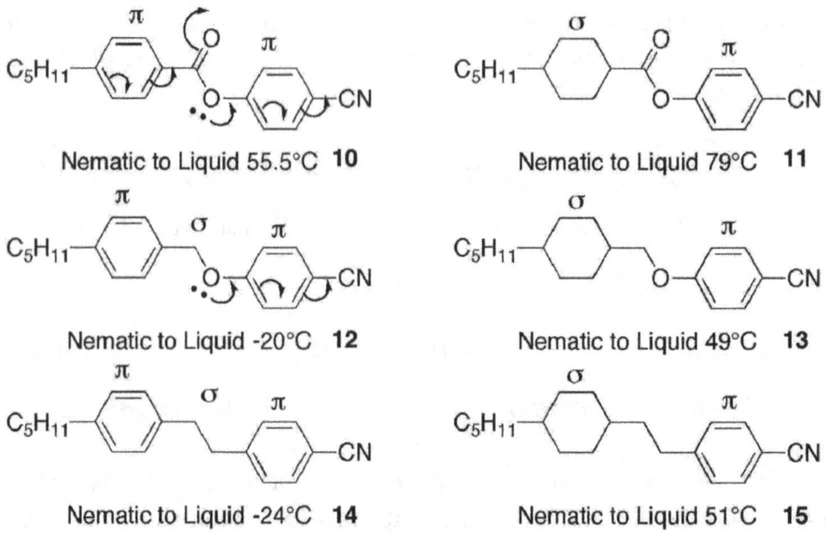

Figure 5.18 The effect of the linking group on the clearing points of a family of nematogens. The curly arrows and pairs of electrons (double dots) show the possible movement of π-electrons.

5.8   LATERAL INTERACTIONS AND BIAXIAL PHASES

In Section 5.5 the lateral breadth to longitudinal length ratio was discussed in terms of the thermal stability of the calamitic nematic phase. If the length to breadth ratio approaches the value of one, the molecules have square shapes, or effectively they are disk-like when all orientations of the molecules are considered. At some point along this path, the length to breadth ratio reaches a point where the molecules have restricted rotation about their long axes. For materials with rod-like or lath-like molecular shapes, their ability to rotate about their long axes coupled with disordering of the short-axes along the director means that there is a C_∞ rotational axis parallel to the director. Conversely, for restricted rotation, the symmetry will be reduced to C_2, assuming that there are as many molecules pointing up as down, or left to right of the director. If the organization of the molecules occurs in the macro-range, the reduced symmetry means that the nematic phase becomes biaxial rather than uniaxial (see Figure 5.19). Consequently, there is also a phase transition between the biaxial and uniaxial variations of the nematic. Furthermore, biaxiality means that the nematic phase has different properties from the calamitic nematic phase; for example, the biaxial nematic phase possesses two directors ($\hat{n}$ and $\hat{m}$) and three optic axes.

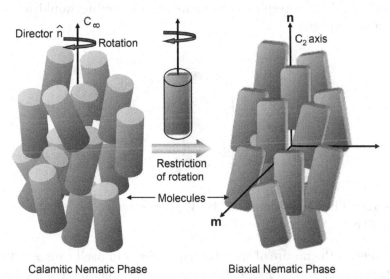

Calamitic Nematic Phase   Biaxial Nematic Phase

Figure 5.19 The structures of the uniaxial calamitic nematic phase (left) and the biaxial nematic phase (right). In both cases the molecules have board-like shapes, but for the calamitic phase the molecules rotate or oscillate relatively freely about their long axes, whereas in the biaxial phase the molecules are restricted. Long-range ordering of the molecules means that the nematic phase is biaxial.

In the design of materials that might exhibit the biaxial phase, there are requirements on the length to breadth to width ratio ($L : B : W$), as shown in Figure 5.20. Theoretical modeling defines an intrinsic biaxiality of shape for molecular biaxiality as λ, which is expressed using the $L : B : W$ ratio for a given molecular architecture as follows:

$$\lambda = \sqrt{\frac{3}{2}} \frac{L(B-W)}{[L(B+W)-2BW]}. \tag{5.1}$$

Eq. 5.1 shows for a rod-like molecular structure, where $B = W$, λ is zero and there is no molecular biaxiality. For disk-like molecules where $L = B$, $\lambda = \sqrt{3/2}$. Between these two uniaxial extremes, there is the possibility for a molecular architecture that has a biaxial shape. Theoretically, the optimum molecular biaxiality has been shown to be a $L : B : W$ ratio of 5:3:1, which corresponds to $\lambda = 1/\sqrt{6}$. For real molecular structures such as a simple aromatic system, the minimum width value W is 0.45 nm, which gives L= 2.2 nm and $B = 1.3$ nm. For a length value L, that corresponds to sexiphenyl, and the breadth value B corresponds to a terphenyl unit, $i.e.$, the structure would mimic a ribbon of graphene.

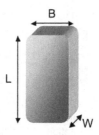

Figure 5.20 The dimensions (L, B, W) for a molecular shape that is board-like.

However, the nature of this size requirement in itself causes difficulties because large rigid molecules have very high melting points. Incorporating flexible chains, while lowering melting points, could also lead to microphase segregation, but this tends to cause self-organization in favor of lamellar or columnar phases. As a consequence, a large number and a wide variety of materials with board-like molecular architectures have been prepared in search of biaxiality, but with no success. The top three compounds in Figure 5.21 are typical examples of materials that have board-like molecular structures, but have been found to exhibit calamitic uniaxial nematic phases rather than biaxial phases. Consequently, there have been many false starts with materials having molecular structures that are X-shaped, T-shaped, board-like twins, rods attached to disks, $etc.$, being claimed to be biaxial, and then found not to be so. Only one material system has been found that has properties that hint at being due to biaxiality, and this is the bent core structures based on oxadiazoles, as exemplified by the bottom compound in Figure 5.21.

The bent core systems have the length to breadth ratio that is re-

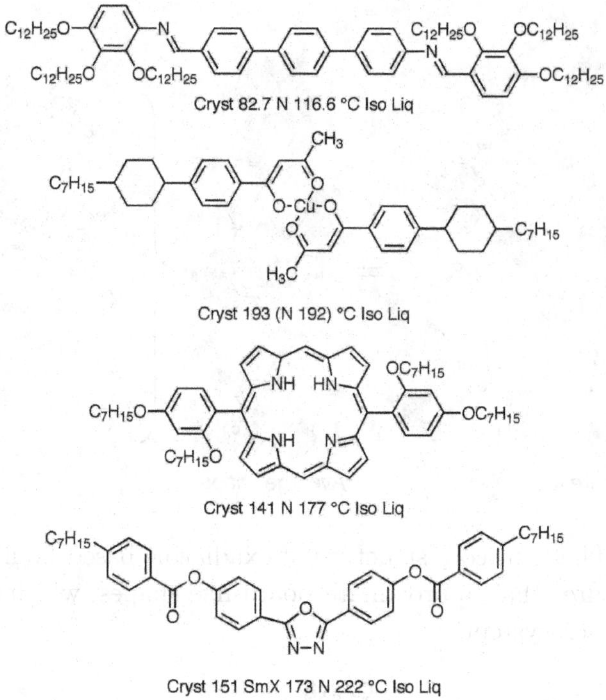

C12H25O
C12H25O
OC12H25
C12H25O
OC12H25
OC12H25
Cryst 82.7 N 116.6 °C Iso Liq
CH3
C7H15
O-Cu-O
C7H15
H3C
Cryst 193 (N 192) °C Iso Liq
NH HN
OC7H15
C7H15O
OC7H15
NH N
C7H15O
Cryst 141 N 177 °C Iso Liq
C7H15
C7H15
N–N
Cryst 151 SmX 173 N 222 °C Iso Liq

Figure 5.21 Molecules with board-like architectures and low aspect ratios. The top three compounds exhibit calamitic nematic phases. The bottom material has been reported to exhibit weak biaxiality.

quired for biaxiality, and because of their relatively open structures, they have low enough melting points and thermal mesogenic stability to exhibit the nematic phase. Figure 5.22 shows the bent-core structure of an oxadiazole. Rather than thinking of it having a bent shape, if both orientations are considered together, the average shape of the combination is board-like. Packing the molecules together in a macrosystem is then biaxial. Retention of this structure probably requires that the molecules strongly interact via dipolar couplings, which is possible through the oxadiazole moieties. The major problem with such materials forming biaxial phases is that the free volume is large and higher densities for molecular packing are required. Therefore the structure of a potential biaxial nematic phase, as shown in Figure 5.23, needs to be one in which the molecules do not dynamically wobble too much in relation to the horizontal plane, as shown by $\hat{l}$ in Figure 5.23, in comparison to the $\hat{m}$ and $\hat{n}$ directors.

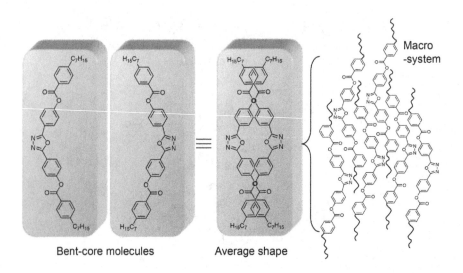

Figure 5.22 The bent-core structure of oxadiazole-based liquid crystals, with structures that approximate board-like shapes, which then form a biaxial macrosystem.

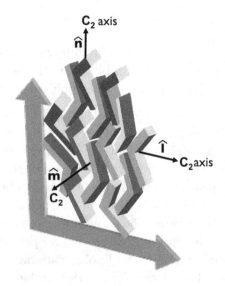

Figure 5.23 The biaxial nematic phase formed by bent-core molecules. The alternate packing of the $\hat{l}$ direction allows for the free volume to be minimized, whereas restriction of rotation along the $\hat{l}$ and $\hat{n}$ directors ensures that the structure is biaxial.

5.9 EFFECT OF ASPECT RATIOS AND LAYER FORMATION

Earlier discussions on aspect ratio focused on the comparison of the thermal stability of the nematic phase for p-sexiphenyl and p-quinquephenyl, shown together in Figure 5.12. In addition to exhibiting a nematic phase, p-sexiphenyl has a smectic A phase, which has a layered structure. In contrast, p-quinquephenyl does not possess a smectic phase even though its melting point is 48°C lower. Further, compound $2',3''''$-dimethylsexiphenyl, which has a rigid rod-like structure like the other two materials, also has two methyl groups attached to the core that reduce the lateral interactions and suppress smectic A formation. These simple comparisons suggest two things: first, the lateral interactions between the molecules for p-sexiphenyl are greater than those of p-quinquephenyl and $2',3''''$-dimethylsexiphenyl, and second, there is a "quasi-lattice" energy to be gained from the formation of layers. As a consequence, greater aspect ratios and tighter lateral packing of the molecules favor the formation of lamellar phases.

As with nematic phases, the next issue to consider is nano-(or micro-)phase segregation. Figure 5.10 shows some simple design characteristics for mesogens that exhibit calamitic mesophases. In all four structures, aliphatic chains are shown attached to the termini of rigid core structures. When the aliphatic chains are long in comparison to the cores, they start to dominate the molecular packing, and hence lamellar phases become more prevalent. This is clearly demonstrated in Figure 5.7 for the 4-alkyl-4'-cyanobiphenyls, where smectic A phases appear first for the octyl homolog and subsequent members of the series.

The 4-alkyl-4'-cyanobiphenyls have a dichotomous, segregated structure consisting of a chain and a rigid core unit, and a smectic phase appears as a layered structure in which the long axes of the molecules are on average orthogonal to the layer planes. The layers themselves are soft and diffuse, and the director of the phase is perpendicular to the layer planes. This soft organization means that the system has one-dimensional sinusoidal ordering as described in Chapter 2.

For a trichotomous segregated system we have two possibilities, either a [chain]-[rigid-unit]-[chain] or a [rigid-unit][chain]-[rigid-unit] structure. The first architecture has the longest history and therefore is the most common, whereas the second design has been utilized for only about the last 30 years. The first option gives property-structure correlations that are generally applicable for layered phases. For at least one short aliphatic chain, orthogonal phases are often found, similar to the smectic A phases of the 4-alkyl-4′-cyanobiphenyls, but sometimes in addition orthogonal hexatic B, crystal B, and crystal E phases are found as one or multiple combinations. When more than one lamellar phase is present, this is termed polymorphism. When the two chains are relatively long ($\sim C_6$ to C_{10}), tilted phases, usually smectic C, are injected into the phase diagram for families of materials. Figure 5.24 shows a typical phase diagram for a family of materials, in this case the alkyl 4′-octyloxybiphenyl-4-carboxylates, commonly known as the nmOBCs, where n is the ester chain length, m is the ether chain length, and BCO is Biphenyl-Oxy-Carboxylate. As previously observed, the small n value generates orthogonal phases, whereas for larger values of n, tilted phases appear in the form of smectic C. Looking at the phase diagram for the octyl family of materials, plotting the value of n on the x-axis and the temperature (°C) on the y-axis, it can be seen that for low values of n, orthogonal smectic A, hexatic B, and crystal E phases are formed. When the value of n reaches four, a tilted smectic C phase appears, with the smectic A phase being retained and the B and E phases being lost. As n is increased the tilted phase disappears, leaving only the smectic A phase.

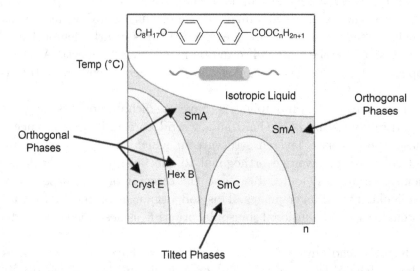

Figure 5.24 Transition temperatures and phase ranges for the smectic phases of the alkyl 4′-octyloxybiphenyl-4-carboxylates, n8OBCs, for which n is the aliphatic chain length (1 to 12 carbon units long).

The reasons the molecules tilt over in their layers in the middle of the phase diagram can be rationalized by considering the potential electrostatic intermolecular interactions, which are allowed by the steric shapes of the molecules. The shapes of the molecules become more important because the density of packing of the molecules in smectic phases is slightly higher than for the nematic phase, and hence the free volume is less. This means that the molecules cannot any longer be treated as simple rods or laths. Consider the series of the alkyl 4'-octyloxybiphenyl-4-carboxylates; for small alkyl chains the molecules need to pack tightly in layers in which there are as many molecules pointing up as down. This means there are π-π, σ-σ, and π-σ interactions, and the molecules have to share interpenetrated space. In the center of the phase diagram the molecules tilt in their layers. This is achieved by staggering the molecules relative to one another, like the stacking of chairs. At longer chain lengths only the smectic A phase is observed as the chains suppress the sharing of space, and the molecules become more like rolling pins. A schematic view of the packing of the molecules in the three stages is shown in Figure 5.25. This observation demonstrates the subtlety of molecular shape in determining mesophase formation in layered structures.

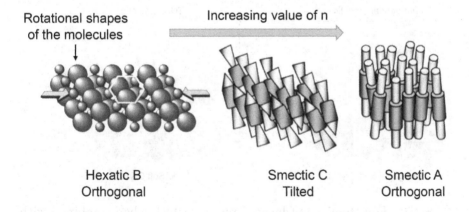

Rotational shapes of the molecules

Increasing value of n

| Hexatic B | Smectic C | Smectic A |
| Orthogonal | Tilted | Orthogonal |

Figure 5.25 The gross shapes and packing arrangements of the molecules in the mesophases of the alkyl 4'-octyloxybiphenyl-4-carboxylates.

As described in Chapter 2, the condensation of smectogens on cooling from the isotropic liquid can result in a number of possible phases being formed via systematic changes in the periodic ordering of the molecules. For the various structures of smectic phases, there is a natural division between those that have long-range periodic ordering and those that have short-range ordering. In this respect the orientation of the molecules relative to the layers, orthogonal or tilted, can be added. Therefore, four differing families of mesophase types can be readily identified, as shown schematically in Figure 5.26. There are tilted and orthogonal groups for smectic phases, which have short-range ordering of the molecules, and tilted and orthogonal soft crystal phases that have long-range ordering. The details of the local organizations of the molecules within the layers are used to determine the mesophase classification, and the finer points of their structures can be found in Figure 2.20. This collection of phases constitutes the smectic state of matter.

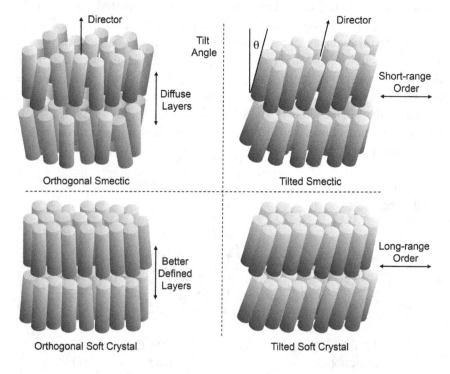

Figure 5.26 Four families of layered phases, which when combined with the hexatic analogs, represent the smectic state of matter.

Some smectogens are found to exhibit only one mesophase, but many others are polymorphic. In some cases smectogens can possess both smectic and soft crystal phases, whereas others may have both tilted and orthogonal phases. Indeed, there is a rich diversity of phases in this state of matter, as demonstrated in Figure 5.27 for examples of three families of materials, which include the N-(4-alkoxybenzylidine)-4-alkylanilines (nOms), the terephthylidene-bis-4-alkylanilines (TB-nAs), and the alkyl 4-alkoxybiphenylcarboxylates (nmOBCs and related nmOBSCs), plus there is a single example of the phenyl-biphenyl carboxylates: 4-(2-methylbutyl)phenyl 4-octyloxybiphenyl-4-carboxylate (8OSI). In the figure the transitions between the phases are shown on cooling from the isotropic liquid, demonstrating how the phase sequences change upon varying aliphatic chain length. It is often the case that more complicated phase sequences are observed for larger aspect ratios, which are related to the size of the central core unit, $e.g.$, three-ring versus two-ring systems.

A transition from an orthogonal smectic phase to a soft crystal orthogonal phase often involves better and tighter lateral packing of the molecules accompanying increasing restriction on the free rotation about their long axes. Consequently, sometimes there are changes in the positional structure of the molecules from hexagonal close packing to, say, rectangular. An example of this process is shown in Figure 5.28 for the orthogonal phases smectic A, hexatic B, and crystal E. In cooling, these processes are in effect associated with the condensing of the material towards the solid state via discrete steps. As there is a transformation in the local lattice structure, there also is an enthalpy evolved at the transition. In this case, it is usual for the transition to be first-order and accompanied by a change in heat capacity between the two phases. Transitions to and from orthogonal to tilted phases are sometimes more difficult to explain. For example, in the transition from the smectic A phase to the tilted smectic C phase, there is no heat evolved at the phase transition. Instead there is usually a change in the heat capacity, so for thermograms of the phase change there is usually a step in the baseline. For related tilted to orthogonal phase transformations, the transition is usually first-order because of a change in local lattice parameters.

Tilted smectic phases are of interest because of the potential to be used in optical devices that exhibit fast response modes, and as a result, there is a fascination as to why tilting occurs. In the following section, tilting in the smectic C phase is explored in more detail.

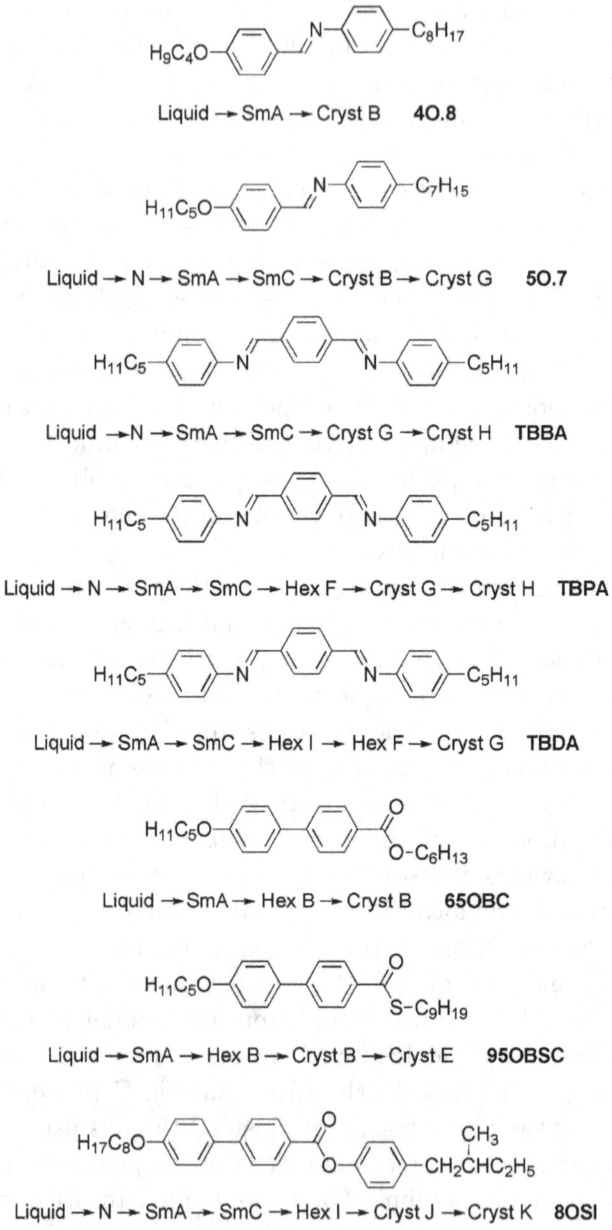
H9C4O
N
C8H17
Liquid → SmA → Cryst B 4O.8
H11C5O
N
C7H15
Liquid → N → SmA → SmC → Cryst B → Cryst G 5O.7
H11C5
N
N
C5H11
Liquid → N → SmA → SmC → Cryst G → Cryst H TBBA
H11C5
N
N
C5H11
Liquid → N → SmA → SmC → Hex F → Cryst G → Cryst H TBPA
H11C5
N
N
C5H11
Liquid → SmA → SmC → Hex I → Hex F → Cryst G TBDA
H11C5O
O
O-C6H13
Liquid → SmA → Hex B → Cryst B 65OBC
H11C5O
O
S-C9H19
Liquid → SmA → Hex B → Cryst B → Cryst E 95OBSC
H17C8O
O
O
CH3
CH2CHC2H5
Liquid → N → SmA → SmC → Hex I → Cryst J → Cryst K 8OSI

Figure 5.27 Examples of families of smectogens that show how the sequences of phase transitions change as a function of aliphatic chain length.

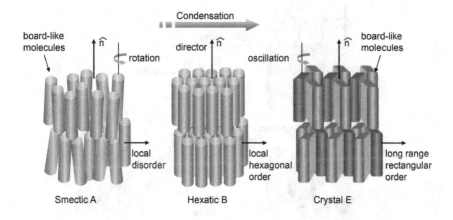

Figure 5.28 A schematic of the condensation process for a material that exhibits the orthogonal phases smectic A, hexatic B, and crystal E.

5.10 TILTING IN SMECTIC PHASES

Two basic theories have been developed in order to explain why molecules tilt within their layers. One depends on dipolar coupling between the molecules, which is exemplified by the McMillan model, and the other depends on the steric packing of the molecules, and is associated with the Wulf model. A schematic representation of the two is shown in Figure 5.29. The tilt angle, usually denoted as θ, can have two forms, one in which the tilt is temperature dependent and has a value between 15° and 30°, and the second in which the tilt is usually invariant with respect to temperature and has a value of around 45°.

The McMillan model focuses on the phase transition from the orthogonal smectic A to the tilted smectic C phase. In the smectic A phase the molecules are assumed to have polar groups that are oriented lateral to the molecular long axis, but the molecules are also rotating rapidly about the vertical axis, as shown in the figure. At the phase transition the rotation is assumed to become restricted, thereby allowing the polar groups to interact giving dipolar and quadrupolar couplings. These couplings create a torque relative to the long axes thereby producing a tilt of the interacting molecules. When the tilts become aligned, a phase transition occurs from the smectic A phase to smectic C. Conversely, the Wulf model assumes that the free rotation of parts of the molecules results in the molecular architecture adopting a zigzag shape at the phase transition from smectic A to smec-

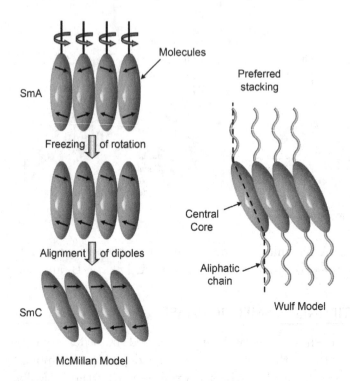

Figure 5.29 Depiction of the McMillan and Wulf theoretical models of the mechanism for tilting of the molecules relative to the layers in the smectic C phase.

tic C. The zigzag conformational structure allows the molecules to be stacked one on top of another, like chairs, to give a tilted layering. Here the molecules are assumed to still be undergoing rapid conformational changes, while still retaining their overall zigzag shapes.

The two models do not solve all of the issues surrounding tilting; however, both have aspects that support such a process to the extent that it is difficult to favor one model over the other. In the following, property-structure correlations are used to compare the two theories.

Considering first the McMillan model, Figure 5.30 shows a structure-property correlation made in relation to the incorporation of polar groups attached to the phenyl-biphenyl-4-carboxylate core unit. Octyl aliphatic chains are located at the termini of the core and separated by the incorporation or not of oxygen atoms. The material at the top of the figure (8O.O8) exhibits a SmA-SmC-SmB phase sequence on cooling. It also has the highest SmC to SmA phase transition temperature, and hence the highest thermal stability. This is followed by 8O.8 and 8.O8, which exhibit the same phase sequences, but at lower temperatures. Thus the sequence for smectic C stability is 8O.O8 > 8O.8 > 8.O8. The bottom structure in the figure is that of 8-8 which has no oxygen atoms between the core and the aliphatic chains. In this case the smectic C phase is totally suppressed and a phase sequence of SmA-SmB-G is obtained, where the B phase is hexatic. This correlation demonstrates that the higher the number of polar groups present in the molecular structure, the higher the thermal stability of the smectic C phase, and with no polar groups, no smectic C phase is found. This supports the hypothesis proposed by McMillan that there are intermolecular interactions between the polar groups, which cause torques to develop and thereby the induction of tilting of the molecules in their layers.

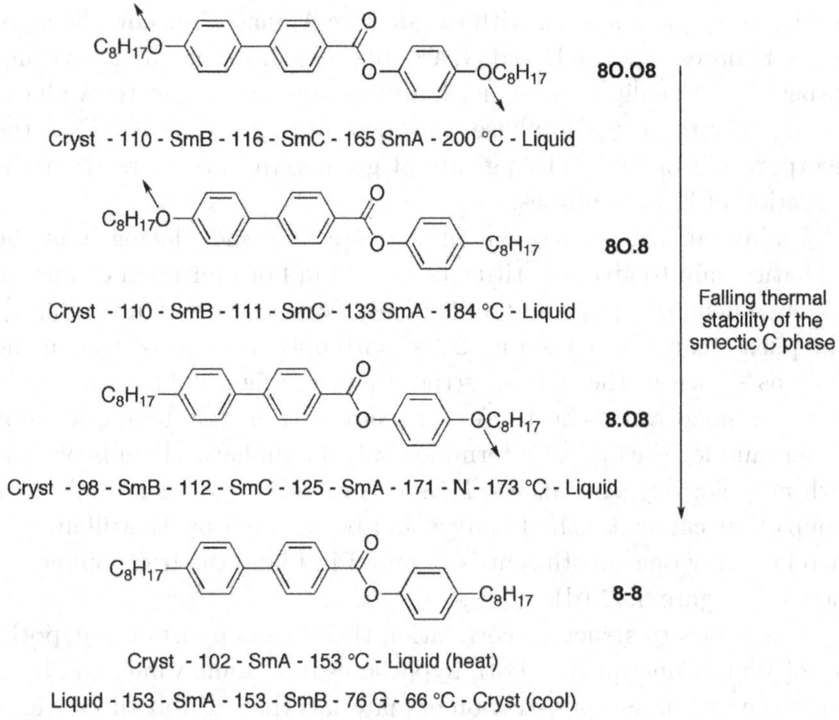

$C_8H_{17}O$ — ... — O — ... — OC_8H_{17} **8O.O8**

Cryst - 110 - SmB - 116 - SmC - 165 SmA - 200 °C - Liquid

$C_8H_{17}O$ — ... — O — ... — C_8H_{17} **8O.8**

Cryst - 110 - SmB - 111 - SmC - 133 SmA - 184 °C - Liquid

C_8H_{17} — ... — O — ... — OC_8H_{17} **8.O8**

Cryst - 98 - SmB - 112 - SmC - 125 - SmA - 171 - N - 173 °C - Liquid

C_8H_{17} — ... — O — ... — C_8H_{17} **8-8**

Cryst - 102 - SmA - 153 °C - Liquid (heat)

Liquid - 153 - SmA - 153 - SmB - 76 G - 66 °C - Cryst (cool)

Falling thermal stability of the smectic C phase

Figure 5.30 A structure-property correlation made in relation to the incorporation of polar groups attached to a phenyl-biphenyl-4-carboxylate core unit.

We turn now to a property-structure correlation that is related to the hypothesis proposed by Wulf. Wulf's model involves the stacking up of zigzag shaped molecules like chairs to form tilted layers, or conversely, molecules tilted within their layers. However, the cross-sectional areas of the terminal aliphatic chains, even in dynamic motion, are much less than those of the rigid central (aromatic) cores, as shown in Figure 5.29. This means that the molecules do not pack snugly together, and that there is free volume between them. Lateral substitution in the molecular architecture, in the form of, say, a methyl group, can be used to probe the free volume and to investigate mesophase stability. Following from the property-structure correlations shown in Figure 5.30, a lateral substituent can be easily positioned in the central core as shown in the chemical structure at the top of Figure 5.31. Locating a methyl group in the core region essentially pushes the molecules apart, see Figure 5.32 (a), and destabilizes the packing. Hence a nematic phase appears along with the smectic A phase, but now the more ordered phases, smectic B and G, are suppressed. Adding a methyl substituent into the aliphatic chain serves to suppress the smectic A phase, leaving only the nematic phase, as shown in Figure 5.32(b). Thus the incorporation of two differing lateral groups appears to restrain the formation of layered phases.

Taking out the substituent in the core unit and placing it in the aliphatic chain to give substituents located in both terminal chains results in a nematic, SmA, and SmC phase sequence (see Figure 5.32(c)). The phase sequence is then repeated with only one substituent in the chain as shown in the bottom structure of the figure. This time however, the monotropic smectic A to smectic C transition temperature is higher, and for the opposite terminus only an aliphatic chain is present with no polar oxygen linkage. Thus this material has no lateral polar groups that can cause the tilt as would be required by McMillan, and therefore only one substituent is required in filling the free volume, as shown in Figure 5.32 (d).

This property-structure correlation therefore supports the hypothesis of Wulf. Consequently both hypotheses have some value, and both aid in the design and preparation of materials that exhibit tilted smectic phases.

When molecules that have rod-like shapes pack together, the possible structural variations are either related to monolayer or bilayer arrangements perpendicular to the layers, or to differences in the in-plane packing. For mesophases that have molecules that are tilted in

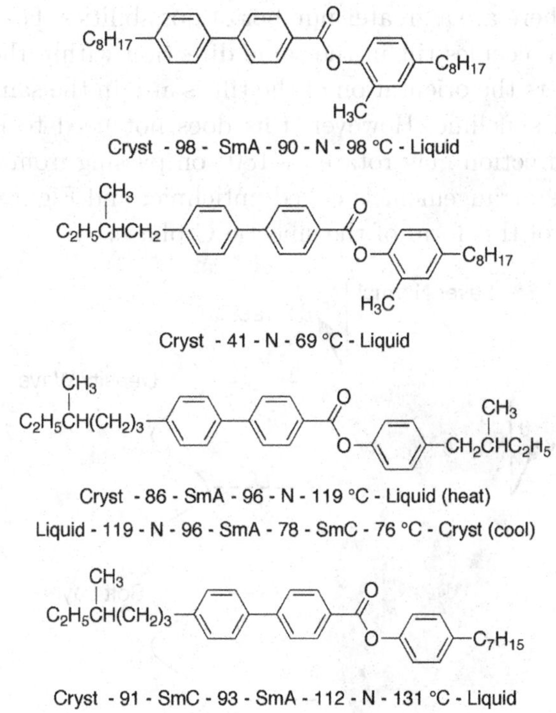

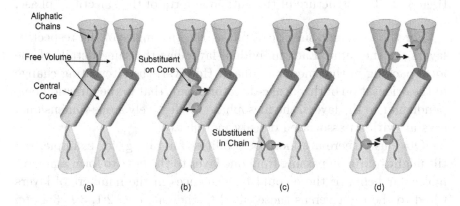

C$_8$H$_{17}$ —[ring]—[ring]— C(=O)—O—[ring with H$_3$C]— C$_8$H$_{17}$

Cryst - 98 - SmA - 90 - N - 98 °C - Liquid

CH$_3$

C$_2$H$_5$CHCH$_2$ —[ring]—[ring]— C(=O)—O—[ring with H$_3$C]— C$_8$H$_{17}$

Cryst - 41 - N - 69 °C - Liquid

CH$_3$

C$_2$H$_5$CH(CH$_2$)$_3$ —[ring]—[ring]— C(=O)—O—[ring]— CH$_2$CHC$_2$H$_5$ (CH$_3$)

Cryst - 86 - SmA - 96 - N - 119 °C - Liquid (heat)

Liquid - 119 - N - 96 - SmA - 78 - SmC - 76 °C - Cryst (cool)

CH$_3$

C$_2$H$_5$CH(CH$_2$)$_3$ —[ring]—[ring]— C(=O)—O—[ring]— C$_7$H$_{15}$

Cryst - 91 - SmC - 93 - SmA - 112 - N - 131 °C - Liquid

Figure 5.31 A structure-property correlation made in relation to the incorporation of methyl groups attached to a phenyl-biphenyl-4-carboxylate core unit, and to the aliphatic terminal chains.

Aliphatic Chains

Free Volume

Central Core

Substituent on Core

Substituent in Chain

(a)   (b)   (c)   (d)

Figure 5.32 Effect of lateral methyl units in property-structure correlations based on the phenyl-biphenyl-4-carboxylate core unit.

their layers, there are a greater number of possibilities. For the smectic C phase the molecules tilt in the same direction within the layer, and out of the layers the orientation of the tilt is also in the same direction. This is called synclinic. However, this does not need to be the case, and the tilt direction may rotate by 180° on passing from one layer to the next. This arrangement is called anticlinic, and Figure 5.33 shows the structure of this form of the smectic C phase.

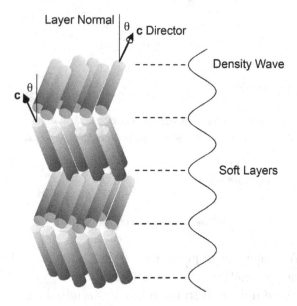

Figure 5.33 The structure of the anticlinic form of the smectic C phase.

For the anticlinic smectic C phase, the tilt angles of the respective layers are the same. Each individual layer has the same details of the local packing in the smectic C phase. However, the alternating change in the orientation of the local c-director means that in the direction perpendicular to the layers, the mesophase is effectively a one-dimensional crystal, with an associated density modulation.

Obviously there are more possibilities for tilting; for example, the tilt might rotate on passing from one layer to the next to form a macromolecular helix, or there could be sequences in the numbers of layers tilted to the right versus those tilted to the left, *e.g.*, 2:1, 3:1, 3:4 *etc.* These modifications are usually associated with systems in which the molecules are chiral (sometimes called optically active). These structures are described later in Chapter 7.

The number of materials that exhibit the anticlinic phase is very small for compounds that have no chirality with respect to their chemcal structures. In comparison, the larger proportion of materials that exhibit the anticlinic phase are chiral, and as a consequence they usually possess antiferroelectric properties, as described in Chapter 7. An example of an achiral, anticlinic smectic C material, 3-4-9OPPBC, is shown in Figure 5.34. The chemical structure of this material is not too dissimilar to those shown in Figures 5.30 and 5.31 for compounds that exhibit the smectic C phase. The major structural difference is that 3-4-9OPPBC has a swallowtail shape based on being a derivative of 4-hydroxyheptane. The bi-forked tail can pack in different ways, and to fill space efficiently, an alternating tilt is needed. However, the transition temperatures for the material are relatively low.

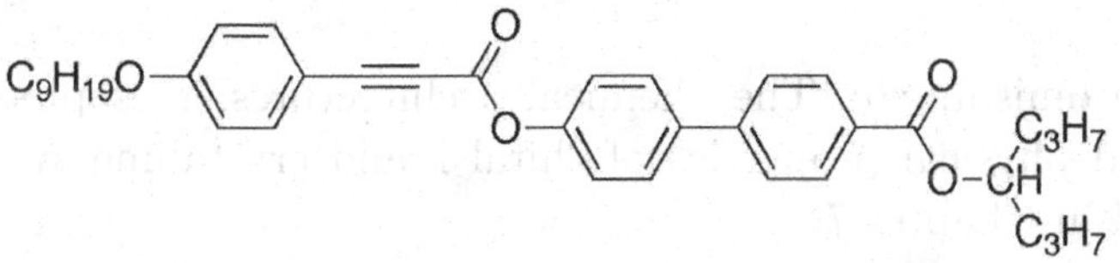

Liquid - 70 - SmA - 55 - Anti - 42 °C - Cryst   3-4-9OPPBC

Figure 5.34 The swallowtail chemical structure of the compound 3-4-9OPPBC, which exhibits an anticlinic smectic C phase.

In the synclinic and anticlinic smectic C phases, the molecules (apart from the tilt) are disorganized in their layers, and possess only short-range periodic order. In the phases that have local hexagonal packing that extends over 20 nm or more, there is a choice for the tilt direction. It can be either to the side of the hexagonal packing array or to the apex, as shown in Figure 5.35. This choice applies to both the hexatic I and F phases and the crystal J and G soft matter phases. For the I and J phases the tilt is to the apex, and for the F and G phases to the side.

It is possible that the hexatic I and F phases may also exhibit anticlinic arrangements, but because of the long-range periodic order in soft crystal phases, it is unlikely that anticlinic arrangements occur for the J and G (plus the H and K) soft crystals.

Designing materials that form tilted structures is of particular interest for compounds that have asymmetric structures because they tend to exhibit non-linear properties, such as ferroelectricity, ferrielectricity, antiferroelectricity, pyroelectricity, electroclinism, flexoelectricity,

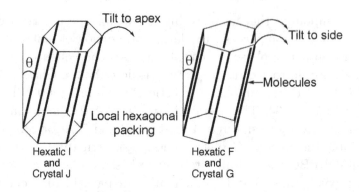

Figure 5.35 The tilt directions in the hexatic I and F phases, and the soft crystal J and G phases.

thermochromism, *etc.* The chemical architectures, mesophase structures, and physical properties of chiral liquid crystalline systems are described in Chapter 7.

EXERCISES

5.1 A powder diffraction pattern of 5OCB in the nematic phase exhibits two scattering rings with d-spacings of 0.35 and 2.38 nm, whereas the diffraction pattern for an aligned sample gives the same spacings, but with diffuse scattering arcs perpendicular to one another. Given that the molecular length of 5OCB is 1.7 nm, use curly arrows and resonance structures to exemplify the molecular interactions and draw a picture of the organization of the molecules in the nematic phase.

5.2 Describe the changes that take place in molecular dynamics and molecular organization at the clearing point of a nematogen, and compare those descriptions with the dynamics in the bulk phase at 20°C lower.

5.3 Compounds 1 to 6 below exhibit a number of calamitic phases including nematic and smectic C phases. Using steric shapes and the relative positioning of polar groups, identify (i) the two materials that do not exhibit a nematic phase, (ii) which compound has the highest thermal stability for the smectic C phase, and (iii) which has the lowest.

5.4 An X-ray power pattern for an unaligned smectic A phase shows two diffuse scattering rings, one associated with the layer spacing and the other with the lateral distances between the molecules. When the same smectic A phase is aligned in a magnetic field, two sets of diffuse scattering arcs are found oriented perpendicular to one another. Upon cooling there is a transition to the smectic C phase. Draw the resulting diffractogram, and describe how the tilt angle of the smectic C phase can be determined.

5.5 The ordering of the polarizabilities of methylene, sulfur, and oxygen is S > O > CH$_2$. Compounds 1 to 4 in the figure are nematogens; with logical reasoning, place them in order of the highest to lowest clearing point.

CHAPTER 6

Discotic Liquid Crystals – Stacking the Dishes and the Bowls

6.1 PILING UP PENNIES AND DISCOTIC LIQUID CRYSTALS

The mesophases of liquid crystals composed of rod-like molecules were given the names of nematic and smectic, derived from the Greek *nematos* and *smectos*, because of their bulk-like appearance. As seen in the previous chapter, it was wrongly thought that a better descriptor was to name the phases after their molecular shapes. The much later discovery of the mesophases formed by liquid crystals composed of disk-like molecules also proved to be problematic when it came to the naming of the new phases. A similar error was made, and the mesophases became named after the molecular architectures as *discotique*, as derived from French, or its English translation of discotic.

With the discovery of discotic phases there came with them a view that there were some analogies with mesophases of rod-like systems. Figure 6.1 gives a brief picture of this concept. Taking a sphere, which has isotropic properties, and by unidirectionally stretching its structure, an anisotropic calamitic phase is formed. Doing the reverse by compressing the sphere, an anisotropic discotic phase is created. In both cases the liquid crystal director field is located along the C_∞ axis of the material. However, the positive properties, such as the optical anisotropy for the calamitic system, become negative for the discotic. Thus discotics can be seen as negative versions of calamitics. As discussed in Chapter 14, a classical example of the opposing properties of discotics and calamitics can be seen in the optical compensation films that are used in LCDs to correct problems associated with poor viewing angle.

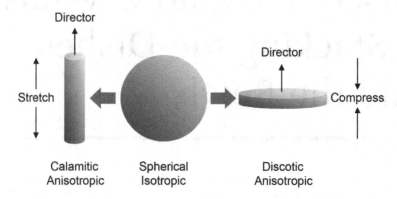

Figure 6.1 Comparison between rod-like and disk-like molecular architectures that form anisotropic mesophases.

Again the issue of molecular shape, as described in this chapter, demonstrates that there are many more types of molecular architecture that can be sculpted to support the formation of discotic mesophases, and indeed the nematic phase appears not only for calamitic (rod-type) systems but also for disk-like ones. For example, Figure 6.2 shows a variety of architectures when used in the design of liquid crystals support the formation of discotic mesophases. Ultimately, these architectures can be nanosculpted and nanoengineered to produce materials that have practical applications. The descriptor discotic was also subsequently redefined into two terms - nematic and columnar, but even these are inconsistent with respect to nomenclature as one describes the bulk and the structure of the respective phases.

In the following, the structures of various types of discotic mesophase are described, and subsequently materials that have the molecular architectures shown in Figure 6.2 are discussed.

6.2 STRUCTURES OF MESOPHASES FORMED BY DISK-LIKE MOLECULES

The phases formed by disk-like molecules fall into two groups; one in which the molecules do not have long-range order, called the nematic

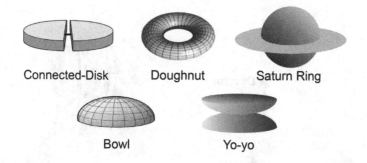

Figure 6.2 Typical shapes of the molecular structures of materials that exhibit discotic mesophases.

phase, and the other in which the molecules stack one on top of another to form a family of modifications that are collectively known as columnar phases.

The discotic nematic phase (N_D), unlike the nematic phases of calamitic materials, has only one modification, plus its chiral form (N_D^*). Like its calamitic analog, the N_D phase is the least ordered liquid crystalline discotic phase, and also the least viscous. As the molecules are disk-like in shape, the flow in the mesophase is anisotropic, with the easiest direction of movement being parallel to the planes of the disks.

The generalized structure of the N_D phase is easy to visualize as shown in Figure 6.3, where the phase structure is analogous to collapsing columns of coins and spreading them out randomly on a table. The disks, like rods, are anisotropic and are able to assemble so that their short axes are, on average, parallel to one another. This direction is the director, $\hat{n}$, of the nematic discotic phase, and for the chiral form, the molecules twist with respect to the local director to form a helical macrostructure as shown in Figure 6.3.

As with the calamitic nematic phase, the N_D phase has only a statistically parallel ordering of molecular orientations with no translational ordering of the molecules. Accordingly, the N_D phase is therefore defined as a true liquid crystal. Recently, a few materials with disk-shaped molecules have been discovered that generate a columnar nematic (N_C) phase; this phase consists of short columns of a few molecules that adopt a discotic nematic packing arrangement.

The calamitic and discotic nematic phases are also differentiated by their optical properties, notably their birefringence, with the calamitic

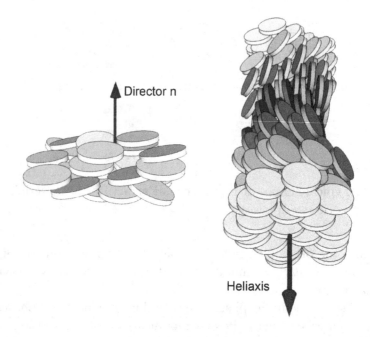

Figure 6.3 Left, the molecular arrangements within the discotic nematic phase. Right, the molecular arrangements within the chiral discotic nematic phase. The local director, which is perpendicular to the planes of the molecules, twists around a perpendicular axis, thereby forming a helical macrostructure.

nematic being classified as a positive uniaxial phase, which is termed N_{u+}, whereas the discotic nematic phase is also uniaxial but the bire-fringence is negative, and hence it is termed, N_{u-}. It has also been predicted that when calamitic and discotic nematogens are mixed, a biaxial nematic phase can be generated. However, as yet no biaxial nematic phase has been produced by this method.

The second class of mesophases formed by disk-like molecules is the collection of columnar phases. These arise because of the differ-ent symmetry classes of two-dimensional lattices of columns of stacked molecules. Further differentiation occurs with the ordering of the molecules along the columnar axes, which may be ordered or disor-dered, and where the planes of the molecules may be tilted or not rel-ative to the column axes. This polymorphism in columnar mesophases is, in many ways, analogous to the polymorphism found in the smectic mesophases generated by calamitic materials, except that the columnar variants do not appear to exhibit short-range or hexatic ordering of the 2D packing of the columns.

The least ordered of these phases, and the one with the highest symmetry, is the hexagonally close-packed phase, which is given the descriptor Col_{hd} (previously this phase was labeled as D_{hd}). The molecules are disordered with respect to the distances between the molecular planes and their relative tilting or not along the column axes. The columns are packed in a hexagonal array, similar to the way in which pencils can be packed together, as shown in Figure 6.4. The columns are relatively soft, and can bend and form hairpins, and the molecules can sometimes translate from one column to the next. Thus, the lattice distance from a molecule to its nearest neighbor is identical whatever the direction (*i.e.*, $a = b$) in the hexagonal array as shown in the figure. The structure of the hexagonal phase has a sixfold rotational axis, a mirror plane perpendicular to this axis, six mirror planes in the plane of the sixfold axis, six twofold rotational axes with respect to the sixfold axis, and a center of inversion; therefore the phase has D_{6h} symmetry, which is the same as for the crystal B phase, and has a 2D space group of P6 2/m 2/m.

Figure 6.4 The structure of the hexagonal disordered columnar phase ($a = b$) where the positions of the disk-like molecules are disorganized along the column axes.

When the molecules become ordered along the columns on cooling, the 2D hexagonal phase undergoes a transition to the 3D hexagonal ordered columnar phase, classified as Col_{ho} (previously called D_{ho}). This phase has the same symmetry elements as the Col_{hd} phase, but because of the long-range ordering within the columns, the phase is really a soft crystal. Thus the columns in the Col_{ho} phase do not bend so easily, and defects form that can be observed in the polarizing microscope.

Squeezing the columns of the hexagonal disordered phase together results in the formation of the rectangular disordered columnar phase (Col_{rd}), which is illustrated in Figure 6.5. Here the two lengths a and b of the 2D lattice are not equal, which results in the structure having three two-fold (C_2) axes as shown, three mutually orthogonal mirror planes, and a center of inversion. Thus the phase exhibits D_{2h} symmetry. Along the columns the molecules are disordered and effectively the planes of their disk-like structures are perpendicular to the column axes with slight tilts being allowed. As the 2D structure has different lattice parameters, as shown by $a \neq b$, the phase exhibits extensive domains as it is formed. The difference in the lattice parameters also means that the phase is biaxial.

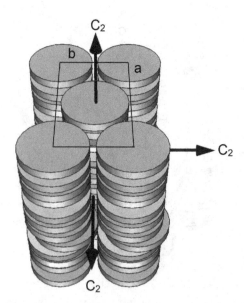

Figure 6.5 The structure of the rectangular disordered columnar phase ($a \neq b$) where the positions of the disk-like molecules are disorganized along the column axes.

As with the hexagonal phase, the rectangular columnar phase can also have the molecules ordered along the column axes, to give a 3D structured soft crystal phase. The phase exhibits the same symmetry operations as the disordered analogs, and thus it is classified as Col_{ro}.

In the above examples the molecules are on average arranged such that the planes of their disk-like structures are perpendicular to the column axes. However, there is another possibility, which is that within a column all of the molecules are tilted in roughly the same direction with respect to the column axis. Thus each column may be termed as being "tilted" and the packing of such columns could be arranged so that the tilt points in the same direction (synclinic) or in opposing directions (anticlinic) or variations in between these two extremes (see Figure 6.6). These structural possibilities are rather similar to those found for "tilted" smectics, and in particular to ferro- and anti-ferroelectric phases.

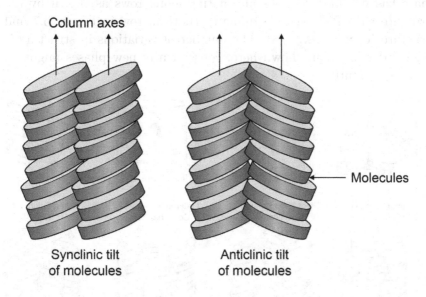

Figure 6.6 Tilted disk-like molecules in the columns of discotic liquid crystals.

The simplest possibility is to have a hexagonal columnar phase in which the molecules are tilted. The symmetry is thereby reduced from D_{2h} to C_2. This mesophase was originally classified as Col_t, where t stands for tilted, and its structure is shown in Figure 6.7 (top left). A similar situation applies to the rectangular phase, but in this case there are more choices for orienting the tilt, *i.e.*, to one or other of the sides of the 2D lattice (thick lines in Figure 6.8). This creates several more complex packing arrangements in which the disk-shaped molecules are tilted with respect to the column axes. Furthermore, classification of such structures by considering their point symmetries is often problematic. Consequently, it is more informative to classify them in accordance with their corresponding two-dimensional space groups. A few representative examples of such packing arrangements, along with each individual 2D space group symmetry, are shown together in Figure 6.7. A simple variant of the tilted hexagonal columnar phases is one in which rows are tilted in one direction, whereas in adjacent rows the columns are tilted in the opposite direction. This phase is called a disordered rectangular columnar phase, and has $P2_1/a$ space symmetry (see bottom row in Figure 6.7 and drawing 6.8(b)). This alternation in the tilt is similar to an anticlinic smectic phase. A similar rectangular phase has its tilt directions alternating along rows as shown by the structure with P_2/a space symmetry (bottom row in Figure 6.7 and structure (c) in Figure 6.8). The number of variations in structure is large and more than shown here; consequently new phases might be found in the future.

Nematic Discotic Disordered Hexagonal Columnar Col$_{hd}$ Ordered Hexagonal Columnar Col$_{ho}$ Tilted Columnar Col$_t$

Disordered Rectangular Columnar Col$_{rd}$ (P6 2/m 2/m) Disordered Rectangular Columnar Col$_{rd}$ (P2₁/a) Disordered Rectangular Columnar Col$_{rd}$ (P2/a) Disordered Rectangular Columnar Col$_{rd}$ (P1)

Figure 6.7 General structures of discotic liquid crystals and soft crystals.

In the above we have considered how the disks might pack together in columns and how the columns pack together in order to minimize the

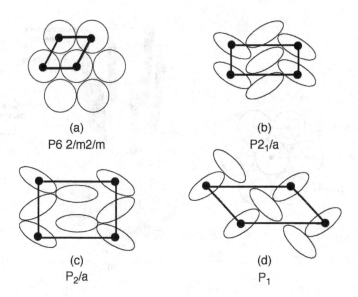

Figure 6.8 Some 2D lattices found in columnar mesophases with; (a) P6 2/m 2/m; (b) P2₁1/a; (c) P₂/a; and (d) P₁ space symmetries. The ellipses indicate that the disk-shaped molecules are tilted with respect to the column axis.

free volume; a bit like packing pencils together. This is a conventional way of thinking, focusing on the solid-like entities. However, there is an alternative view, instead looking at the space between the columns. If the disks are hard, then clearly there is space between the columns, and therefore an associated lattice is possible, as shown in Figure 6.9(a). For actual materials, the space is filled by flexible, fatty, aliphatic chains, which can bend through conformational variations to fill the voids. Therefore rigid parts of the disks effectively sit periodically in a sea of fat, in a form of nanophase segregation. In Figure 6.9(b) the columns possessing disks of different sizes exemplify this. This arrangement is similar to having a stack of coins of different value, where the column width is determined by the largest coin, but then not all of the centers of the coins are located along the column axes. In a real material, again the fatty chains occupy the space between the columns, resulting in the defects having long-range periodicity.

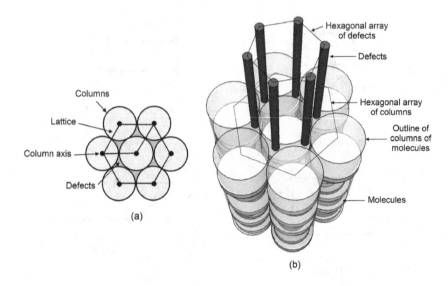

Figure 6.9 Hexagonal packing of columns showing the defects between the columns (a), and the hexagonal structure composed of disks of different sizes in which the defects are still arranged in a hexagonal lattice (b).

6.3 CHEMICAL MOIETY SELECTION — HARD DISKS

We now turn to discussing the types of molecular structures that support the formation of discotic phases. For example, we investigate the structural moieties that constitute suitable molecular architectures for the generation of discotic mesophases, starting with hard disk systems. This allows us to make direct comparisons with similar studies on rigid-rod systems described in Chapter 5 for polyphenyls, such as sexiphenyl.

There are very few studies on the phase behavior of materials possessing hard disk-like structures such as triphenylene, truxene, coronene, *etc.* However, nematic phases are found in oil sludge and tar when heated to high temperatures. These phases are called carbonaceous mesophases, and photomicrographs of their textures confirm that they are nematic. Most of the materials in the mixtures of the oil sludge and tar are found to be polyaromatics.

Table 6.1 shows the melting behavior of a few polyaromatic hydrocarbons in comparison to p-sexiphenyl. All of the materials that possess disk-like molecules have melting points lower in temperature than p-sexiphenyl, but similar to that of p-quinquephenyl (in the case of the larger disks). However, no mesophases are found for the discoidal materials; this contrasts with the calamitic analogs that are found to be nematic. For example, truxene and coronene exhibit high melting points (384 and 433°C, respectively), and even upon supercooling from the liquid, no monotropic mesomorphic properties are observed, whereas p-quinquephenyl melts at 386°C, and has a nematic to liquid transition at 415°C. Therefore these results appear to indicate that hard disk materials are non-mesomorphic.

Table 6.1 The chemical structure and melting points for a variety of polyaromatics.

Compound	Chemical Structure	Melting Point (°C)
Triphenylene		197.3
Pyrene		150.2
Perylene		274.9
Truxene		384.7
Corenene		432.9
p-Sexiphenyl		439.9

A way to investigate the potential for achieving mesomorphic behavior for the hard disk systems is through creating low melting mixtures. This was done by constructing binary phase diagrams to determine the virtual discotic liquid crystal to liquid transition for a hard disk compound that exhibited no discotic phases (stable or monotropic). In these studies, they mixed amphiphilic discotic liquid crystal materials with triphenylene and pyrene to determine the virtual transitions for the two polyaromatics. Thus the virtual nematic to liquid transitions for the hard disk materials were determined by extrapolation across the binary phase diagrams. (A virtual transition in this case means a transition that would occur below the recrystallization temperature provided that recrystallization could be suppressed by supercooling.) Their work demonstrated that hard-disk materials had much more inferior tendencies to form mesophases than rod-like analogs.

Overall these results may be an indicator of the π-π face-to-face interactions being too strong, even at high temperatures, to allow the molecular disks to flow past one another in forming a nematic phase. Figure 6.10 shows the space-filling structures of triphenylene and coronene, which are effectively flat, and truxene, which has a slightly twisted structure. Triphenylene and coronene stacking in a face-to-face arrangement have strong π-π interactions even though the disks are offset with the carbon atoms of one disk positioned over gaps in the π-cloud of the structure of the adjacent disk, as shown in Figure 6.11. Truxene on the other hand has a slightly twisted structure, which would be expected to push apart adjacent molecules in order to weaken the interactions, as shown in the figure. But even this does not enhance relative mesomorphic properties.

As noted above, there is a tendency for aromatic carbon atoms to be positioned over the centers of other aromatic rings, and as a consequence stacking in columns occurs without the disks being positioned directly above and below adjacent disks. Figure 6.11 shows this effect for triphenylene in which adjacent disks might be rotated relative to a chosen disk. It has been proposed for some materials that such packing is capable of producing a spiraling columnar structure. Thus it appears that the columns themselves could have superstructures, particularly when they have lateral substitutions, which may be chiral.

If we now turn to molecular design, as with calamitic materials, hard disks are unlikely choices for development. Therefore, the disks are usually substituted along their edges, using peripheral units that are similar to those employed in calamitic design. The disk-shaped central core unit is usually benzene, a heterocyclic, or a polyaromatic such as phthalocyanine or triphenylene, but columnar phases have been generated with alicyclic cores such as cyclohexane, carbohydrate, and crown ether. In order to maintain the disk-like structure, the central core is usually symmetrical and the peripheral dendritic units are present in

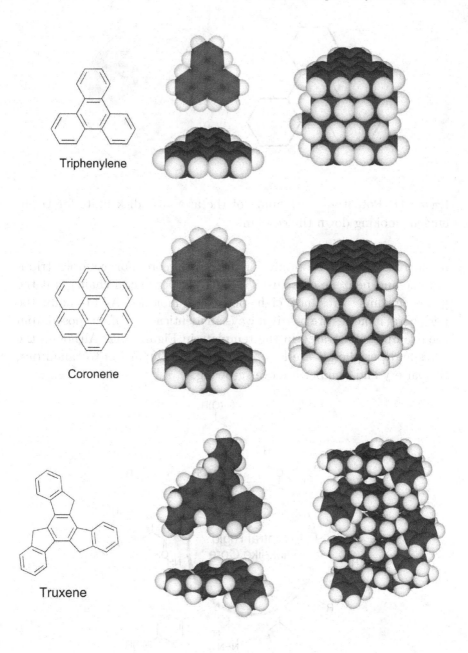

Figure 6.10 The space-filling structures and face-to-face interactions of triphenylene, coronene, and truxene.

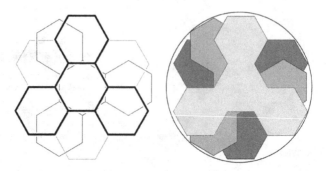

Figure 6.11 Potential overlapping of the aromatic disk units for triphenylene looking down the column.

numbers that are appropriate for the central core. For example, triphenylene and benzene cores usually have six peripheral chains, but the phthalocyanine cores have eight peripheral moieties. Additionally, the peripheral moieties are nearly always all identical and often chosen from the examples illustrated in the template in Figure 6.12. Although the cores and peripherals are less in number in comparison to calamitics, the variety and number are ever growing.

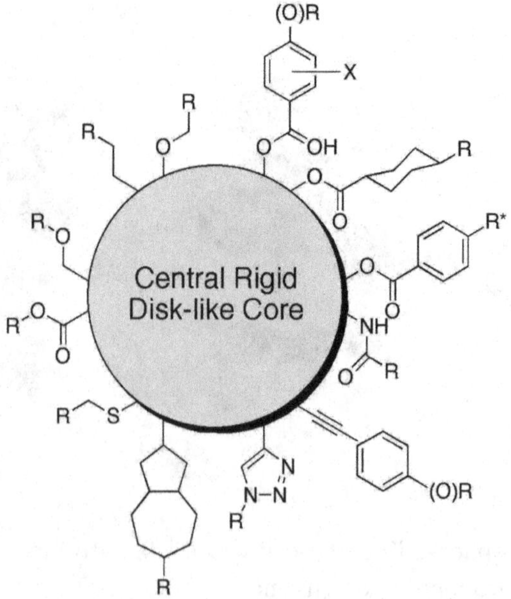

Figure 6.12 A general structural template for discotic liquid crystals.

6.4   CHEMICAL MOIETY SELECTION — DISK PERIPHERY

The majority of discotic materials exhibit columnar phases. Some additionally show the nematic discotic phase, but very few exhibit only the nematic phase. If we start now with the simplest rigid disk-like core unit, benzene, it has six possible substitution points for attachment of aliphatic units. In addition, having all the peripheral units identical helps to maintain the overall discotic architecture of the structure. As with calamitic systems, the core unit provides the rigidity and the relative flexibility of peripheral moieties serve to reduce melting points so that mesophases are exhibited. Thus, the hexaalkanoyloxybenzenes were the first discotic materials in which mesomorphism was observed. Hexahexanoyloxybenzene is an example of this homologous series; it exhibits a rectangular disordered phase (see Figure 6.13), and from the simulation in the figure, it shows that the benzene unit sits in the middle of an aliphatic sea around it. This begs the question: to what extent should the peripheral space along the edges of the disks be fully filled?

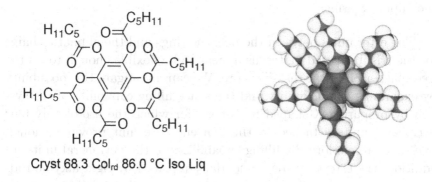

Cryst 68.3 Col$_{rd}$ 86.0 °C Iso Liq

Figure   6.13 Chemical   and   simulated   structures   of   hexahexanoyloxybenzene, which forms a rectangular disordered columnar phase.

It is generally accepted that a fairly large, planar, and rigid core that has six or eight peripheral moieties attached is required for the generation of "discotic" mesophases. Surprisingly therefore, the nematic discotic phase was thought to be exhibited, albeit monotropically, by one of the very few discotic compounds that has only three peripheral moieties, as shown in Figure 6.14. In the case of this material, there is a lot of free space around the core that reduces the ability of the molecules to pack in a columnar fashion, but it appears that intermolecular interactions are still sufficient to enable the exhibition of the nematic phase.

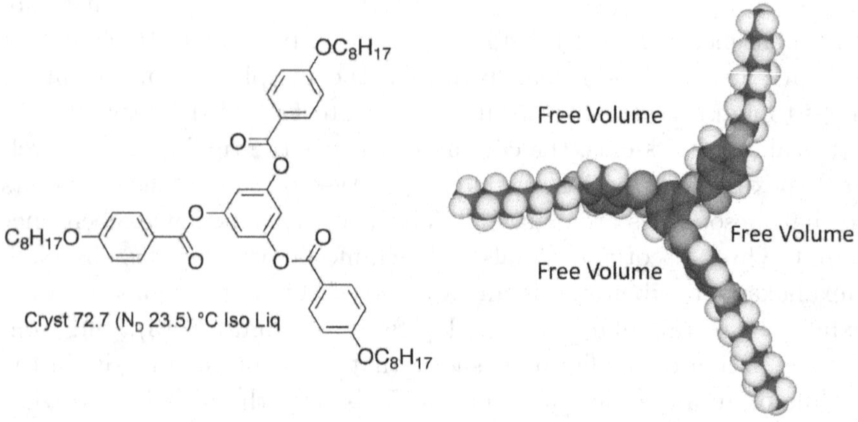

Figure 6.14 A trisubstituted mesogen, showing the free volume between the aliphatic arms.

The ester units between the benzene rings and the aliphatic chains for the trisubstituted materials appear to be flexible enough to fill the free space around the disk-like core. We can investigate this possibility by examining the flexibility and the space filling capability of the system by preparing X-shaped tetramers, for which the use of only four peripheral units attached to the benzene core unit provides enough free space to explore the filling capabilities of the peripheral units. In addition, the esters can be sequentially replaced by rigid alkynic rod-like units, such that the arms become more rigidly placed around the periphery, as shown in Figure 6.15, and in the space filling models in Figure 6.16.

Figure 6.15 shows the transition temperatures for the materials in this systematic study, with most of the materials exhibiting nematic and smectic C phases. The presence of the calamitic smectic C phase indicates that the materials are behaving as though they have rod-like architectures, suggesting that the arms are coming together through conformational changes to minimize the free space available, as shown in Figure 6.17. Consequently the nematic phase is also calamitic rather than discotic.

Thus the first property-structure correlation for designing and

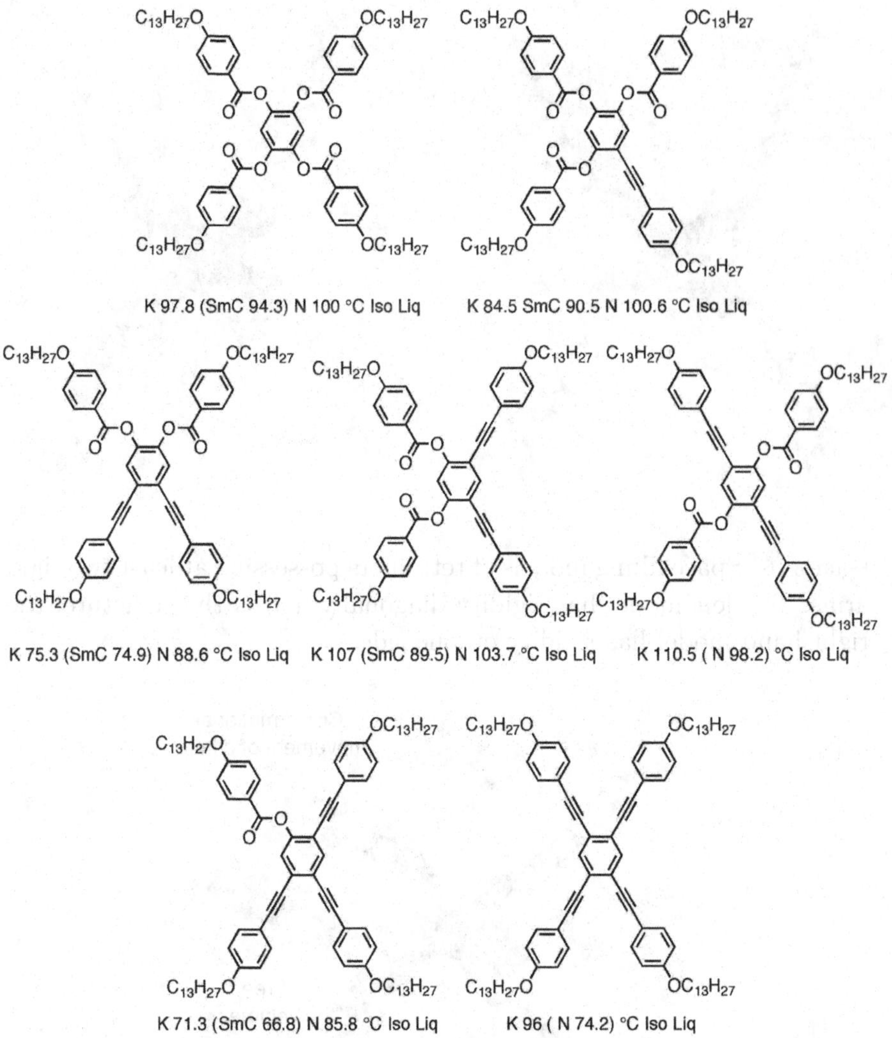

K 97.8 (SmC 94.3) N 100 °C Iso Liq   K 84.5 SmC 90.5 N 100.6 °C Iso Liq

K 75.3 (SmC 74.9) N 88.6 °C Iso Liq   K 107 (SmC 89.5) N 103.7 °C Iso Liq   K 110.5 (N 98.2) °C Iso Liq

K 71.3 (SmC 66.8) N 85.8 °C Iso Liq   K 96 (N 74.2) °C Iso Liq

Figure 6.15 The effect of sequential replacement of esters with alkynic units in tetramer liquid crystals.

preparing materials to exhibit discotic nematic and/or columnar phases is to ensure that the edge of the central disk is as densely packed with substituents as possible in order to produce a disk-like hierarchical architecture.

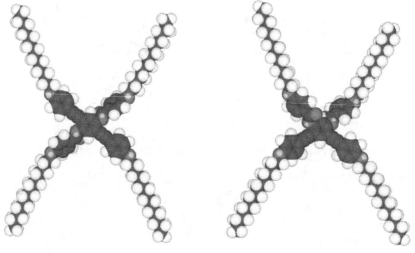

Figure 6.16 Space filling models of tetramers possessing at least two rigid arms. The left model has rigidity diagonally across the structure; the right-hand model has rigidity on one side.

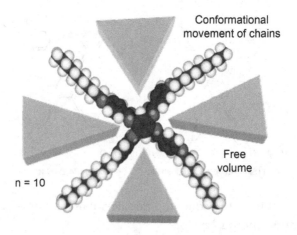

Figure 6.17 The free space available in X-shaped molecular architectures for a tetraester material. In this case the esters allow the arms to undergo conformational changes so that the arms come together to give a rod-like structure.

6.5 CHEMICAL MOIETY SELECTION — DISK THICKNESS

As with calamitic phases, there are small design changes that can be made to molecular architectures in order to control phase formation. For example, it can be seen from the previous section that the density of substituents along the disk edge can be used to enhance mesophase formation and temperature range. In this section we examine the effect of disk thickness and inter-disk separation on the formation of nematic and columnar phases. The first set of examples we consider are derivatives of triphenylene, simply because it was the second family of discotic mesogens that were discovered. The triphenylene core consists of three benzene rings conjugatively joined to give a planar aromatic unit that enables six peripheral units to be symmetrically attached with ease, and because the core is much larger than benzene, the mesomorphic tendency of such materials is much higher.

Both of the hexaalkoxy-substituted triphenylene compounds illustrated in Figure 6.18 exhibit columnar phases. Interestingly, the material to the left (HAT5) exhibits a hexagonal ordered columnar (Col_{ho}) phase, whereas the hexaoctanoyloxy-substituted triphenylene to the right possesses higher mesophase stability, but exhibits a rectangular columnar phase (Col_{rd}) phase. The preference for columnar phases is possibly due to the polar oxygens combining with the large delocalized π-electron core to facilitate column formation. However, the carbonyl groups in the right-hand structure have the ability to cause some repulsion between the disk-like units, and hence this phase has a disordered columnar structure.

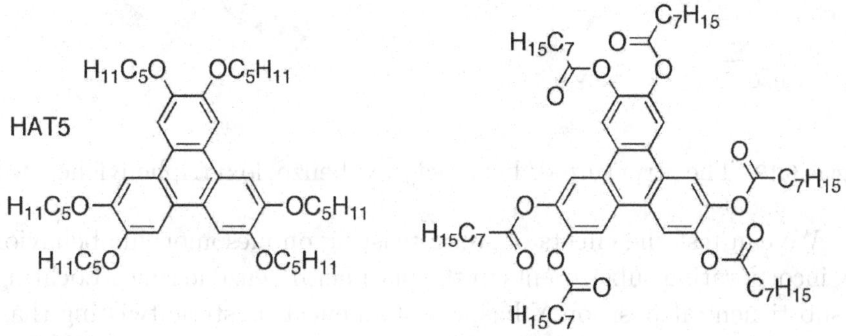

HAT5

Cryst 69.0 Col_{ho} 122.1 °C Iso Liq   Cryst 66.0 Col_{rd} 126.0 °C Iso Liq

Figure 6.18 Simple aliphatic substituted derivatives of triphenylene.

If we now expand the size of the core of the disk to give the substituted hexabenzoyloxytriphenylenes, we have a much larger core unit that is extended by six ester-linked benzene rings. Accordingly, these materials have a much higher mesomorphic temperature range than the triphenylenes based on simple ester or ether chains as shown in Figure 6.18. The hexa-octyloxybenzoyloxytriphenylene compound, for example, is shown in Figure 6.19. It melts at 152°C into a rectangular columnar phase, which then transforms into a nematic discotic phase at 168°C, and finally to the liquid at 244°C. The liquid crystal temperature ranges for these materials are over 100°C greater than for the simple aliphatic ester variants, simply because the core unit is bigger and ultimately possesses a larger number of delocalized electrons. However, a question arises as to why there is the formation of a nematic discotic phase. This is probably because the exterior benzene rings have the potential to twist out of the plane of the central triphenylene disk, thereby causing at high temperature steric interference, which dislocates the molecules along the columnar axis. It is almost like a column of pennies being tipped over.

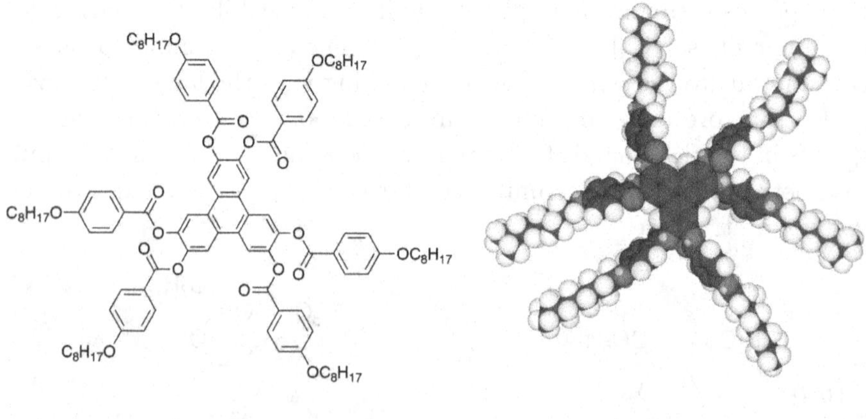

Figure 6.19 The structure of hexa-octyloxybenzoyloxytriphenylene.

We can test the effects of steric twisting on mesomorphic behavior by incorporating substituents in the peripheral benzene rings. Locating a substituent at position Y has less of an effect on steric twisting than a substituent at position X in the structure in Table 6.2. Thus, the table shows systematic comparisons for X and Y both being a methyl unit, and the only variable being the length of the chains attached to the edge of the extended aromatic disk. The first observation of note is that the unsubstituted parent materials have higher transition temperatures than the substituted analogs. The second observation is that the parent materials are more likely to exhibit columnar phases in comparison to the analogs, which do not exhibit columnar phases at all. Third, the substituted materials are all nematogens with relatively low melting points, which is caused by the lateral substituents disrupting the packing of the cores, as shown in comparative models in Figure 6.20. Fourth, the greater the twist (inner > outer) the lower the transition temperature.

Table 6.2 Lateral substitution (X and Y) in the exterior benzene rings of the hexa-alkyloxybenzoyloxytriphenylenes. Temperatures are in °C.

C_nH_{2n+1}	X or Y	Cryst		Col$_{rd}$		Nematic		Liquid
C_6H_{13}	X=H, Y=H	·	186	·	193	·	274	·
C_8H_{17}	X=H, Y=H	·	152	·	168	·	244	·
$C_{10}H_{21}$	X=H, Y=H	·	142	·	191	·	212	·
$C_{12}H_{25}$	X=H, Y=H	·	146	·	174	·	-	·
Outer		**Cryst**				**Nematic**		**Liquid**
C_6H_{13}	X=H, Y=CH$_3$	·	126		-	·	233	·
C_8H_{17}	X=H, Y=CH$_3$	·	134		-	·	223	·
$C_{10}H_{21}$	X=H, Y=CH$_3$	·	103		-	·	192	·
$C_{12}H_{25}$	X=H, Y=CH$_3$	·	99		-	·	163	·
Inner		**Cryst**				**Nematic**		**Liquid**
C_6H_{13}	X=CH$_3$, Y=H	·	172		-	·	230	·
C_8H_{17}	X=CH$_3$, Y=H	·	126		-	·	198	·
$C_{10}H_{21}$	X=CH$_3$, Y=H	·	109		-	·	165	·
$C_{12}H_{25}$	X=CH$_3$, Y=H	·	105		-	·	129	·

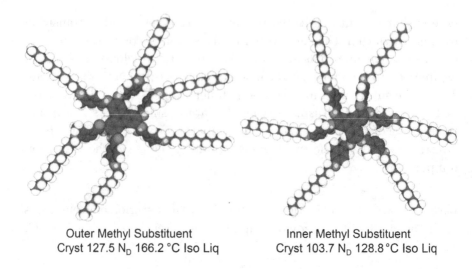

Outer Methyl Substituent
Cryst 127.5 N_D 166.2 °C Iso Liq

Inner Methyl Substituent
Cryst 103.7 N_D 128.8 °C Iso Liq

Figure 6.20 Space filling models for the methyl substituted hexa-dodecyloxybenzoyloxytriphenylenes.

We can investigate to some degree the effect of the out-of-plane twisting and steric shape by controlling the size of the lateral substituents. This has been done via a classical approach by systematically varying the substituent from methyl to ethyl, isopropyl and tert-butyl while retaining consistency in the remainder of the molecular structure. Table 6.3 shows the results on the mesomorphic behavior of the laterally substituted hexa-decyloxybenzoyloxytriphenylenes.

As predicted, the only material of the inner and outer substituted families to exhibit a columnar phase is the parent unsubstituted compound. The results show that all of the substituted materials are nematic discogens, and that the inner substituted systems have the lower transition temperatures in comparison to the outer. The larger the substituent, the lower the clearing point. Thus the steric twist coupled with the size of the substituent appears to reduce the interdisk interactions, thereby suppressing formation of columnar organization and favoring nematic mesophase formation. The steric twist is demonstrated clearly in the blow-up of the central disk region of the inner tert-butyl substituted and unsubstituted forms of benzoyloxytriphenylene shown in Figure 6.21.

These observations have been utilized in the nanoengineering of nematic discogens for applications in thin optical retardation films that are now used extensively in wide viewing angle displays, and also as lubricants used in all computer disk drives. Both of these applications are described in Chapters 14 and 15.

Table 6.3 Effect of the size of the lateral substituent (X and Y) in the exterior benzene rings on the transition temperature (°C) of the hexa-4-decyloxybenzoyloxytriphenylenes.

X or Y	Cryst		Col_{rd}		Nematic		Liquid
Y=H, X=H	·	142	·	191	·	212	·
Y=CH$_3$, X=H	·	102	·	-	·	192	·
Y=C$_2$H$_5$, X=H	·	129	·	-	·	206	·
Y=CH(CH$_3$), X=H	·	161	·	-	·	202	·
Y=C(CH$_3$)$_3$, X=H	·	194	·	-	·	225	·
X or Y	**Cryst**				**Nematic**		**Liquid**
X=H, Y=H	·	142	·	191	·	212	·
X=CH$_3$, Y=H	·	107	·	-	·	162	·
X=C$_2$H$_5$, Y=H	·	117	·	-	·	131	·
X=CH(CH$_3$), Y=H	·	93	·	-	·	70	·

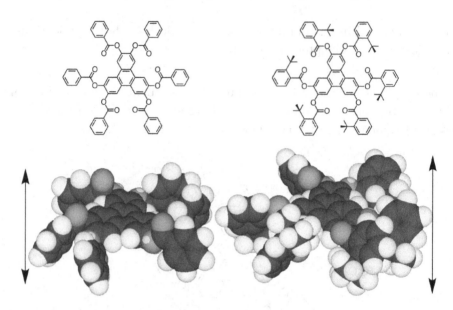

Figure 6.21 Exemplification of steric twist around the central disk region of the unsubstituted (left) and inner substituted tert-butyl (right) forms of benzoyloxytriphenylene.

The steric repulsion produced by varying the lateral substituents is achieved by controlling the shape/volume of non-polar groups. Further studies using polar substituents, however, produce slightly different results. Taking the structure of the discogen in Table 6.3 and replacing the substituents X and Y with halogens, produces materials with much stronger polar disk-like core units. For the inner-substituted materials, neither the fluoro- nor chloro-substituted compounds are mesomorphic, with both melting at temperatures greater than 300°C, upon which decomposition starts to occur. The outer-substituted analogs prove to be mesomorphic with the fluoro-substituted material melting at 159°C and exhibiting a nematic discotic phase up to a temperature of 194°C. The chloro homolog melts at only 80°C to a nematic phase and then to the liquid at 105°C. Thus the same principles still apply: (1) the introduction of a lateral substituent favors the formation of nematic discotic phases over columnar phases, (2) lower melting and clearing points are found for larger lateral substituents; only for certain architectures do the melting points rise to a level whereby no mesophase behavior is observed.

6.6 CHEMICAL MOIETY SELECTION – CHANGING THE DISK-LIKE CORES

After the discovery in 1978 of the first discogen based on hexa-substituted benzene, the natural tendency for designing and synthesizing discotic materials was to select chemical motifs that were disk-like for the core unit. This approach led almost immediately to reports on discogens based on triphenylene. Quickly following this, truxene was exploited using the same approach.

The truxene core is larger than the triphenylene core and consists of four benzene rings. The three radial rings are symmetrically attached to the central ring in two ways; first, by a conjugative single bond, and second, through a methylene spacer that locks in an approximately planar structure by preventing dynamic inter-annular twisting, as shown in Figure 6.10. Accordingly, the mesomorphic tendency of the compounds based on the hexa-substituted truxene core is relatively high. For example, the simple hexadecyloxytruxene, as shown in the left-hand structure of Figure 6.22, exhibits a wide temperature range Col_{ho} phase up to 260°C.

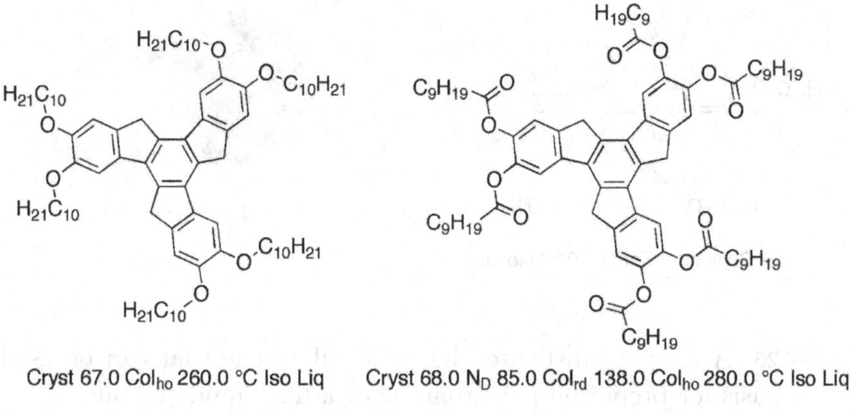

Cryst 67.0 Col_{ho} 260.0 °C Iso Liq   Cryst 68.0 N_D 85.0 Col_{rd} 138.0 Col_{ho} 280.0 °C Iso Liq

Figure 6.22 Two examples of hexa-substituted truxenes.

Like order-disorder-order transitions in smectic phases, the phase sequences for discotic materials can exhibit disordered phases at lower temperatures changing to ordered ones at higher temperatures. For example, the right-hand material shown in Figure 6.22 exhibits an inverted phase sequence for which the nematic discotic phase is exhibited at a lower temperature than the rectangular disordered columnar (Col_{rd}) and the hexagonal ordered columnar (Col_{ho}) mesophases.

This type of behavior usually relates to changes in the ability of the molecules to pack with respect to temperature, often caused by the conformational arrangements and dynamics of the peripheral chains decreasing on cooling, whereas at higher temperatures the chains possibly adopt all trans conformations that allow the discotic cores to pack into columns.

Following on with the concept that the central core units are made up of smaller ring structures that are fused together, we enter into the area of polyaromatics. Polyaromatics have been widely studied for their potential applications in lubricants, organic electronics, light emission, *etc.*, and have been described as supermolecular nanographenes. Their syntheses usually start with a smaller discotic liquid crystal such as a substituted hexaphenylbenzene, as shown in Figure 6.23, and then fuse the outer rings together. This approach to making discogens gives easy access to coronenes, and ultimately to graphenes and graphene ribbons, as shown in Figure 6.24.

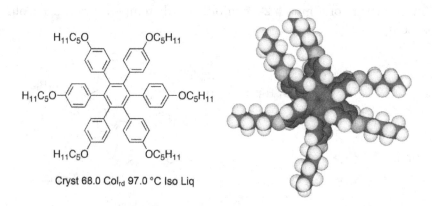

Cryst 68.0 Col$_{rd}$ 97.0 °C Iso Liq

Figure 6.23 A simple substituted hexaphenylbenzene that can be used as the basis for preparing polyaromatic disk-like liquid crystals.

The materials shown in Figure 6.24 are compared to analogs with the same aliphatic chain substitution (R = $C_{12}H_{25}$). All three melt at temperatures just above 100°C into columnar hexagonally ordered mesophases which then persist to temperatures above 400°C, at which point they start to decompose. The ordered columnar organizations are probably due to the strong π-π interactions between the polyaromatic cores coupled with a limited influence of the aliphatic chains, which are not great in number around the edges of the disks. It is interesting to note that even though some of the materials are more board-like in shape, they still appear to form hexagonal phases rather than rectangular.

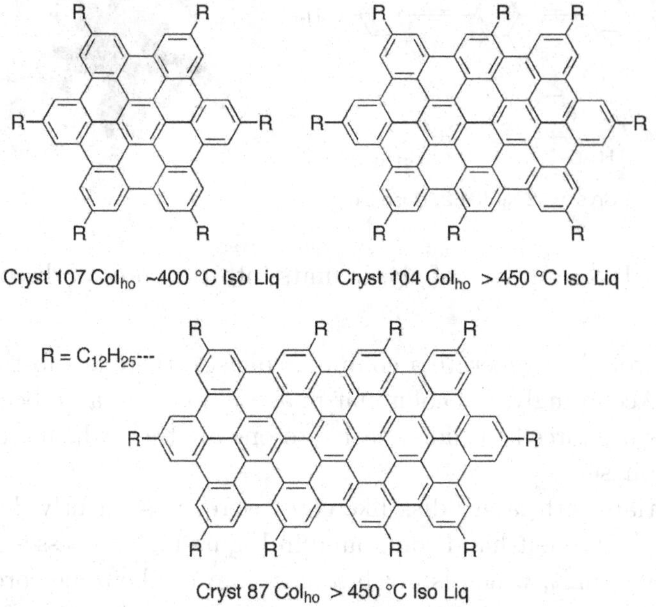

Cryst 107 Col$_{ho}$ ~400 °C Iso Liq   Cryst 104 Col$_{ho}$ > 450 °C Iso Liq

R = C$_{12}$H$_{25}$---

Cryst 87 Col$_{ho}$ > 450 °C Iso Liq

Figure 6.24 Supermolecular disk-like materials based on coronenes.

Having examined disk-like materials in which the π-electrons are delocalized across the central core units, we now turn to systems in which the regions of the core are not as freely delocalized, and investigate how this affects the formation of various mesophases. This can be achieved by incorporating sp^1 hybridized carbon atoms into the disk-like core in addition to those that are sp^2 hybridized, i.e., through the incorporation of alkynic bonding, as shown in Figure 6.25.

The first thing to note about materials that possess alkynic groups is that there is a greater tendency to form nematic discotic phases, as shown by the phase transitions exhibited by the material in the figure. The incorporation of the sp^1 hybridized acetylene linkages removes the steric interactions between the aryl rings and allows the rings to be easily twisted at 90° with respect to each other. This arrangement of benzene rings probably prevents the molecules from adopting a columnar structure. Additionally, the acetylene units are of localized polarizability, and as the molecules approach each other face-to-face

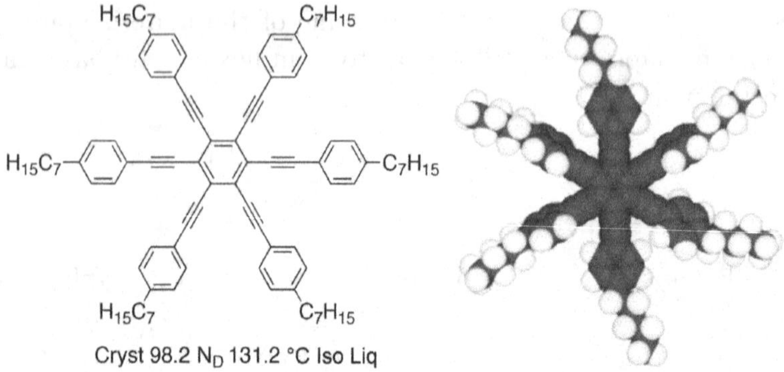

Cryst 98.2 N_D 131.2 °C Iso Liq

Figure 6.25 Incorporation of alynic units into the cores of discogens.

(as they would to generate a columnar phase), there is substantial repulsion. Accordingly, the columnar phase is not generated because the molecules are forced to slide relative to one another, which creates the nematic phase.

Materials with larger disk-like cores were subsequently developed; for example, the left-hand compound in Figure 6.26 possesses a central triphenylene unit, which is much larger than the benzene core in Figure 6.25. Hence its transition temperatures are much higher. However, strange behavior is seen for the analogous acetylene-linked naphthalene discogen shown to the right in the figure. This compound exhibits a nematic phase that is slightly lower in phase stability, but this is to be expected because of the slight asymmetry of the core. However, there is an inverted phase sequence with columnar phases being generated at higher temperatures. A similar effect was described previously for truxene discogens. This strange behavior is also sensitive to small changes in molecular architecture, because the slightest increase in the lengths of the peripheral chains suppresses columnar mesophase formation.

All of the above descriptions of the structures of discogens have been made on carbon based core units. However, it is also possible to create cores possessing hetero-atoms that may be conjugated or not. For example, discotic mesomorphism is supported in materials that have a tricycloquinazoline polyaromatic heterocyclic core, as shown in Figure 6.27. This type of central unit was investigated because it has strong core-core attractions and a low ionization potential. Such materials are known to be useful in electronic and photoelectronic applications such as electron carriers.

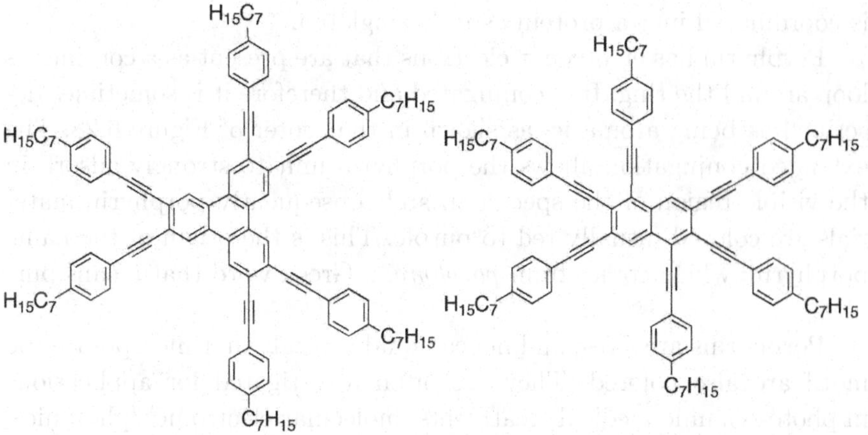

Cryst 1215 N_D 175 °C Iso Liq   Cryst 60 N_D 113 Col_{ro} 137 $Col_{ho} \sim 200$ °C Iso Liq

Figure 6.26 Nematic phases are exhibited by materials with larger disk-like core units, but anisotropy of shape can introduce columnar phases in a strange inverted phase sequence.

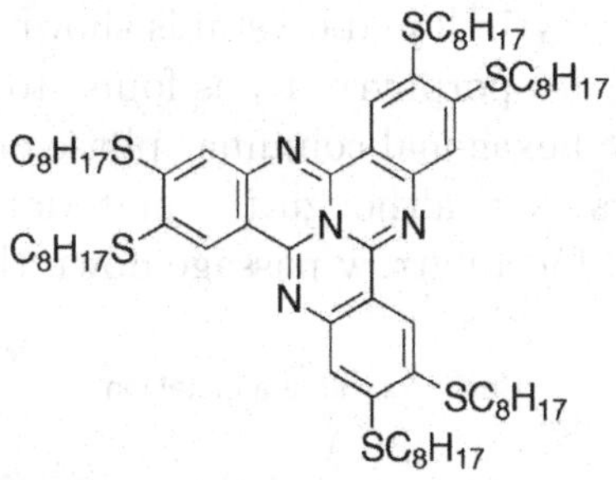

Cryst 80.7 Col_{ho} 207.4 °C Iso Liq

Figure 6.27 Example of a discogen that has a tricycloquinazoline poly-aromatic heterocyclic core.

Another heterocyclic system, which has been used in the investigation of discotic liquid crystals, is that based on the porphyrin motif as the core unit. Porphyrins are common in biology as the hemes in chlorophyll and hemagolobin. In biosystems, hemes are effectively coordination complexes of metal ions to the porphyrin unit, which acts as a tetradentate ligand. There are also axial ligands when the porphyrin is coordinated into a protein as in hemaglobin.

Porphyrin has eighteen π-electrons that are present as a continuous loop around the ring. It is conjugated and therefore it is sometimes described as being aromatic, as shown in the center of Figure 6.28. The extended conjugation allows the porphyrin unit to strongly adsorb in the visible region of the spectrum, and consequently porphyrin materials are colored, usually red to purple. This is the origin of the name porphyrin, which comes from *porphyra*, a Greek word that means purple.

Porphyrins are dyes, and hence liquid crystals that incorporate the motif are also colored. They are often investigated for applications in photodynamic medical treatments, molecular electronics/photonics, and dye-sensitized solar cells.

The porphyrin motif is capable of having eight peripheral groups associated with the edge of the core, and with its extensive conjugation, it is more likely to induce columnar phase formation than nematic. Moreover, the potentially strong inter-disk interactions are more likely to produce high melting points and hence monotropic phases. The first observed discotic porphyrin liquid crystal is shown in Figure 6.28. The dodecyl derivative of uro-porphyrin I was found to melt at 107°C, and to form a monotropic hexagonal columnar phase on cooling at 96.8°C. The figure also shows that, although the material is not coordinated, there is the potential for a narrow passage down the column axes.

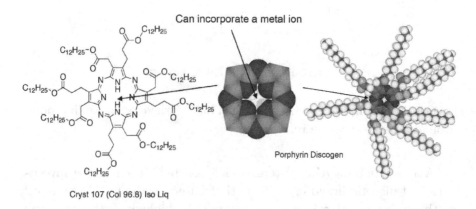

Figure 6.28 The structure and phase transitions for the octa-dodecyl derivative of uro-porphyrin I.

Extending the concept of using the porphyrin unit as a central core for discogens, phthalocyanine was the next step along the path of developing dye-based columnar liquid crystals. Phthalocyanines have similar applications to porphyrins in that they are used for dyes, electrochromics, photodynamic therapy, catalysts, and non-linear electrooptic materials.

Unlike porphyrins, phthalocyanines possess potentially sixteen peripheral locations at which different groups can be attached. However, usually only eight are used, and the substituents are often the same, as shown in the two structures in Figure 6.29. Phthalocyanine discogens with outer substitutions (left structure) nearly always exhibit wide temperature range hexagonal columnar (Col_{ho} and/or Col_{hd}) mesophases, whereas for materials with the non-outer eight peripheral sites available for substitution (right structure) tend to form rectangular columnar (Col_{rd}) phases.

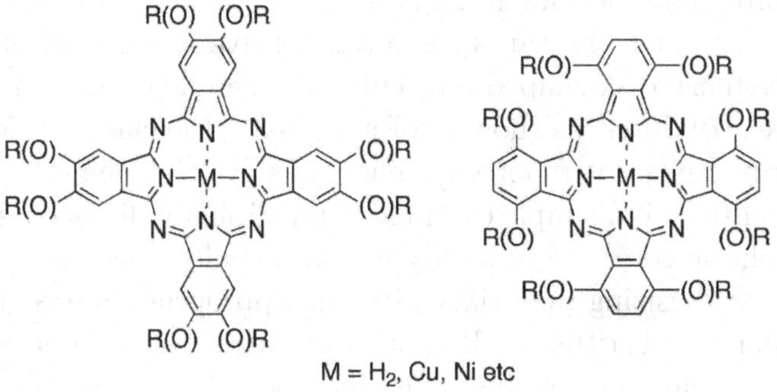

M = H_2, Cu, Ni etc

Figure 6.29 Possible substitution positions for phthalocyanine discogens.

The phthalocyanine core, like porphyrin, is able to hold a metal in the center which is often copper or nickel. The coordinated metal has the effect of increasing the columnar mesophase stability, but this usually results in decomposition before the clearing point is reached.

Other dye chromophores are useful sources of central cores for producing colored discogens. A very common chromophore is based on anthraquinone, which can produce calamitic materials with disubstitution and columnar discotic phases with hexa-substitution. For example, the hexa-substituted anthraquinone shown in Figure 6.30 exhibits a monotropic columnar phase. This result is unusual because the material exhibits columnar mesophases even though the molecular architecture is not quite disk-like. In this case the high polarity of the carbonyl units within the central core aids in the generation of the necessary intermolecular forces of attraction that support the formation of a columnar structure.

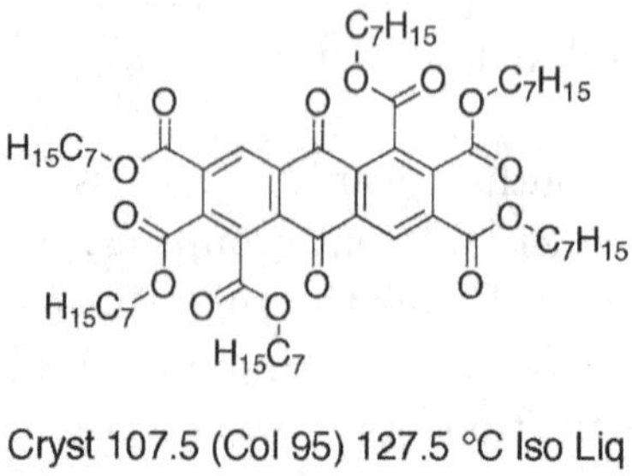

Cryst 107.5 (Col 95) 127.5 °C Iso Liq

Figure 6.30 A dye discogen based on an anthraquinone core unit.

The core systems described so far have all been based on aromatic or conjugated heterocyclic units, and indeed most discogens are in fact aromatic. Although the variety is not extensive in comparison to conjugated systems, disk-shaped molecules also can be generated from alicyclic core structures, as shown in Figure 6.31 (the classification of the columnar phase is not reported). The trans cyclohexane ring (equatorial substitution) is a simple example of an aliphatic discogen, however, larger aliphatic cores are probably not so prevalent because of the difficulty in synthesizing materials with the appropriate stereochemistry to give planar substitution. It is also interesting in comparison that calamitic systems tend to favor the incorporation of cyclohexyl units in their core segments.

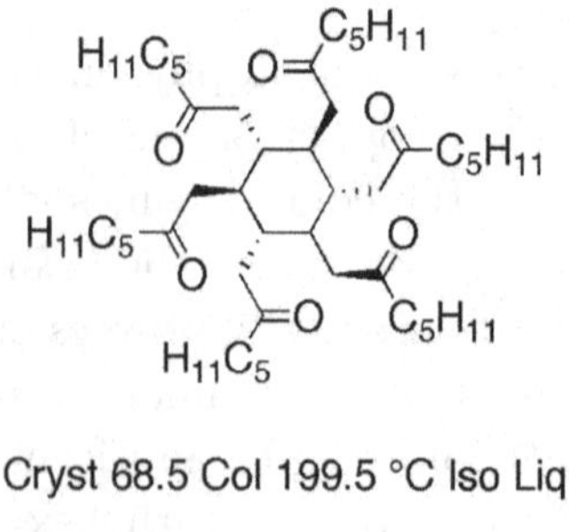

Cryst 68.5 Col 199.5 °C Iso Liq

Figure 6.31 An all aliphatic discogen.

6.7   DISCOTICS WITH CONNECTED DISKS

Disk-like core units can also be designed by connecting sub-units together to form a disk. Often the connection is made via H-bonding. The first observed discotic phase formed by connected molecular entities was di-isobutylsilanediol as shown in Figure 6.32. Pairs of molecules H-bond together to form a disk, and the disks stack together via H-bonding perpendicular to the disk to form columns. An interesting outcome from this study was realized when other analogs were prepared. It turns out that the isobutyl unit is the only moiety that supports mesophase formation, with the butyl, propyl, ethyl, isopropyl, and tert-butyl substituents forming H-bonded structures that do not generate columnar phases. This result indicates that the disk-like structure has to be just the right shape to fill the space around the core, thereby indicating the sensitivity of discotic phase formation. Moreover, di-isobutylsilanediol is also very unusual in terms of its chemical constitution; it is not conjugated, and does not have any carbon atoms in the disk-like core. Thus it is very different from the conventional design structures for discogens. It might also be expected that the H-bonding along the column axes would lock-in the structure, but the columns still tend to curve and fold, as can be seen by optical microscopy.

H-bonding can also be achieved easily between aromatic systems, but because of the geometry of benzene rings with angles of 120°, and similar for substituents, the formation of disk-like structures is usually between three units. This structural arrangement is demonstrated by the trimer in Figure 6.33.

Earlier a columnar phase was described for an equatorial hexa-substituted cyclohexane. Other cyclohexane systems based on inositol give an opportunity to form connected disks through hydrogen bonding for which the bonding is dependent on the number of hydroxyl units available and their stereochemistry. Inositol, which is effectively hexa-hydroxycyclohexane, has a variety of stereochemical forms as shown in Figure 6.34, where the hydroxyl units are either equatorial or axial with respect to the cyclohexane ring.

In Figure 6.35 three discogens that form various connected disks are selected to exemplify the following: (1) the effect of axial versus equatorial stereochemistry on mesophase formation; and (2) the effect of having multiple H-bond forming large disk-like entities on transition temperatures. The left-hand and center inositols have four ap-

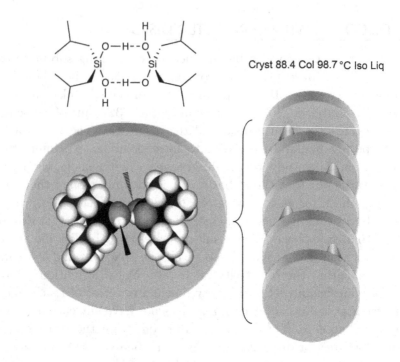

Cryst 88.4 Col 98.7 °C Iso Liq

Figure 6.32 The structure of di-isobutylsilanediol and the structure of the columnar mesophase it forms.

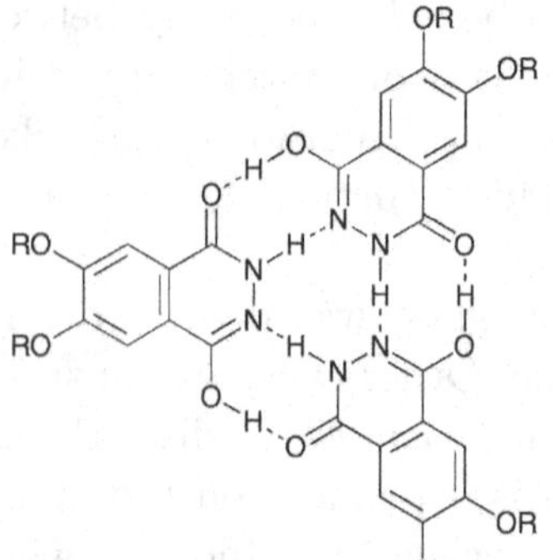

Figure 6.33 A hydrogen bonded discogen with three sub-units bonded at 120°.

pending groups that are essentially aliphatic chains, and two hydroxyl groups, that are available for H-bonding. For the left-hand structure there is one axial OH and one equatorial, whereas for the center structure there are two equatorial. The axial-equatorial combination gen-

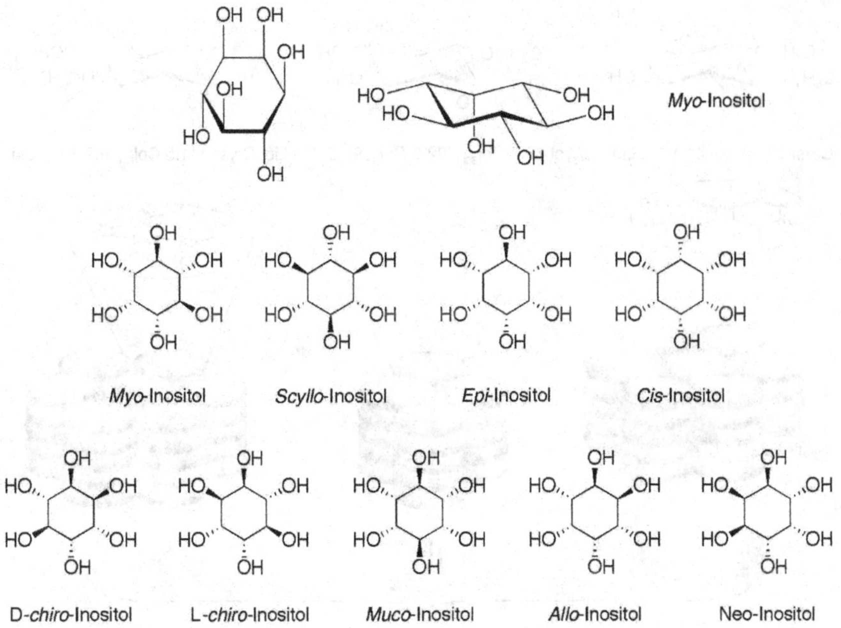

Figure 6.34 Stereochemical structures associated with the various forms of inositol.

erates a hexagonal columnar phase, whereas the equatorial-equatorial combination produces rectangular ordered columnar phases, with the axial-equatorial combination having much lower transition temperatures. Thus, the out-of-plane of the disk OH appears to interfere with the creation of the columns, which destabilizes mesophase formation.

For the di-substituted inositol on the right-hand side of the figure, which has four OH groups available for H-bonding, there is the opportunity for creating much larger central disk structures. In this case investigations show this could be as many as five sub-units, with the aliphatic chains being in an equatorial arrangement. The resulting mesophase has a hexagonal columnar structure, but this time the column has a disordered structure along its axis. As a result of the larger disk structure, the transition temperatures are higher than both of the other examples discussed.

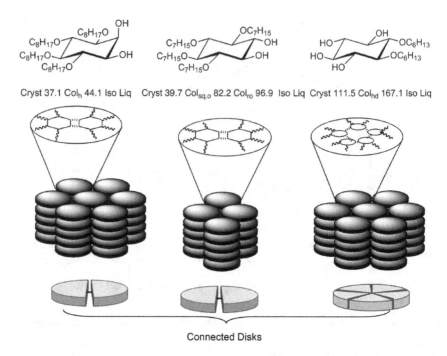

Cryst 37.1 Col$_h$ 44.1 Iso Liq   Cryst 39.7 Col$_{sq,o}$ 82.2 Col$_{ro}$ 96.9 Iso Liq   Cryst 111.5 Col$_{hd}$ 167.1 Iso Liq

Connected Disks

Figure 6.35 Columnar structures formed by various connected derivatives of inositol that are substituted by octyl chains.

6.8 DISCOTICS WITH DONUT RINGS

So far we have only looked at discotic liquid crystals formed by materials that have disk- or oblong-shaped central core units. However, there are many more possibilities for disk-shaped core units that are not ones one might immediately think of. For example, instead of a filled disk, consider what could happen if there is a hole in the middle.

In phenylacetylene macrocycles, acetylene-linking units have been employed in the construction of a conjugated central core unit to give a disk-like architecture. However, the core is not of the usual type but has a hollow center surrounded by alternating benzene rings and acetylene-linking groups; conventional ether and ester units have been used with peripheral aliphatic chains to provide nano-segregative moieties. The materials are designed to exhibit columnar mesophases that self-organize to give molecular or ion channels that can be used for the transport of entities for applications in molecular wires and membranes. However, as shown by the example in Figure 6.36, these materials tend to exhibit nematic discotic phases. The central core unit modeled in the right-hand structure of the figure shows that the aromatic rings can twist out of the plane of the disk. The twisting of course weakens the intermolecular interactions between the donut shaped cores, thereby supporting the formation of nematic discotic phases. This view is supported by the studies on the equivalent ester ($-OCOC_7H_{15}$), which has an extremely wide nematic temperature range (melting point of 121°C, clearing point of 241°C).

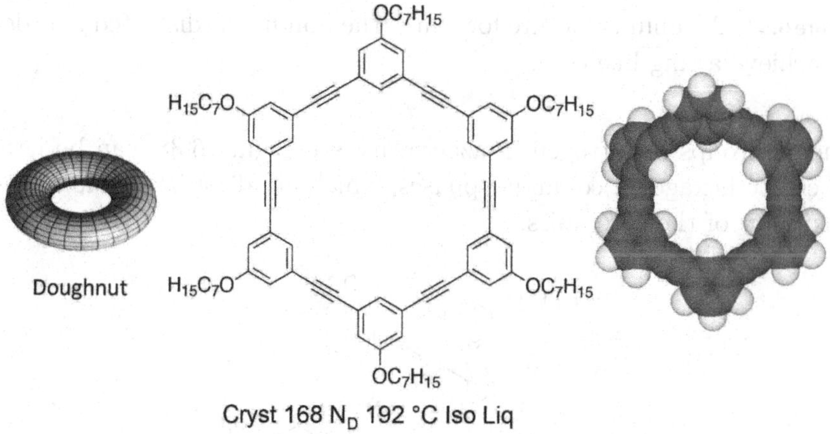

Cryst 168 N_D 192 °C Iso Liq

Figure 6.36 Donut ring nematic discogen, with the central core modeled to the right.

A similar effect is seen for the compound shown in Figure 6.37. In this material there are two triphenylene units held together by bridging diacetylene moieties. The diacetylene moieties are linear in terms of their bonding, and so the overall structure is sterically stressed, as the triphenylene motifs are flat. Steric twisting/bending therefore affects the face-to-face disk interactions, thereby favoring the formation of nematic discotic phases.

The fixed structures of the donut shaped conjugated core units can, in principle, be changed to flexible ring structures based on aliphatic moieties. For example, crown ethers are well known to have flexible ring structures that can accommodate various metal ions. In the case of the alicyclic azacrown systems, the molecules may assemble in a columnar fashion, and due to the ring nature of the molecules, the mesophase has been denoted as tubular. Accordingly, such systems that assemble into columns offer the potential for ion transport. Substitution with

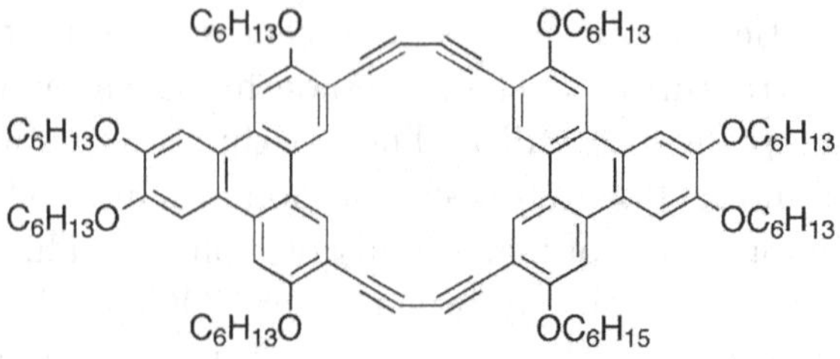

Cryst 209 N_D >280 °C Iso Liq

Figure 6.37 A donut structure for which the bonding is distorted in order to achieve a ring-like core.

phenyl groups at nitrogen in azacrowns, see Figure 6.38, can be used to create hexagonal columnar phases, which are disordered due to the flexibility of the core units.

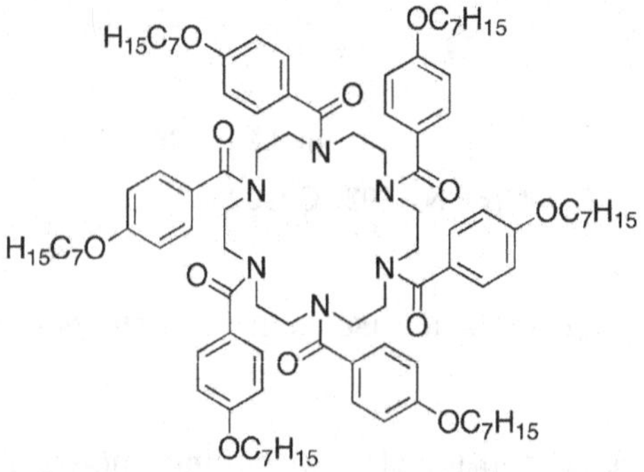

Cryst 121.5 Col 141.5 °C Iso Liq

Figure 6.38 An azacrown columnar discotic liquid crystal with a flexible donut central core.

6.9 DISCOTICS WITH BOWLIC CORES

Another example of the effect that the central core of a discogen has on mesophase formation and structure, and hence properties, is for the core to be bowl-like. Unlike a disk, a bowl has reduced rotational symmetry perpendicular to the vertical axis of rotation through the center of the bowl. Consequently, when bowls are stacked into columns, the symmetry of a column is reduced to $C_{\infty V}$. In effect this means that columns have directionality, *i.e.*, stacking up or down, as shown in Figure 6.39. If the charge distribution is different, up or down, for a molecular bowl, then the columns also are polar, as shown in the figure.

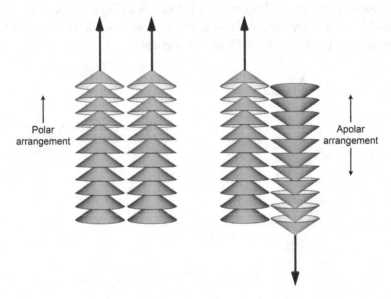

Figure 6.39 Stacking of bowlic discogens one on top of another. The columns can then be arranged either to be parallel (left) or antiparallel (right).

The types of bowl-like core units that can be constructed have two architectural variations, one for which the bowlic structure is relatively rigid and another for which the bowls have more flexibility and the potential to flip inside out. In the first case for parallel columns, the structure is polar, but for an antiparallel arrangement the structure is apolar. However, if the bowls are able to flip inside out, the structure is reversible under an applied electric field, changing from pointing up to pointing down. Hence the structure is ferroelectric, with the possibility of an antiferroelectric intermediary state if the flipping is energetically difficult to achieve.

An example of a relatively stiff bowlic system is shown by the tribenzocyclononatrienes in Figure 6.40. Although the structure of tribenzocyclononatriene is only weakly polar, substitution along the edges of the bowl (R groups) significantly increases the polarity along the vertical rotational axis, and therefore the properties of this system are dependent to some degree on the substituents. From the figure it can be seen that for simple alkoxy substituents, the melting points are relatively low and the clearing points are under 100°C. Moreover, the materials are reported to exhibit hexagonal columnar phases, as shown schematically in Figure 6.41 for the polar system. Of course, in the real system there are defects along and across the columns, which make perfect orientations difficult to generate or retain.

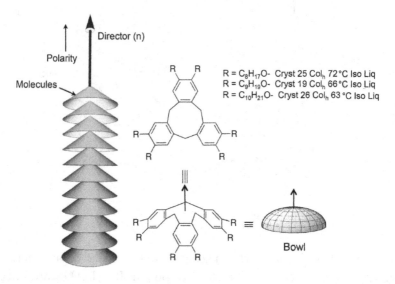

Figure 6.40 Structure and stacking of a bowlic system based on tribenzocyclononatriene.

Figure 6.41 A perfectly oriented and defect-free hexagonal columnar bowlic structure.

Flexible rings are usually achieved by expanding the cyclic ring structure. This is the case for the tetrabenzocyclododecatetraene (left-hand structure), and the metacyclophane (right-hand structure) in Figure 6.42. For materials based on these two systems, derivatives of the tetrabenzocyclododecatetraenes tend to support the formation of rectangular and hexagonal columnar phases, whereas the metacyclophanes exhibit ordered hexagonal columnar phases, with the aliphatic esters melting below 50°C and the clearing points just below 70°C.

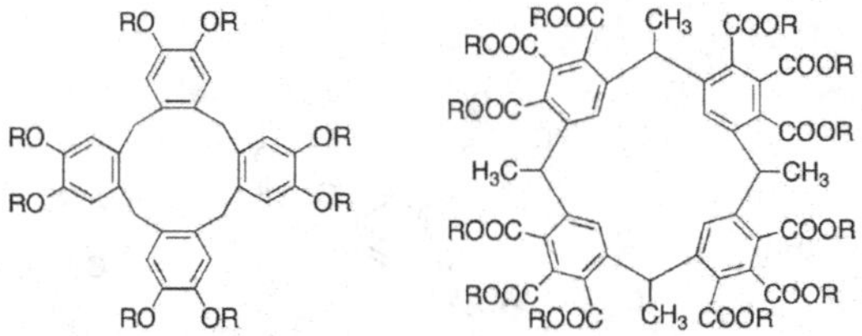

Figure 6.42 Columnar liquid crystals based on substituted tetrabenzocyclododecatetraene (left) and metacyclophane (right).

6.10 DISCOTIC SUPERMOLECULES WITH HYBRID COLUMNAR AND CALAMITIC STRUCTURES

Supermolecules are a slightly different concept compared to supramolecular systems. Supramolecular materials are effectively sub-units bound together via various interactions, other than covalent bonding, to create a large-scale definitive entity. In discotic systems, a supramolecular entity could be one in which several disk-like core units are bonded together in a defined way to give a larger disk-like unit.

For example, a central triphenylene core with six points of substitution could have six peripheral triphenylene units attached, as shown in Figure 6.43. This compound uses flexible spacers to attach peripheral triphenylene units to the central discotic core in a star-like manner. Hence such supramolecular materials are commonly called "star-like" liquid crystals, with the resulting macro-disks stacking to form columnar mesophases. The mesophase formed by the compound shown is identified as hexagonal.

The structure of the example described is effectively uniform (monodisperse), composed as it is by the same disk-units. But this type of arrangement is not necessarily the only form for discotic supramolecular materials. Supramolecular discogens are formed with calamitic peripheral units and wedge-like moieties as well as disk-like entities, and at the center there can be spheres, cubes, flexible chains, alicyclic

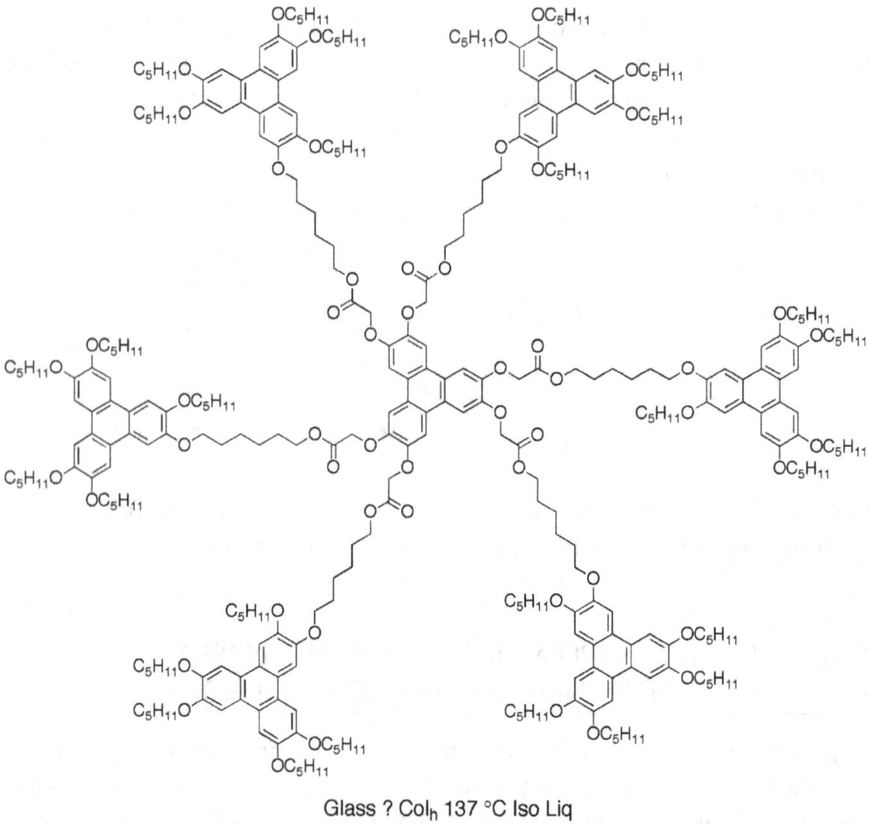

Glass ? Col$_h$ 137 °C Iso Liq

Figure 6.43 A supermolecular discotic mesogen comprising seven disk-like core units.

rings, *etc.* In many of these cases the peripheral arms, at least in the gas phase, spread out to give a spherical form. However, in the mesomorphic state the spherical forms are squashed down into disks in order to minimize the free volume. When the center of the supermolecule is relatively rigid, the overall structure is analogous to a saturn-like shape, as shown in Figure 6.44 for a C_{60} fullerene core spherically substituted with wedge-like moieties. The material shown, for $R = OC_{12}H_{25}$, exhibits a hexagonal columnar phase.

Supermolecular materials of course have some analogies with oligomers and polymers, but clearly when there are no repeats with respect to the peripheral groups, then a material is not an oligomer or polymer, as described later in Chapter 10 on polymeric liquid crystals.

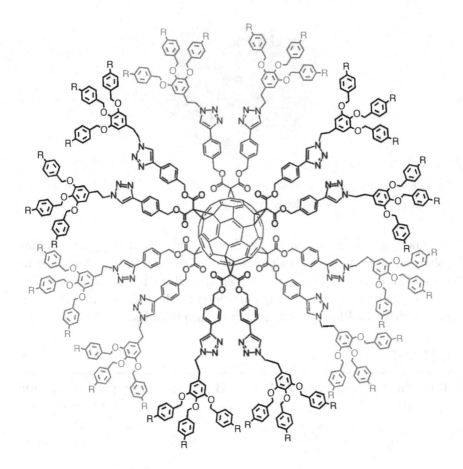

Figure 6.44 A supermolecular saturn-like discogen where the spherical structure is squashed into a disk.

So far, all of the material types that have been described have disk or board-like central core units, even when they are connected systems. However, there is the possibility that columnar phases may be created through the associations of wedge-like entities, which do not form disk-like structures. For example, Figure 6.45 shows some wedge-like components that form a hexagonal columnar phase, but for which the individual components are disordered along the axes of the columns, i.e., none of the components come together to form a disk or board; they just pack together to minimize the free volume and maximize the molecular density via the formation of flexible pillars, as shown by the disorganized arrangement of the wedges in the figure.

Figure 6.45 Hexagonal columnar phase, which is formed by materials with no disk- or board-like core units.

There is a large set of materials that exhibit these types of pillar columnar mesophases. This particular set of materials is described as being amphitropic. The materials are based on lipids, or fats, which have polar or ionic head groups attached to them. This class of materials is described later in Chapter 9 on amphiphiles.

EXERCISES

6.1 The two compounds **A** and **B** shown below possess discotic phases. Giving explanations, predict the phases that they exhibit.

$H_{19}C_9$ C_9H_{19} $H_{19}C_9$ C_9H_{19}

$H_{19}C_9$ C_9H_{19} $H_{19}C_9$ C_9H_{19}

$H_{19}C_9$ C_9H_{19} **A** $H_{19}C_9$ **B** C_9H_{19}

6.2 The two polar compounds **C** and **D** have the potential to exhibit discotic mesophases; however, one is non-mesogenic, whereas the other possesses a monotropic phase. Giving explanations, identify which compound is non-mesogenic. For the other compound choose the mesophase that is formed, and comment on why a monotropic mesophase is observed.

6.3 Phasmids are unusual materials that exhibit columnar liquid crystal phases. The compound shown below possesses a disordered hexagonal columnar phase. Draw at least two potential structures for this phase showing the arrangements of the molecules.

6.4 The drawing below shows four possible molecular structures when R is replaced by groups **E** to **H** resulting in these groups being attached to the triphenylene unit.

Of the four compounds that are possible, with reasoning, identify the following, (a) compounds that are non-mesogenic, (b) compounds that only exhibit nematic phases, (c) compounds that possess columnar phases but are not nematogenic, (d) compounds that exhibit columnar and nematic mesophases, and (e) a compound that has the largest negative birefringence.

6.5 The compounds **I** and **J** have identical aliphatic chains, R. Explain why material **J** has a higher clearing point than **I**, and also a higher birefringence. Elucidate why both materials exhibit columnar phases whereas the parent truxene analog is nematogenic.

I   J

CHAPTER 7

Chiral Liquid Crystals – Twisted, Frustrated, and Sometimes Defective

7.1 HANDED LIQUID CRYSTALS

We break now with the descriptions of liquid crystal phases to describe the features of objects that are said to be chiral, which is Greek ($\chi\epsilon\rho\iota$, kheir) for "hand", and how they self-organize into beautiful, complex structures of reduced symmetry, that often have fascinating physical properties.

According to Lord Kelvin (1884), an object is called chiral when it has a non-superimposable mirror image. Many natural systems are composed of mixtures of materials that have left-handed and right-handed antipodes, called enantiomers, that are present in unequal proportions. Such systems are also said to be chiral.

In chemical systems, the topic of chirality has reached a pinnacle of importance in recent years, particularly with the development of new asymmetric reactions, which allow chemists to design and create chiral materials almost at will. Nowhere is this success more apparent than in the synthesis of biologically and pharmacologically active substances; life, after all, is "handed". The ability to synthesize chiral compounds, with high enantiomeric purities, i.e., those compounds that have very large differences between the amounts of their left- and right-handed enantiomers, has also influenced the development of "advanced materials" through the preparation of novel polymers, liquid crystals, non-linear optoelectronic materials, and compounds used in molecular recognition processes.

For liquid crystals, the availability of chiral starting materials has led to the design and synthesis of many chiral compounds that exhibit mesophases, which are themselves chiral. This simple statement implies that there are a number of levels of chirality in relation to self-assembly and self-organization. Thus, we can define chirality in liquid crystals in a few simple ways: first, by the point symmetry or asymmetry of molecular architecture (usually labeled with an *); second, through the space symmetry or asymmetry of the local structure of the mesophase; and third, through the form chirality of the macrostructures that many phases exhibit. Point symmetry is itself described by the spatial configuration rules of Cahn, Ingold, and Prelog, space symmetry by group theory, and form chirality by helicity or "handedness". Thus chiral liquid crystals are also identified with the asterisk label, *e.g.*, N*, SmA*, SmC*, *etc.* Two other important factors that are also coupled to the asymmetry in such systems are the molecular biaxiality and the conformational structure(s) of the molecules. In the molecular design of chiral liquid crystals, all of these factors have to be taken into account. In the next sections we examine the various levels of symmetry.

7.2 MOLECULAR ASYMMETRY AND DISSYMMETRY IN LIQUID CRYSTALS

The concepts of molecular symmetry and asymmetry (point symmetry) are used to describe the spatial configuration of a single molecular structure, inasmuch as they describe the geometric, conformational, and configurational properties, *i.e.*, the stereochemistry of a material. An asymmetric atom can be described as an atom that has a number of different atoms or functional groups bonded to it for which their arrangement in 3-D space has no superimposable mirror image.

Using a tetrahedral atom (*e.g.*, carbon, silicon, germanium, *etc.*) to create an asymmetric architecture, Figure 7.1 shows the spatial configuration about two asymmetric atoms, which are related to one another as mirror images (enantiomers). The two atoms are labeled by the absolute configuration system introduced by Cahn, Ingold, and Prelog as being either *R* (*rectus*, right) or *S* (*sinister*, left). The absolute spatial configuration designation of an asymmetric atom is obtained by first determining the relationship of the group priorities about the asymmetric atom based on oxidation number. The group with the lowest priority (4) is positioned, according to the conversion rule, towards the rear of the tetrahedral structure as shown in the figure. The other three groups are viewed from the opposite side of the asymmetric center. If the remaining groups are arranged in descending order (1 to 2 to 3) of priority in a clockwise direction about the asymmetric center, the spatial configuration is designated *R* (left-hand part of figure). For the reverse situation, when the priority order descends in an anti-clockwise direction (right-hand part of figure), the spatial configuration is given the absolute configuration label *S*. For the vast majority of materials, the asymmetric center is a carbon atom, and rather than being called a chiral center, it is now known as a stereogenic center.

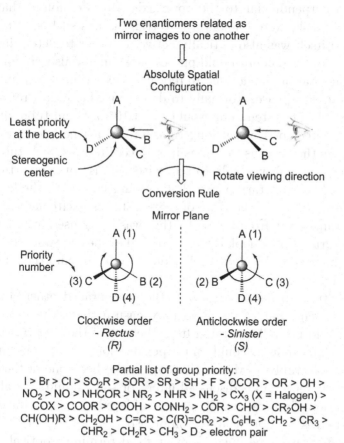

Figure 7.1 Cahn, Ingold, and Prelog rules for the classification of asymmetric centers in molecules, and a partial list of group priority in descending order.

The history of the discovery of chirality in molecular materials involves the rotation of the direction of linearly polarized light as it passes through a solution of a potentially chiral solute in a solution of a non-chiral solvent. The rotation of the direction might be clockwise or counterclockwise for an observer watching the light emerge from the sample. For a clockwise rotation, the material is called dextrorotatory and given a sign of (+); for counterclockwise rotation, it is called levorotatory and labeled as (−). Dextrorotatory and levorotatory materials are said to be optically active or possess optical activity.

Biot, in 1812, discovered optical rotation, sometimes called optical rotatory power, while experimenting with sections of quartz disks that were cut perpendicular to the optic axis. He later found that when the quartz disks were immersed in certain liquids such as turpentine, the liquid itself was also optically active. In order to determine if the structures of the solid or liquid phases of the materials were important to optical rotation, he decided to repeat his experiments on turpentine while in its gas phase. Normally to determine the optical rotation for a solute/solvent system, one would use a 10 to 20 cm tube, but for a gas Biot opted for a 30 m long tube into which he would introduce the gas. As the gaseous form was introduced into his 30 m tube, Biot, sitting at one end was able to see, with the use of a polarizer, that the gas exhibited rotary power that was indeed due to the individual molecules and not to the state of matter. In his excitement, he called his assistants away from the boiler that was being used to produce the gaseous turpentine to look through the end of the tube and polarizer to confirm his result. The boiler left alone caught fire and the laboratory burned down.

In writing up his observations, Biot encouraged others to repeat his experiment, but in doing so they ensure that "the furnace and boiler should be well-separated from the rest of the apparatus, or the explosion of vapor ... could cause persons located at some distance to perish miserably."[1] Maybe this could have been one of the earliest safety documents! As is discussed later, for some liquid crystal phases the detection of optical activity can be achieved for certain wavelengths over distances as short as a micron or two.

The standard experimental set up for determining optical rotation is shown in Figure 7.2. This involves a monochromatic light source, usually from a sodium lamp, a polarizer to give a linearly polarized beam,

[1]J. Applequist, *American Scientist* **75**, 58 (1987).

a tube usually a decimeter long, or multiples thereof, with optically transparent entrances to the tube, and a rotatable analyzer for observation. The tube is filled with a solvent and a soluble chiral compound of known concentration, and the rotation of the linearly polarized light through the medium is determined via rotation of the analyzer. When this experiment is performed at constant temperature using monochromatic light, the specific optical rotation for the chiral compound is determined. The optical rotation is given by $\alpha = kcl$, where c is the concentration of the chiral material, l is the length of the tube, and k is a constant.

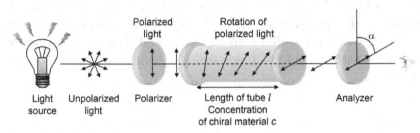

Figure 7.2 The experimental set up for the determination of specific rotation for a chiral compound.

When c is in g/ml and l is in decimeters, then $k = [\alpha]_\lambda^T$, which is called the specific rotation for a defined material at temperature T, usually $25°C$, and at a selected wavelength λ, which is usually the sodium D line (589 nm). Thus for a chiral material, the optical rotation will also be defined as dextrorotatory (+) for a clockwise rotation of the analyzer or levorotatory (−) for a counterclockwise rotation, for an observer looking at the light emerge from the tube. For a sample of a known chiral material but of unknown spatial configuration, this is another way of determining if the material in excess is of the R or S type, irrespective of optical purity.

For a chiral compound for which the specific rotation has not been standardized for concentrations of 100% R or S, the measured value cannot be used to determine the relative proportions of R or S.

More complications in the labeling of materials can arise from historical labeling of chiral compounds derived from biological systems, particularly sugars. Although the labeling of dextro and levo chiral materials can sometimes be synonymous as both (+) and d can be used for dextro, and (−) and l can be used for levo, there is a further complication due to the use of capital D and L, which are related to the stereochemistry of sugars rather than the experimental determination of the physical property of d or l. The D and L labels are a configurational notation that was introduced by Emil Fischer to designate the configurations of various sugars relative to the $(+)$ and $(-)$ forms of glyceraldehyde. The D and L labels are still in use today for sugars and amino acids, and particularly for reaction substrates used in the preparation of chiral liquid crystals. For example, lactic acid is used as a substrate in the preparation of many varieties of chiral smectic liquid crystals. Its structure in terms of the absolute spatial R and S configurations of Cahn, Ingold, and Prelog is shown in Figure 7.3. Using the other labels, the two enantiomers could be designated as R-D-$(-)$ and S-L-$(+)$. However, it should be noted that R and S are not necessarily synonymous with $(+)$ and $(-)$, nor with the helical twist directions in liquid crystals.

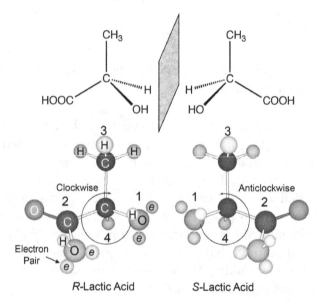

Figure 7.3 Cahn, Ingold, and Prelog configurations for the enantiomers of lactic acid.

For a material with one stereogenic center, there are two enantiomers. The two forms are expected to have the same values of their physical properties, such as melting point, boiling point, *etc.* However, what happens when there is more than one stereogenic center? Take for example cholesterol. There are many derivatives of cholesterol that exhibit liquid crystallinity, with a large number of these possessing helical macrostructures that selectively reflect light. The architecture of cholesterol is shown in Figure 7.4; it has eight stereogenic centers shown by the * symbol in the structure. The number of stereogenic isomers is given simply by $2n$, where n is the number of stereogenic centers. For cholesterol, the number of stereogenic isomers is 256. Pairwise, some of the relationships between isomers are enantiomeric, that is, the two isomers are mirror images of each another. However, many isomers are not mirror images of one another; these are diastereoisomers. Unlike enantiomers, diastereoisomers do not have the same physical properties, and for diastereoisomers of cholesterol, they exhibit selective reflection for different wavelengths of light.

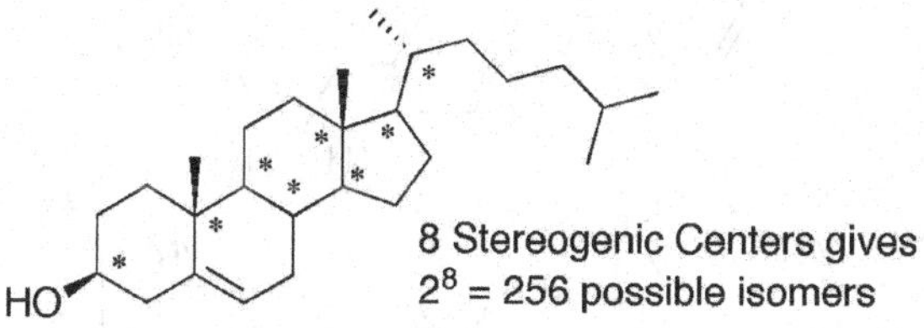

8 Stereogenic Centers gives
$2^8 = 256$ possible isomers

Figure 7.4 Structure of cholesterol showing the locations of eight stereogenic centers.

Consider for example a simple structure in which a molecule has two stereogenic centers. The isomer that has RR configurations for its stereogenic centers is the enantiomer of the one with an SS configuration. However, the RR isomer is the diastereoisomer of the RS isomer. Conversely, the RS isomer is the enantiomer of the SR variation. These relationships are shown in the chemical structure formats in Figure 7.5. A mesoform has a mirror plane dividing two parts of the molecules, each of which has at least one stereogenic center. Mesoforms are not chiral even though they possess two or more sterogenic centers.

So far we have only discussed chirality in molecules that contain an asymmetric atom. However, some molecules are optically active even when they do not possess an asymmetric atom, *e.g.*, substituted allenes, spirocyclobutanes, *etc.* For molecules with a dissymmetric structural grouping, the R or S configuration can be found by assigning priority 1 to the higher priority group in front, 2 to the lower priority group in front, 3 to the higher priority group in the back, *etc.*, then examining the path $1 \rightarrow 2 \rightarrow 3 \rightarrow 4$. For example, the absolute spatial configurational assignment for the simple molecule 1,3-dimethylallene is shown

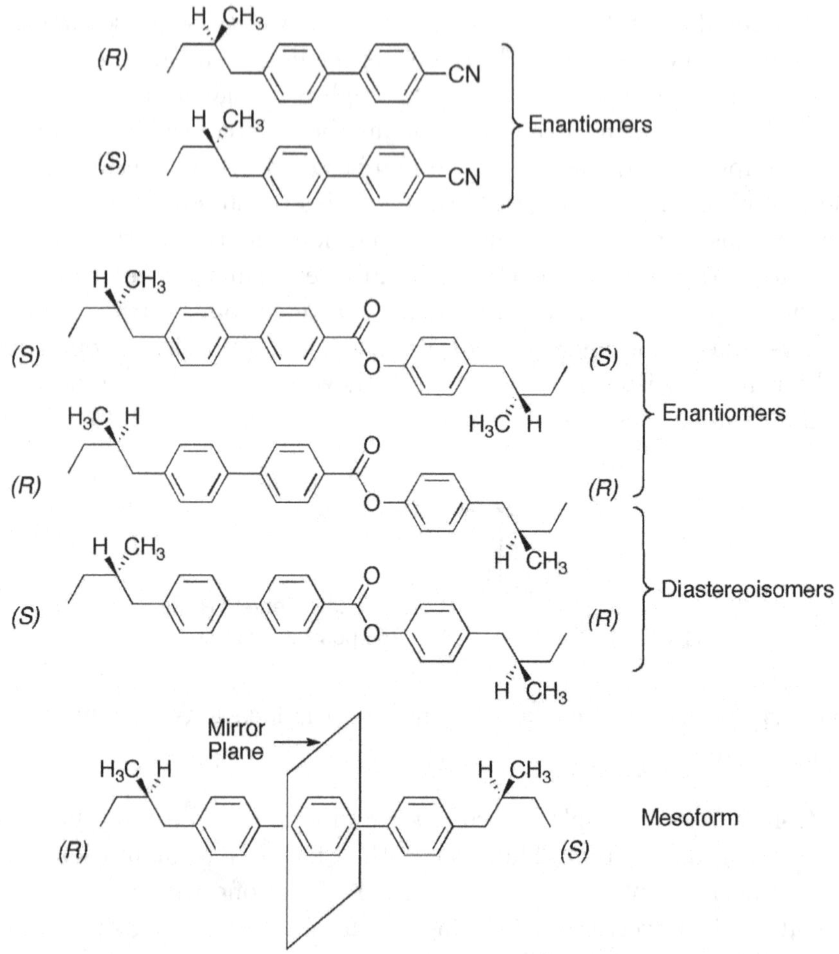

Figure 7.5 Enantiomers, diastereoisomers, and mesoforms of various materials associated with calamitic liquid crystals and liquid crystal technology.

in Figure 7.6. In the figure the R-enantiomer is shown by the structure on the left. Rotating this structure by $90°$ so that we are looking down the main molecular axis, represented by the circle, gives the center-left structure in the figure. The two methyl groups are to the top (front) and to the left (back) thereby creating a clockwise path of group priority and hence an R configuration. Flipping the horizontal methyl group from position 3 to position 4 gives a counterclockwise path of group priority, and hence a change to the S enantiomer. As the two structures are related to one another as non-superimposable forms, they are chiral.

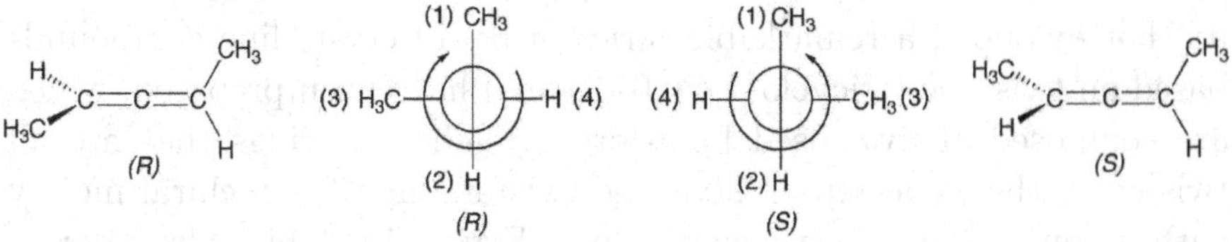

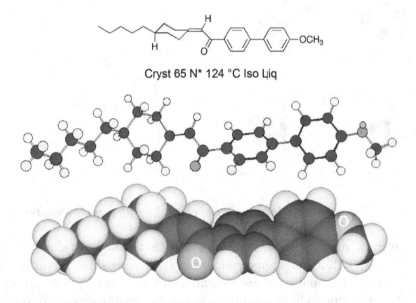

Figure 7.6 Molecular dissymmetry showing the enantiomers of 1,3-dimethylallene. The two structures on the left are R-enantiomers, whereas the two structures on the right are S enantiomers.

This type of dissymmetric structural concept is used as a template in the creation of a range of chiral nematic liquid crystals based on the chiral cyclohexylidene ethanone unit, as shown in Figure 7.7. This material exhibits a chiral nematic phase and its structure is described in a later section. However, the broken symmetry, which is located in the core section of the molecule, does not have a particularly strong effect on the physical properties that are associated with chirality.

Cryst 65 N* 124 °C Iso Liq

Figure 7.7 A chiral nematic liquid crystal derived from 1-biphenylyl-2-cyclohexylidene ethanone, which has a dissymmetric structure.

Dissymmetry can also occur for materials that possess rotational symmetry elements. Typically dissymmetric materials that have two- or threefold rotational axes of symmetry can also exhibit optical activity. For example, a remarkable series of liquid crystalline compounds based on twistane (tricyclo(4,4,0,0)decane) have been prepared, which are composed of five fused boat-form cyclohexane rings that are all twisted in the same sense. Thus the twistane motif is a chiral moiety with unique D_2 symmetry, as shown in Figure 7.8. This chiral moiety exists in two enantiomeric forms, (a) and (b), as shown in the figure in which the rotational axis is pointing into the page. The twist is exemplified by the positions of the two hydrogen atoms in the center at the top of the cage, just above the two carbon atoms located on the rotational axis. The pair in the left-hand structure is rotated to the left, whereas the opposite is the case for the right-hand structure. The chirality of twistane is therefore due to the space asymmetry of the molecule as a whole and does not depend on the presence of asymmetric atoms. Again, the chiral effects are relatively weak because the asymmetry is located in the core of the molecular architecture.

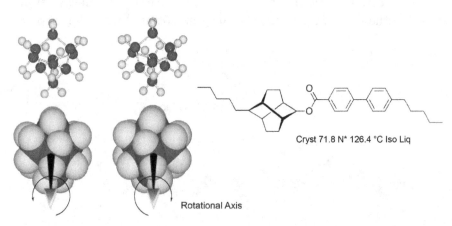

Cryst 71.8 N* 126.4 °C Iso Liq

Rotational Axis

Figure 7.8 Enantiomeric forms of tricyclo(4,4,0,0)decane (twistane).

Overall, dissymmetric molecules commonly have a simple axis of symmetry but no stereogenic centers, whereas for asymmetric molecules, rotational axes are absent. However, both species are optically active due to having no superimposable mirror images. In liquid crystal systems, both types of materials are capable of exhibiting chiral properties. Table 7.1 summarizes the relationships between optical activity, molecular structure, and rotational symmetry operations.

Table 7.1 Relationships between optical activity, molecular structure, and rotational symmetry operations.

Term	Alternating Axis	Simple Axis	Optical Activity
Symmetric	Present	Present or Absent	Inactive
Dissymmetric	Absent	Present or Absent	Usually Active
Asymmetric	Absent	Absent	Usually Active

The search for new chiral effects in liquid crystals, such as planar chirality, prompted the investigation of unsymmetrically 1,3-disubstituted ferrocene-containing compounds, as shown in Figure 7.9. Planar chirality occurs when a chiral molecule lacks a stereogenic center(s), but possesses two non-coplanar rings that are each dissymmetric and cannot easily rotate about the chemical bond connecting them. Having different substituents at the 1- and 3-positions in ferrocene-containing compounds thereby generates structures that create the planar chirality.

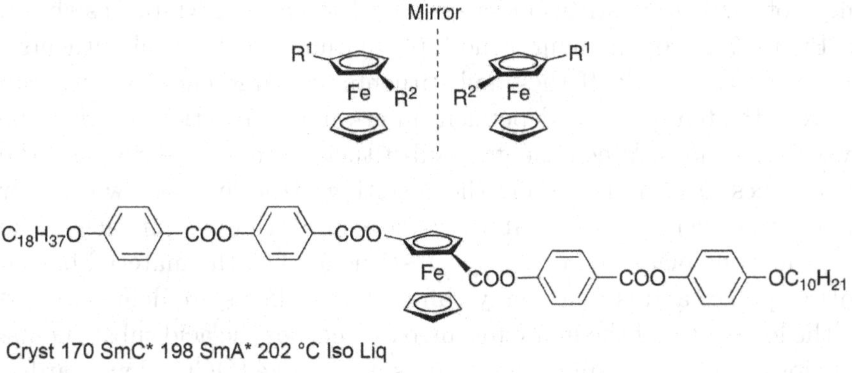

Cryst 170 SmC* 198 SmA* 202 °C Iso Liq

Figure 7.9 Planar chirality in asymmetrically 1,3-disubstituted ferrocene derivatives.

A typical example is illustrated in Figure 7.9 by a compound that has different terminal alkyl chain lengths. It exhibits smectic C and smectic A phases, denoted by the * symbol to indicate that the phases are chiral. Various homologs of this material also give compounds that exhibit smectic C and smectic A phases, and in some cases additional nematic phases. The packing constraints of the molecules, within the layers, induces restricted rotation of the molecules about their long axes.

Dissymmetric structuring, and hence chirality, can be achieved through rotational entrapment of conformers. A conformer has certain orientations of the atoms in a molecule that differ from all the other possible orientations by rotations about single bonds, unless they are held rigid by small ring structures or double bonds. Thus a molecule can have an infinite number of conformers, but only one configuration. Consequently, conformers have the potential for having chiral structures, but these structures are rapidly interconverting between one another even at very low temperatures. For example, derivatives of ethane can interconvert millions of times per second at room temperature.

Rotational entrapment of conformer interconversion is mostly achieved via stereochemical interactions. For example, derivatives of biphenyls, which we see in many forms of liquid crystals and their applications, are candidates for trapped material design. Indeed many chiral biphenyls, which are not themselves liquid crystalline, are utilized in liquid crystal mixtures to induce various forms of chirailty.

Substitution adjacent to the bond between the two phenyl units can interfere with rotation between the two. In doing so, judicious placement of further substitution can render the material chiral, as shown in Figure 7.10. In the figure, the biphenyl unit has four substituents - two gray, two black. If these substituents are large enough, they can prevent the temperature dependent inter-annular rotation between the two phenyl rings. When the gray substituents are on one ring and the black ones are on the other, the resulting structure has two mirror planes as shown, and the material is not chiral. However, if each of the rings has one black and one gray substituent, then the material has no mirror planes and is potentially chiral. These effects are demonstrated in the lower part of the figure for nitro and carboxylic acid substituents. Obviously, when the substituents are small in size (OCH_3 downwards), the rotation is not trapped and the material is not chiral irrespective of the locations of the substituents.

7.3   OPTICAL PURITY AND ENANTIOMERIC EXCESS

Up to this point in the discussion, it has been presumed that the molecules in each example are of one handedness, *i.e.*, they are defined as right-handed (R) or left-handed (S) by application of the Cahn, Ingold, and Prelog rules. However, this is rarely the case and most chiral materials are composed of unequal mixtures of left- and right-handed enantiomers. The relative proportions of left- to right-handed

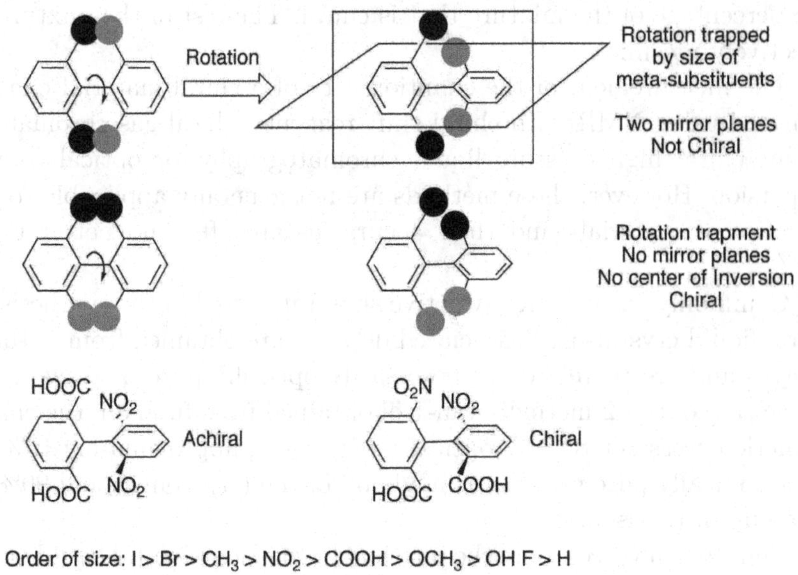

Order of size: I > Br > CH₃ > NO₂ > COOH > OCH₃ > OH F > H

Figure 7.10 Rotational entrapment in meta-substituted biphenyls that leads to chiral structures.

enantiomers in a mixture is termed the enantiomeric excess (*ee*), and when these proportions are equal, the mixture is called a racemate. The percentage enantiomeric excess can be defined as

% enantiomeric excess =
$$\frac{\#\ \text{moles enantiomer A} - \#\ \text{moles enantiomer B}}{\#\ \text{moles both enantiomers}}. \quad (7.1)$$

In terms of materials with molecular structures possessing one stereogenic center of spatial configuration R or S, the enantiomeric excess is given by

$$ee = \frac{[R] - [S]}{[R] + [S]} \times 100 \quad \text{or} \quad ee = \frac{[S] - [R]}{[R] + [S]} \times 100. \quad (7.2)$$

The enantiomeric excess is sometimes called the optical purity (OP). Both the *ee* and OP can be expressed as follows:

$$\text{OP} = 2(\%\ \text{of the enantiomer that is in excess}) - 100\%. \quad (7.3)$$

Another way of thinking about *ee* and OP is to realize that they equal the percentage of the mixture that is chiral. The rest of the mixture is effectively racemic.

The measurement of the enantiopurity of a chiral material can be achieved using NMR with chiral shift reagents, chiral gas chromatography, chiral high pressure liquid chromatography, or optical rotary dispersion. However, these methods are not generally applicable to all varieties of materials and their accuracies are often no better than ±2%.

Commonly, most optically active substrates used in the synthesis of chiral liquid crystals and associated dopants are obtained from natural sources and are therefore not necessarily optically pure. For example, in the case of (S)-2-methylbutan-1-ol obtained from fusel oil, the enantiomeric excess is 0.8 (80% optical purity); *i.e.*, approximately 10% of the chemically pure material is made up of the (R)-isomer, and 90% is made up of (S)-isomer.

Thus, when we compare the physical properties of chiral liquid crystals, the results obtained for each individual material should be normalized to take into account the optical purity. Linear relationships between enantiomeric excess and the physical properties dependent on the chirality of a material are to be expected. For example, for a chiral liquid crystal that exhibits a helical macrostructure in a condensed mesophase, *e.g.*, a chiral nematic phase, the pitch length of the helix is inversely proportional to the enantiopurity. At zero enantiomeric excess, the pitch length is infinite, and at 100% *ee* the pitch reaches a minimum value. Assuming an inverse linear relationship, the inverse pitch of the helix is proportional to the enantiomeric excess (*ee*), the proportionality constant of which is sometimes defined as the helical twisting power (HTP).

Conversely, if the properties of chiral liquid crystals are known, then they can be used to determine structural features of a chiral material such as spatial configuration (R or S), etc. In addition, since liquid crystal systems can amplify physical properties, they are capable of producing highly accurate evaluations of material properties. These possibilities not only apply to liquid crystals, but also to non-mesogenic materials that can be dispersed in a liquid crystal.

7.4 SYMMETRY BREAKING FOR LOCAL SPACE SYMMETRY AND FORM CHIRALITY

We start this section with smectic phases that have their molecules tilted in their layers because they provide exemplary symmetry breaking that leads to local chiral organization, or space asymmetry, as it is sometimes known. In this analysis we assume that there are as many molecules pointing in one direction as in the opposing direction, thus $\hat{n} = -\hat{n}$ in terms of the director.

Let us consider the structure of the smectic C phase, in which achiral molecules have rod-like central rigid core units, to which are attached aliphatic chains located in the terminal positions, as shown in Figure 7.11(a). Remembering dynamic motion, the figure shows two molecules, one pointing up, the other down, which represents the local environmental structure of the phase. Examining the symmetry elements for this arrangement, there is a mirror plane, a center of symmetry (or inversion), and a twofold axis of rotation. If a stereogenic center is now introduced into the aliphatic chain of each molecule, as shown in Figure 7.11(b), the smectic C phase becomes chiral, and the mirror plane is lost along with the center of symmetry, leaving just the twofold axis. The coupling of the stereogenic structure with the overall molecular architecture results in the twofold C_2 axis being polar.

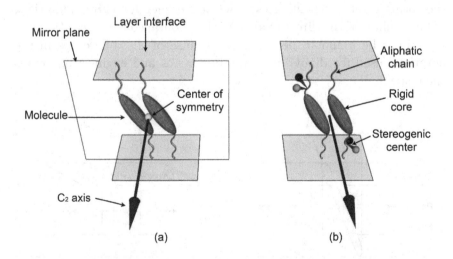

Figure 7.11 (a) The symmetry elements in the local environment of the achiral smectic C phase, and (b) the introduction of stereogenic centers in the terminal chains breaks the symmetry of the smectic C* phase.

The introduction of a stereogenic center into the molecular architecture therefore results in a reduction of symmetry from C_{2h} in the smectic C phase to C_2 in the smectic C* modification, and the spontaneous introduction of a polarization along the C_2 axis. On a larger scale beyond the "two molecule" environmental arrangement, for molecules in domains with their tilts in the same direction, the spontaneous polarization generates ferroelectricity. A similar argument also applies to other tilted smectic phases such as hexatic phases I and F, and soft crystal phases such as G, J, H, and K. Thus the presence of a spontaneous polarization in tilted smectics and crystal phases stems from a time-dependent coupling of the lateral components of the dipoles of the molecules with the chiral environment. As a consequence, only the time-averaged projections of the dipoles along the C_2 axis contribute to the spontaneous polarization (P_s). Reorientation of the structure, via flipping of the tilt directions, can be achieved upon the application of a DC electric field. This effect has consequences for fast switching optical shutters and displays.

One outcome of the analysis of the symmetry operations is that the spontaneous polarization is directional, and that the direction, $P_s(+)$ or $P_s(-)$ is as shown in Figure 7.12. The direction of the polarization relative to the tilt plane of the molecules is dependent on the stereochemistries of the structures of the molecules, which includes all of the forms of stereochemistries described earlier for chiral materials. (The labeling of the directionality of the spontaneous polarization can sometimes be found to be opposite because of the differences in the notations used for dipole moments in physics and engineering versus chemistry.)

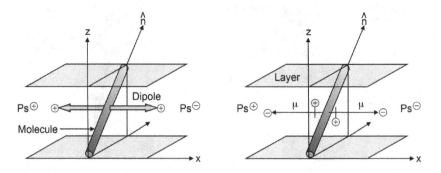

Figure 7.12 The direction of the spontaneous polarization in vector form (left), and using the chemical form for dipole moments (right).

In the bulk 3D smectic C* phase, there is a further aspect of chirality that needs to be considered, namely, the formation of a helical macrostructure, as shown in Figure 7.13. In the bulk C* phase the molecules do not usually tilt in the same direction on passing from one layer to the next, but instead they do so in a circular mode that leads to the formation of a helical structure with the orientation of the tilt and local polarization rotating around the heli-axis. The pitch of the helix of the smectic C* phase is determined via one full 360° rotation and translation of the tilt around the heli-axis. Right-handed and left-handed helical macrostructures are possible as shown in Figure 7.14. For investigations of the helical twist in the following studies, the handedness of the helical structure is defined by pointing the helix away from you, and if the twist is clockwise on moving away, it is considered to be right-handed, otherwise it is left-handed, as shown by the eyes looking down the heli-axis in the figure.

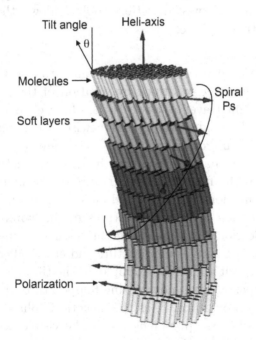

Figure 7.13 The structure of the chiral smectic C* phase showing the spiraling of the twist and polarization around the heli-axis.

The twist sense of the helical structure can be investigated by transmitted polarized light microscopy, in a similar way to how polarimetry is used to determine the specific rotation for a chiral compound. Thus,

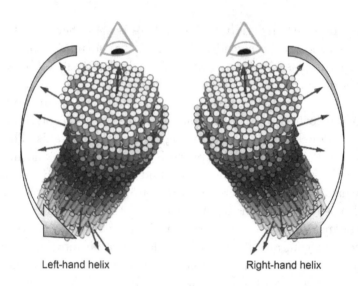

Left-hand helix   Right-hand helix

Figure 7.14 The helical twist directions looking down the heli-axis for the chiral smectic C* phase.

the optical rotation can be detected for homeotropically, aligned samples that are a few microns thick by rotation of the analyzer of the microscope for light traveling towards the observer as in polarimetry. A color change observed by either a clockwise or counterclockwise rotation indicates if the rotation is dextro (d) or levo (l), respectively.

The left- or right-handed nature of the helical structure appears to be dependent on the molecular architecture of the material in terms of spatial configuration (R and S), and molecular dissymmetry (P and M), and the location in the architectures of the symmetry breaking structure (see Section 7.9), and the conditions under which the experiments are performed such as temperature and how it affects mesophase structure, such as tilt angle. For example, as the tilt angle in smectic C phases usually increases with decreasing temperature, the pitch length decreases with temperature. Thus the smectic C phase is potentially thermochromic when the pitch length is in the visible range. As a consequence, similar studies can be performed on mixtures between an achiral host and chiral dopant in order to exclude external effects.

In addition to the spiraling tilt, the chiral smectic C* phase possesses a spiraling spontaneous polarization, which in the bulk is zero due to cancelation. However, the helical structure can be unwound by an applied electric field due to coupling of the spontaneous polarization with the field, and after unwinding of the helix, ferroelectric switching can be observed. Therefore the smectc C* phase can be described as being helielectric.

The minimization of the structural effect of the tilt and polarization means that the chiral smectic C* phase has form chirality. Thus the smectic C* phase has three levels of chirality: configurational stereo-chemisty, environmental space chirality, and form chirality of the condensed macrostructure. These three levels do not have to apply to all liquid crystal phases. For instance, the smectic A phase and the soft-crystal phases possess the first two because they do not have helical macrostructures, whereas the hexatic I and F phases do since the molecules in these two phases are tilted in layers.

7.5 HELICAL STRUCTURES OF VARIOUS LIQUID CRYSTAL PHASES

Probably the most well-known chiral liquid crystal is the chiral nematic phase (N*). It comes in two forms, one based on chiral rod-like (calamitic) molecules, and the other based on chiral disk-like molecules, as shown in Figure 7.15. In the chiral nematic phase, although the molecules have local orientational order, they have no long-range positional ordering, thus the phase has D_∞ symmetry. In both forms it is the local director that is rotated perpendicularly on passing along the heli-axis. The repeat in the resulting helix structure occurs for a rotation of 180°, and the pitch length for the 360° twist can vary from 0.1 μm to infinity. When the pitch length is similar to the wavelength of light, both nematic variants selectively reflect circularly polarized light, and so appear iridescent and colored. The form based on rod-like molecules was first discovered in derivatives of cholesterol, and so for the early studies on chiral nematics the phase was labeled the cholesteric (Ch) phase (now called the N* phase as shown for cholesteryl benzoate in Figure 7.16. Consequently, there is mixed notation in the literature.

Materials that generate the calamitic chiral nematic phase are very common and in many respects parallel those that give the nematic phase except that a chiral unit is present. The chiral unit is most often found in the terminal chain because of the relative ease of synthesis. The inclusion of a stereogenic center in the chain usually produces a sizeable steric effect, which inevitably depresses the clearing point. Chiral unit inclusion into cyanobiphenyls demonstrates this ef-

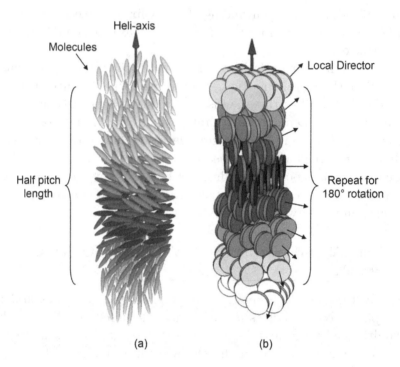

Figure 7.15 Structure of the chiral nematic phase, (a) the calamitic chiral nematic, and (b) the discotic chiral nematic.

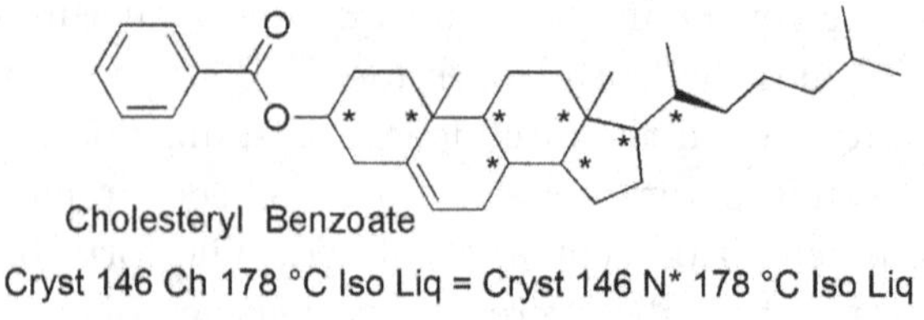

Cholesteryl Benzoate

Cryst 146 Ch 178 °C Iso Liq = Cryst 146 N* 178 °C Iso Liq

Figure 7.16 The structure and melting properties of cholesteryl benzoate.

fect with the clearing points being reduced to the point that the chiral nematic phase is rarely seen. Figure 7.17 demonstrates the effect of mesophase suppression in comparison to the phase behavor of achiral pentyl cyanobiphenyl (5CB).

Discotic systems can be made chiral by incorporating a chiral unit into one or more of the peripheral units that surround the discotic core. Typically symmetrical substitution is employed because of the ease of synthesis. Thus, early chiral discotic liquid crystals have struc-

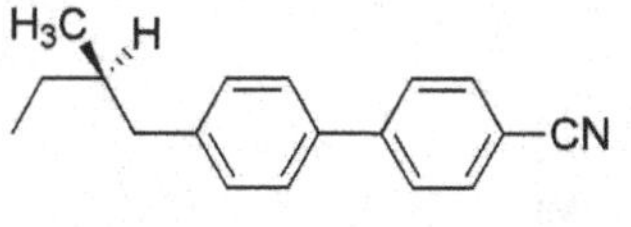

Cryst 4 (SmA -50 N* -30) °C Iso Liq

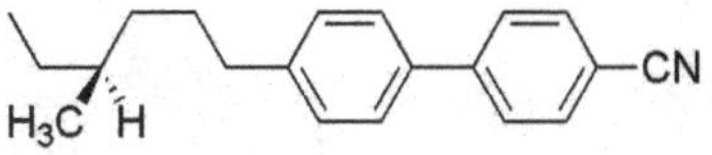

Cryst 28 (SmA -20 N* -10) °C Iso Liq

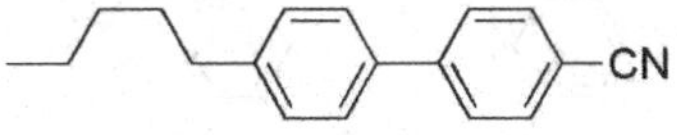

Cryst 24 N 35 °C Iso Liq

Figure 7.17 The effect on the phase transitions by the incorporation of a stereogenic center in alkyl-cyanobiphenyls in comparison with achiral 5CB.

tures where all of the peripheral units are chiral, as shown in Figure 7.18(a). The hexa-substituted triphenylene shown exhibits only a chiral nematic discotic phase (N_D^*) because the steric effects of the branched chains at the stereogenic centers disrupt the ability of the molecules to pack into columns. The large size of the planar aromatic core ensures a high clearing point, but the liquid crystal tendency depends critically on the type of chiral peripheral chain employed. Hexa-substituted phenylacetylenes, on the other hand, were discussed in Chapter 6, and described as preferentially exhibiting the N_D phase. Not surprisingly, when one of the peripheral acetylene units possesses a stereogenic center, a chiral nematic discotic phase is exhibited as shown in Figure 7.18(b).

More recently, chiral discotic materials that exhibit tilted columnar mesophases have attracted attention because of their potential for ferroelectric switching. A spontaneous polarization is generated because of the restriction in rotation of the peripheral chiral chains.

The calamitic and discotic chiral nematic phases do not exhibit a polar structure; however, under applied electric fields, dielectric coupling can induce the helix to unwind to give a structure similar to that of the achiral nematic phase. In addition, it is also possible for the twist direction of the chiral nematic phase to invert as a function of temperature, thereby passing through an unwound nematic phase. For calamitic systems which have molecules that are pear-shaped or slightly bent, there is the possibility for the chiral nematic phase to exhibit flexoelectricity, and as a consequence the phase responds to electric fields.

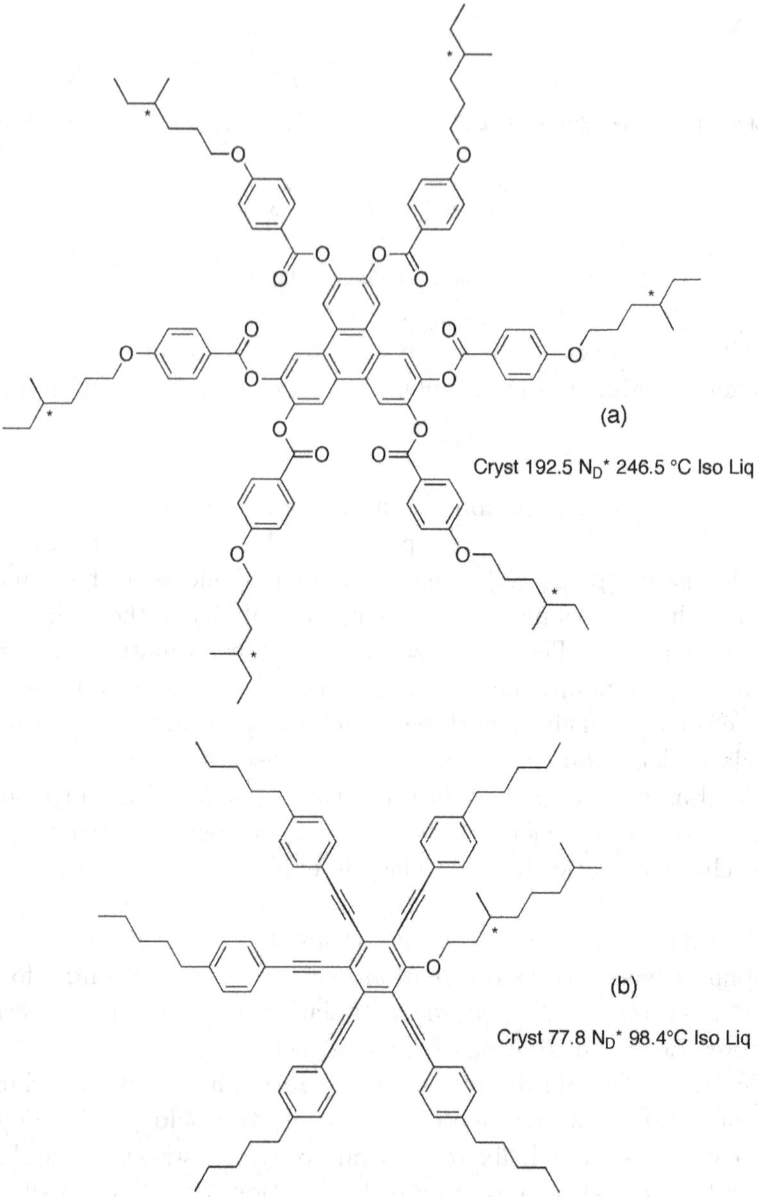

Figure 7.18 Examples of chiral nematic discotic materials, one symmetrical with six stereogenic centers (a), and the other with a single stereogenic center, which can induce form chirality (b).

In terms of other mesophases that form helical structures, there are not many. Like the calamitic chiral nematic, the discotic chiral nematic exhibits a helical structure, as described above, but the columnar mesophases do not. For the layered phases of rod-like systems, those in which the molecules are on average perpendicular to the layer planes, SmA*, Hex B*, Cryst B*, and Cryst E*, and those for which the molecules are tilted but have long-range positional order, soft crystal phases, G*, H*, J*, and K*, do not support the formation of helical macrostructures. Of the remainder, the synclinic tilted smectic C phase and hexatic phases I* and F* form helical structures as described above. In addition the anticlinic variants of these phases, plus the ferrielectric modifications, exhibit helical macrostructures.

The local structure of the anticlinic smectic C* phase has rod-like molecules that are tilted in their layers in a similar way to the smectic C* phase, but on passing from one layer to the next the tilt direction is rotated by 180°. When the molecules have chiral architectures, the anticlinic phase has C_2 axes that oppose one another parallel and perpendicular to the tilt planes of the molecules, as shown in Figure 7.19. As in the smectic C* phase, the C_2 axes are polar, and because they oppose one another, the overall polarization of the phase averages out to zero, hence the phase exhibits antiferroelectricity. Thus under an applied electric field, coupling can occur and the molecules reorient. Unlike the smectic C* phase, the antiferroelectric smectic C* phase has a threshold to switching, which makes it of interest in bistable display devices.

Like the smectic C* phase, the chiral anticlinic phase also exhibits a helical macrostructure. The packing of the molecules, which are shaped like bent cranks, twists the packing and tilt directions between the layer interfaces. The interpenetration between the layers and the continuation of the orientational ordering of the tilt across the layers results in the formation of a helical structure, as shown in Figure 7.20. The definition of the twist direction is the same as for the smectic C* phase, and is related directly to the stereochemistries of the molecules. The repeat for the helical twist occurs for a 180° rotation of the tilt in passing from one layer to another, which is due to the opposing tilts of the molecules in adjacent layers.

In terms of the materials that exhibit anticlinic phases, it appears from property-structure correlations that the location of the asymmetric center needs to be adjacent to the rigid aromatic core of the molecular architecture. In this position the rotation about the center

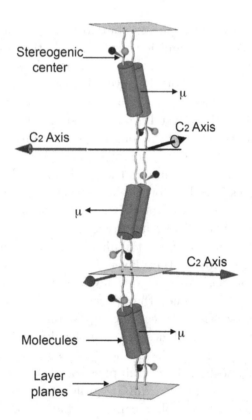

Figure 7.19 The structure of the chiral anticlinic (antiferroelectric) smectic C phase.

is restricted, thereby fixing to some degree the angle of the terminal aliphatic chain (C_6H_{13}) relative to the core unit, as shown in Figures 7.20 and 7.21. In Figure 7.21 the structures also demonstrate the sensitivity of mesophase formation to electron flow in the core unit, exemplified by the orientation of the ester unit (COO) in the core segment (see Figure 7.21(a)). Naturally, the structure associated with the stereogenic center is also important in determining the formation of anticlinic phases (see Figures 7.21(b) and 7.21(c)). Because of the three-dimensional structure produced by rotational damping about the stereogenic center, the propagation of the two terminal chains of the molecules across the interfaces of the layers causes a twist as shown in Figure 7.20. For the analogous synclinic materials, the stereogenic centers have greater rotational freedom and there is not the same need to form a twist from one layer to the next.

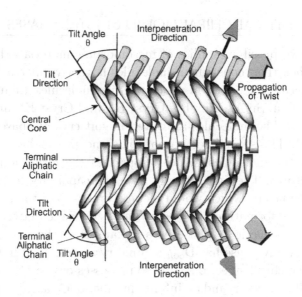

Figure 7.20   Helical structure of the chiral anticlinic smectic C* phase.

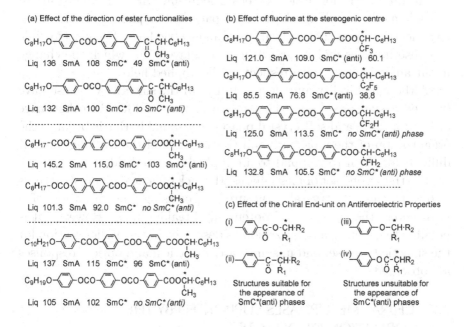

Figure 7.21   Materials that exhibit synclinic versus anticlinic mesophases. Sections (a) and (b) indicate the changes in the structures; the phase sequences and the temperatures (°C) are shown on cooling.

7.6 NON-HELICAL CHIRAL LIQUID CRYSTAL PHASES

There are two possible families of mesomorphic materials that exhibit chirality when their constituent molecules have asymmetric or dissymmetric structures: the ones in which rod-like molecules are upright in a lamellar system, SmA*, Hex B*, Cryst B*, and Cryst E*, and another where the rod-like molecules are tilted in soft crystal phases (G*, H*, J*, and K*). The second set of materials and phases have long-range ordering of the molecules, which suppresses the formation of helical macrostructures. The first set only have the capability to twist in the planes of the layers, which would mean a break-up of the layer organization and the formation of defects. This is discussed later.

Using the smectic A phase as an example of the first group, the achiral smectic A phase has $D_{\infty h}$ symmetry as described in Chapter 5. However, the chiral smectic A* phase possesses only a C_∞ axis that is parallel to the director, and an infinite number of C_2 axes perpendicular to the director, resulting in D_∞ symmetry. The reduced symmetry can result in the formation of electroclinism, which is analogous to the soft-mode switching observed in paraelectric phases of solid-state ferroelectrics upon the application of an electric field. The electroclinic response involves molecular tilting in the layers, for which the change in tilt angle is linear with respect to the applied field.

Although chirality and electroclinism in the smectic A* phase have been extensively explored, the same cannot be said for the other more ordered hexatic and soft crystal phases. Presumably this is because the organization of the molecules and the strength of the layering make it difficult for them to respond to an applied electric field.

For the second set of mesophases, the lack of helical macrostructures is due to the soft crystal structures suppressing the formation of a twist across the layering and potentially breaking it up. However, it is possible to generate a polarization as described in the earlier section for the smectic C phase, and in some cases electrical and optical responses are observed.

7.7 CHIRAL MESOPHASES SUPPORTED BY THE FORMATION OF DEFECTS

If we examine how a twisted structure forms starting from a rod-like molecule, because of the local molecular chirality the packing of two molecules together automatically forms a twist as shown in Figure

7.22(a). If we define one of these as the primary object, then the other can pack in any direction around it. If another chiral object is added, there is only a limited number of packing positions it can take. As more objects are added to the nucleating site, they spread outwards from the primary object to give a cylinder as shown in Figure 7.22(b). This arrangement is called a double twist cylinder, and we see that as its diameter is increased in size, the chiral objects start to rotate towards being perpendicular to the main axis of the cylinder. At this point the packing on the outer surface can change direction when it meets up with other adjacent nucleating sites, as shown in Figure 7.23. At the meeting point, the objects on the outer surface can transfer without change in orientation to adjacent cylinders as shown by the director fields on the adjacent double twist cylinders. If we focus on the crossover point, we see that the orientations of the objects become random and a defect forms. As the space is filled with more objects, the defects become disclination lines. The lines then meet together to form a cubic lattice. In the situation for which the chiral objects are now replaced by molecules, the molecules have no positional ordering, but the resulting phase is cubic based on the repeating arrangement of the defects.

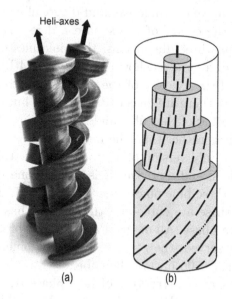

Figure 7.22 (a) The packing of two chiral objects together to form a helical structure, and (b) the molecules packing together to form a double twist cylinder.

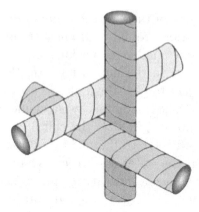

Figure 7.23 The crossing point between three double twist cylinders. The director field moves easily and continuously from one cylinder to the next. At the center there is a defect.

The simplest way to envisage this is first to look at how the helical structure can develop in two dimensions, as shown in Figure 7.24(a). The molecules can twist around the edges of a square lattice with the molecules moving from one twist cylinder to the next, but the molecules cannot move continuously and diagonally across the square, and therefore at the center of the square a defect is formed. Taking this discussion into three dimensions, the molecules can twist around the twelve edges of a cube as shown in Figure 7.24(b). The twists are continuous around the exterior of the cube, but not diagonally across the cube. Therefore there are defect (disclination) lines from one corner to the opposing corner of the cube, and the resulting structure is cubic based on the defects rather than the molecules.

This is not the only version of a cubic structure that has a pattern of twisted cylinders, and moreover it is not the one that is observed. There are in fact three blue phases identified as BPI, BPII, and BPIII. The first two of these are cubic and the last one has sometimes been called the blue fog phase. All of them are termed "blue" because when they were first observed, they tended to reflect blue light due to the lattice dimensions (~500 nm) of the cube.

If we look first at the structure of blue phase II, it is easier to see the relationship between the organization of the double twist cylinders and the line disclinations, which are shown in Figure 7.25. The twist cylinders are packed together in a cubic array (7.25(a)), and the line disclinations are shown as dotted lines in a tetrahedral format as shown together with the cubic cell in Figure 7.25(b). Although the depicted structures show space in the structure, in reality all of the space is filled with molecules.

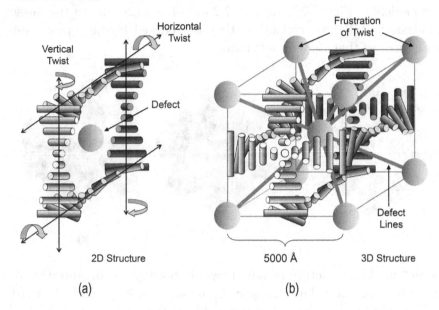

Figure 7.24 (a) A two-dimensional arrangement for helical structures, and (b) the extended structure into three dimensions. In this case the structure has defect lines from corner to corner.

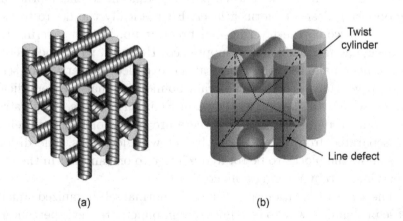

Figure 7.25 The structure of blue phase II showing the organization of the helices (a), and the unit cell for the cubic structure with locations of the line disclinations superimposed (b).

Blue phase I has a similar structure to that of blue phase II, as shown in Figure 7.26(a) for the packing arrangements of the twist cylinders. The cylinders themselves interpenetrate one another as shown in the models in Figure 7.26(b) and 7.26(c). The molecules in the phase continue to fill the structure as they are packed further upon these structures so that they fill all space.

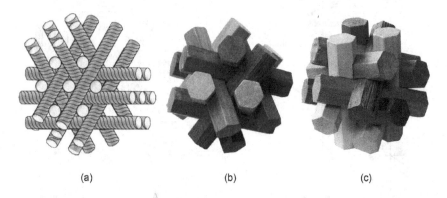

(a)   (b)   (c)

Figure 7.26 The structure of blue phase I showing the organization of the helices (a), and the local space filling structures in which the twist cylinders interpenetrate one another (b) and(c).

We now turn to layered mesophases. For chiral molecules when the layers are relatively soft or weak, there is a competition between the molecules wanting to twist when packing adjacent to one another and the opposing desire to form a layer. Energetically the desire to twist tries to overcome the layering and break it up, but for layering that is strong enough, the twist is suppressed. However, when the energetics are near balanced, the twist can be expelled into arrays of screw dislocations that assemble into grain boundaries that are periodically located within the layers, as shown in Figure 7.27. The overall structure then becomes one in which there are blocks of layers of defined size separated from other blocks by screw dislocations. The dislocations allow the blocks to be rotated relative to one another in the same direction, thereby forming a macro-helix.

The screw dislocations punctuate the normal self-organized lamellar phase in a similar way to how lines of flux punctuate the superconducting phase of the Abrikosov vortex state found in type II superconductors, thereby linking the physics of superconductors with that of phase transitions in liquid crystals. This defect stabilized phase is called the

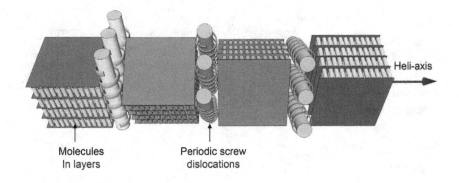

Molecules   Periodic screw
In layers   dislocations

Figure 7.27   The structure of the TGBA phase showing the periodic array
of screw dislocations.

twist grain boundary (TGB) phase, and for molecules organized in layers similar to those in the smectic A phase, the phase is denoted as the
twist grain boundary A phase (TGBA) phase.

The TGB family of phases can have variants based on the layering
structures of smectic phases, so in addition to the TGBA phase, there
are the TGBC synclinic, TGBC* synclinic, and the TGBC* anticlinic
phases, and other phases for which the layers are undulating, called
the UTGBC* phase, or if the layer blocks are very large, called giant
block twist grain boundary phases GBTGBs. For all of these materials,
the helical structure can be commensurate, in which the blocks rotate
and line up through 360° of rotation around the heli-axis, or incom-
mensurate in which the ordering of the blocks does not repeat every
360°.

The first compound to exhibit the TGB phase was in the
enantiomeric forms of 1-methylalkyl 4'-(4-n-tetradecyloxybenzoyloxy)
biphenyl-4-carboxylate (14P1M7). This material is remarkable in the
phases that it exhibits. On cooling from the isotropic liquid, it exhibits
a TGBA phase, which goes on to form ferroelectric, ferrielectric, and
antiferroelectric phases on cooling, thereby demonstrating that these
phases may be interlinked through their property-structure correlations. Most of these phases are described in this section except for the
ferrielectric phase. The ferrielectric phase in Figure 7.28 is shown as a
part-way structure between the SmC and SmC$_A$ phases, in which there
are repeats of anti- and synclinic layering.

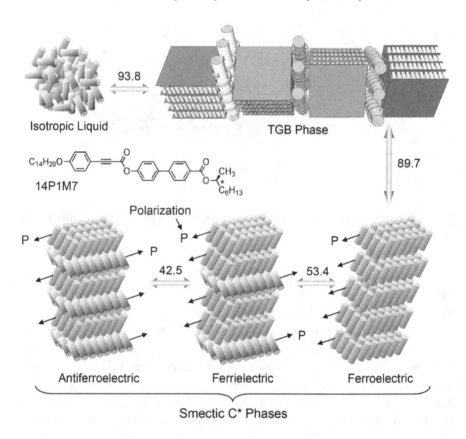

Figure 7.28 The phases and associated transition temperatures (°C) for 1-methylalkyl 4'-(4-n-tetradecyloxy-benzoyloxy)biphenyl-4-carboxylate (14P1M7).

7.8 CHIRAL MESOPHASES AND CONGLOMERATES FORMED BY ACHIRAL MATERIALS

A racemate is a compound, the individual crystals of which contain equal numbers of $(+)$ and $(-)$ enantiomers. It has a sharp melting point and physical properties that usually differ from the pure enantiomer, but it has zero optical rotation determined by polarimetry. A small amount of an enantiomer added to the racemate always lowers the melting point. A conglomerate is like a racemate, but each individual crystal contains a single enantiomer. The conglomerate contains a 50:50 mixture of $(+)$ and $(-)$ crystals, and when a small amount of an enantiomer is added, the melting point always goes up. This means that a racemate acts like a single crystal, whereas a conglomerate behaves more like a eutectic mixture.

Most of the materials discussed so far contain molecules that are rod-like in shape, and to a lesser extent disk-like systems have been mentioned. In both cases the symmetry is broken with the introduction of chiral stereochemistry. Now consider the chiral structural forms that can be produced by changing the shape of the molecules to being bent or banana-like. For bent architectures, the molecules are polar along an axis between the two halves of a molecule, as shown in Figure 7.29. Assuming now that the molecules are in layers and that a plane containing the bent architecture is tilted with respect to the layer planes, then the mirror image of an individual molecule is not superimposable upon the original molecule as shown in the figure. The polar axis in this structural organization is a C_2 axis, and the structure has no mirror plane or center of symmetry, and is therefore chiral without the need for the molecules to be asymmetric or dissymmetric. Extension of this local structure into three dimensions and assuming that the molecules are relatively fixed in their positions, a material could exhibit a conglomerate.

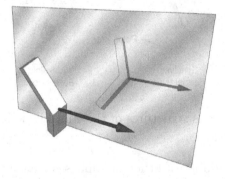

Figure 7.29 A tilted bent-shaped molecule in a layer and its non-superimposable mirror image. The polar axis is C_2.

There are four combinations for the arrangements of tilted bent-core molecules in layers based on synclinic or anticlinic tilts between layers, and the directionality of the polar axes, parallel or antiparallel. The four pairs of combinations, synclinic/ferroelectric, synclinic/antiferroelectric, anticlinic/ferroelectric, and anticlinic/antiferroelectric, are shown together in Figure 7.30. For each combination there is an associated tilt to the left or to the right of a molecule pointing out of the page, as shown in Figure 7.29. The two tilts effectively form the basis for the arrangement of domains and the formation of a conglomerate structure. The molecules are of course not chiral; it is only the packing arrangements that produce the chirality. Therefore, the molecules are not able to flow as in a normal liquid crystal phase in which diffusion and rotations are possible. Thus, the lamellar phases behave as soft crystals. Examples of the chemical structures of bent-core materials are given in Chapter 8, along with details of the phases they exhibit.

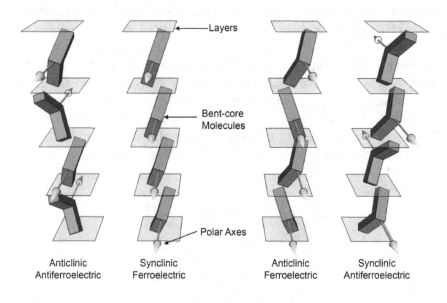

Figure 7.30 The various structures of the chiral lamellar phases of tilted bent-core materials.

There are a number of other mesophases that exhibit chirality effects without the individual components being chiral; these include the so-called "dark conglomerate phase", and the "twist bend phase". Like the bent-core systems, these phases have a balance between two domains of opposite handedness. The phases have local space asymmetry and form chirality, but not molecular asymmetry. This is in contrast to phases such as smectic A* that has molecular asymmetry and local asymmetry but not form chirality, or smectic C* that has the three levels of chirality.

7.9 CONNECTING LOCAL CHIRALITY TO MACROSCOPIC CHIRALITY

There is clearly a relationship between molecular stereochemistry and helical macrostructures. A reasonable question to ask is whether or not there is a relationship between the labeling of stereogenic centers (R and S) with optical rotary direction (dextrorotatory and levorotatory), and hence helical twist, for chiral nematic phases? For simple molecular systems with one stereogenic center, there are relationships called the Gray-McDonnell rules, which relate the stereogenic label to the optical rotation via the position of the stereogenic center in the overall molecular architecture as shown in Figure 7.31. In the figure, a chiral molecule is shown as possessing a common architecture of two terminal chains linked to a rigid central core unit. The stereogenic center is located within a terminal chain, at a distance from the core unit by a number of atoms n, called the linking chain. Thus in the figure the total number of atoms the stereogenic center is from the core is $n + 1$. When this combination is even, the parity (e) is even; when it is odd, the parity is odd (o). Therefore, the Gray and McDonnell rules indicate for a stereogenic center R and an even parity (e), the optical rotation for a chiral nematic phase is levo. The reverse is the case for the helical twist sense if either R or e is inverted. The list of the relationships for the chiral nematic phase are given in the figure. These relationships hold for most simple molecular architectures, but this is not necessarily the case for materials with more complicated molecular structures.

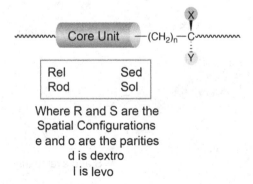

| Rel | Sed |
| Rod | Sol |

Where R and S are the
Spatial Configurations
e and o are the parities
d is dextro
l is levo

Figure 7.31 The Gray and McDonnell relationships between the spatial configuration, parity, and helical twist sense for chiral nematogens.

The rationale for these relationships is probably due to the gross shapes of the molecules and their packing arrangements. The lowest energy level for the molecules is when they are in their all trans conformations. Overall the molecules have zigzag shapes, and predominantly the substituents at the stereogenic centers are on one side or the other of the molecular architecture even though the molecules are undergoing many conformational changes. This is illustrated by three analogs in Figure 7.32, in which the stereogenic centers in all three are (S) and the length of the methylene chain to the stereogenic center, n, is sequentially increased, thereby flipping the parity from even to odd. The energy-minimized structures of the molecules are shown in their space-filling formats, and it can be seen via the arrows in the figure that the lateral methyl group at the stereogenic center flips back and forth with respect to the parity, and consequently the helical twist sense also alternates.

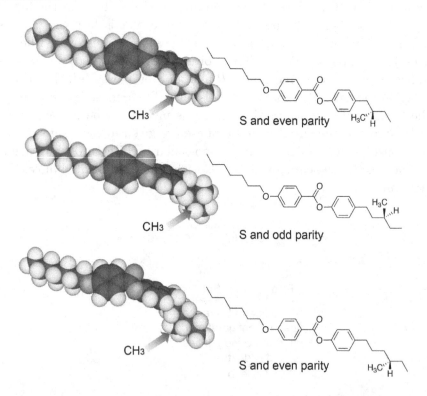

Figure 7.32 The depiction of the Gray-McDonnell rules relating the stereochemical structure to the parity of the center from the aromatic core, and hence to the helical twist sense.

This model is somewhat supported by the measured values of the pitch length, as shown in Table 7.2. As the stereogenic center is moved away from the rigid core section, the terminal chain exhibits a greater variety of conformers and hence increased disorder. Therefore, the pitch length increases, as shown in the table.

Table 7.2 The Gray-McDonnell rules applied to a homologous series of chiral cyanobiphenyls. The pitch measurements are determined in a host liquid crystal and normalized with respect to concentration.

$$NC-\langle\bigcirc\rangle-\langle\bigcirc\rangle-(CH_2)_n\overset{*}{-}\overset{\underset{\displaystyle CH_3}{|}}{C}-C_2H_5$$

n	Parity	Opt. Rot,	Pitch (nm)
1	e	d	0.15
2	o	l	0.30
3	e	d	0.40

Although the Gray-McDonnell rules have been applied to chiral nematogens, there is the possibility that they may also apply to the helical structures of the chiral smectic C* phase (Goodby-Chin rules). Furthermore, there is also the possibility that there might be a connection with the directionality of the spontaneous polarization. A wide variety of smectogenic materials have been investigated and indeed this is found to be the case, as shown in Table 7.3. In addition to the issues of the steric packing associated with the stereogenic center, there is also the direction of the polarity at the center. Because of the σ-bonding at the stereogenic center, we can consider the relative inductive effects (I). For the materials shown in Figure 7.32, the methyl group is donating towards the center, and therefore is labeled a +I effect. Substituents like chlorine pull electrons away from the stereogenic center and therefore are labeled as -I. These two effects are directly related to the polarization and the C_2 axes. Changing the inductive effect thereby inverts the Gray-McDonnell rules as shown in the table. Further, as with helical pitch, the spontaneous polarization and the pitch of the smectic C* phase fall with the increasing value of n in the terminal chain.

The effects of the structure associated with a stereogenic center on physical properties, in particular the spontaneous polarization, are shown in Figure 7.33. The figure shows two sections, (a) and (c), in which the freedom of movement/rotation of the stereogenic center is

Table 7.3 The stereochemical Goodby-Chin rules applied to materials that exhibit ferroelectric smectic C* phases.

Inductive Effect	Spatial Config.	Parity	Opt. Rot.	Ps Direction
+I	S	e	*d*	Ps(-)
+I	R	o	*d*	Ps(-)
+I	R	e	*l*	Ps(-)
+I	S	o	*l*	Ps(-)
-I	S	e	*l*	Ps(+)
-I	R	o	*l*	Ps(+)
-I	R	e	*d*	Ps(-)
-I	S	o	*d*	Ps(-)

affected by steric hindrance. For section (a) the stereogenic center is brought close to the aromatic core, and in doing so the molecular polarization is increased by about an order of magnitude compared with a linking group of one carbon atom longer. For section (c) rotational entrapment occurs as the external aliphatic chain to the stereogenic center is increased in length, which dampens rotational freedom. Thus the spontaneous polarization is increased two to three times for a four-carbon atom chain length increase. A change to the polarity of the groups at the stereogenic center also has a marked effect, as shown in section (d). For a relatively nonpolar hydrocarbon unit at the center, there is an order of magnitude increase in the spontaneous polarization when the hydrocarbon is transferred into a fluorocarbon unit. Conversely, when the stereogenic center is not changed, but the length and number of functional groups in the core are, there is virtually no change in the spontaneous polarization, as shown in section (b). Thus, it appears that the polar nature of the stereogenic center and its steric environment strongly influence the magnitude of the spontaneous polarization. It is not surprising therefore that when the number of stereogenic centers in rotationally trapped environments is increased, the spontaneous polarization is also increased, as shown in section (e) in which a further improvement of an order of magnitude is achieved. These predictive design and synthesis concepts are in effect "nanoscale molecular engineering".

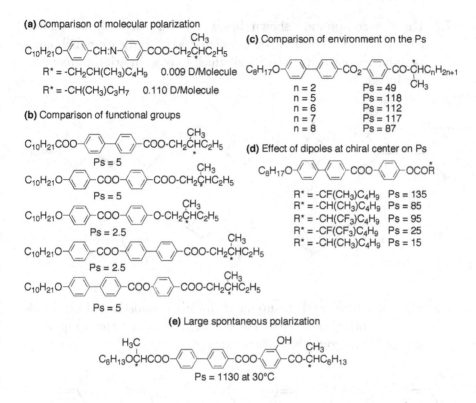

(a) Comparison of molecular polarization

$C_{10}H_{21}O$—⟨⟩—CH:N—⟨⟩—COO-CH$_2$$\overset{*}{C}HC_2H_5$ (CH$_3$)

R* = -CH$_2$CH(CH$_3$)C$_4$H$_9$ 0.009 D/Molecule

R* = -CH(CH$_3$)C$_3$H$_7$ 0.110 D/Molecule

(b) Comparison of functional groups

$C_{10}H_{21}COO$—⟨⟩—⟨⟩—COO-CH$_2$$\overset{*}{C}HC_2H_5$ (CH$_3$)

Ps = 5

$C_{10}H_{21}O$—⟨⟩—COO—⟨⟩—COO-CH$_2$$\overset{*}{C}HC_2H_5$ (CH$_3$)

Ps = 5

$C_{10}H_{21}O$—⟨⟩—COO—⟨⟩—O-CH$_2$$\overset{*}{C}HC_2H_5$ (CH$_3$)

Ps = 2.5

$C_{10}H_{21}O$—⟨⟩—COO—⟨⟩—⟨⟩—COO-CH$_2$$\overset{*}{C}HC_2H_5$ (CH$_3$)

Ps = 2.5

$C_{10}H_{21}O$—⟨⟩—⟨⟩—COO⟨⟩—COO-CH$_2$$\overset{*}{C}HC_2H_5$ (CH$_3$)

Ps = 5

(c) Comparison of environment on the Ps

$C_8H_{17}O$—⟨⟩—⟨⟩—CO$_2$—⟨⟩—CO-$\overset{*}{C}$HC$_n$H$_{2n+1}$ (CH$_3$)

n = 2	Ps = 49
n = 5	Ps = 118
n = 6	Ps = 112
n = 7	Ps = 117
n = 8	Ps = 87

(d) Effect of dipoles at chiral center on Ps

$C_8H_{17}O$—⟨⟩—⟨⟩—COO—⟨⟩—O$\overset{*}{C}$OR

R* = -CF(CH$_3$)C$_4$H$_9$ Ps = 135
R* = -CH(CH$_3$)C$_4$H$_9$ Ps = 85
R* = -CH(CF$_3$)C$_4$H$_9$ Ps = 95
R* = -CF(CF$_3$)C$_4$H$_9$ Ps = 25
R* = -CH(CH$_3$)C$_4$H$_9$ Ps = 15

(e) Large spontaneous polarization

$C_6H_{13}O\overset{*}{C}HCOO$—⟨⟩—⟨⟩—COO⟨⟩—CO-$\overset{*}{C}HC_6H_{13}$ (H$_3$C, OH, CH$_3$)

Ps = 1130 at 30°C

Figure 7.33 Effect of nanoscale molecular engineering on the helical twist and spontaneous polarization (nC/cm^2) in ferroelectric smectic C* liquid crystals.

EXERCISES

7.1 Applying the Cahn, Ingold, and Prelog sequence rules for the compounds shown below, designate the stereogenic centers as either (R) or (S) .

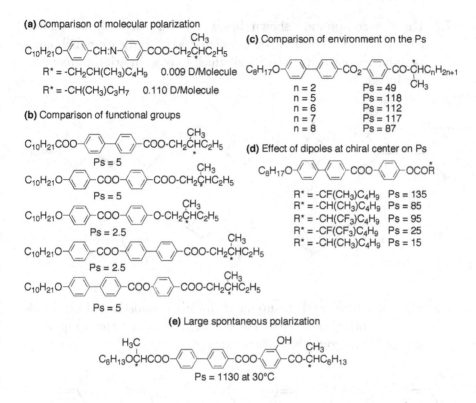

7.2 For the compounds shown below, identify any mirror planes, centers of inversion and/or rotational axes. Indicate which compounds are optically active and which have a superimposable mirror image.

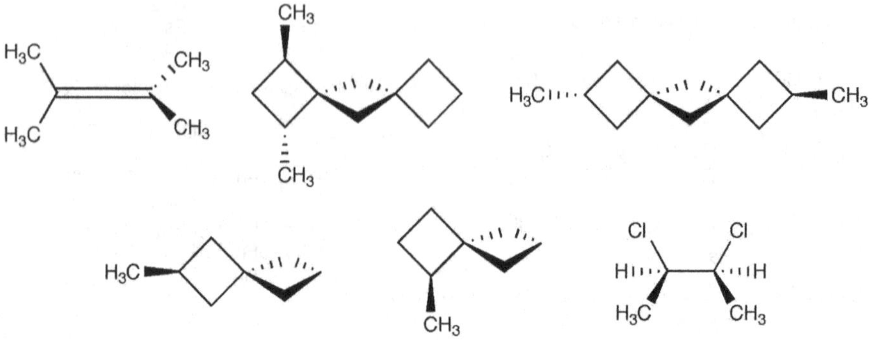

7.3 State the structural requirements for ortho-substituted biphenyls to show optical isomerism. Discuss whether or not the compounds shown below could be optically active.

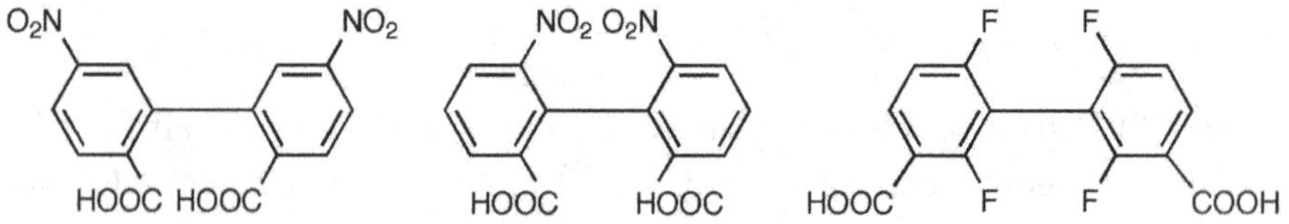

7.4 The picture below shows two aligned ferroelectric liquid crystals, A and B, sandwiched between two pieces of conducting glass. A DC battery is attached to the glass cell with the top glass plate being positive. The cell has been placed between crossed polars and rotated so that compound A is in a dark state, whereas B is in a bright state. Upon rotation B becomes dark and A becomes bright.

(i) What are the polarization directions of compounds A and B?
(ii) If both of the compounds have even parities, what are the possible relationships between their absolute spatial configurations?
(iii) Upon rotation, when A is dark, B is bright and vice versa. Comment on the tilt angles for both of the compounds?
(iv) If A and B have the same chemical constitutions, what is the relationship between A and B?

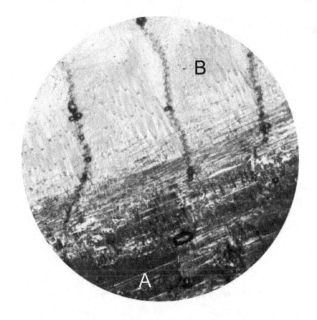

7.5 It has been suggested that ferrielectric phases of chiral mesogens are composed of defined ratios of anticlinic and synclinic tilts, *e.g.*, 2:1,3:1, *etc.* Explain the observation that when a mesogen is in its racemic form it does not exhibit an analogous achiral form of the ferrielectric phase.

CHAPTER 8

Bent-Core Liquid Crystals – It's Bananas

8.1 MOLECULES OF UNCONVENTIONAL STRUCTURE

As we have seen from the earlier chapters, many of the materials that exhibit calamitic phases possess molecules with prototypical shapes, and this is somewhat the case also with discotic liquid crystals. However, with the design of molecular architectures increasing in size and complexity, new molecular forms that support mesomorphic properties have been produced. The development of "unconventional" mesogenic materials can be traced through the various chapters exemplified by bowlic, dendritic, oligomeric, and phasmidic forms, which produce many unusual phases. As the sizes of the molecules have increased due to advanced synthetic methodologies, the reorientation and diffusion times for discrete molecular entities have slowed, such that many of the new phases formed are more like soft crystal phases than "conventional" liquid crystals that easily flow. Although there are many individual examples of such materials, there are few that comprise families of mesophases as found for calamitics. There is one example of multiple discoveries of new phases that are linked together to give a large family of related phases; this family when first found was entertainingly called "banana" liquid crystals and later described as "bent-core" liquid crystals. The name was changed partly because banana liquid crystals were labeled by the letter B, which clashed with the labeling of some calamitic liquid crystals, such as the hexatic B and crystal B modifications. Moreover, this new family of phases also included columnar and frustrated phases in addition to lamellar phases, thereby combining most types of liquid crystals. However, there are some crossovers and confusion in the labeling systems for these phases, many of which were discovered as little as 20 years ago.

The bent-core (BC) family of soft materials is usually exhibited by molecular systems that have dichotomous architectures for which the overall shape is bent. Three examples of the molecular shapes that support the formation of BC phases are shown in Figure 8.1. These include a bent rigid central entity attached to flexible units, the inverse of this in which the flexible unit is in the center of the structure and to which are attached rigid entities, and then a mixture of the two, which can only be described as being a hockey stick structure in which a rod-like moiety has a kink at one end that creates the bent structure.

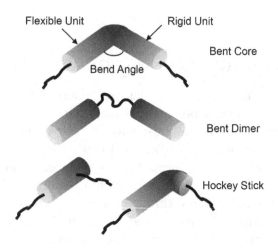

Figure 8.1 Examples of the molecular shapes that support the formation of "bent-core" mesophases.

In the following, the mesophases formed by materials that have central rigid bent-core architectures are described. The systems based on bent dimers are relatively new, and knowledge of their structures is not well developed, so materials of this type are not described. Lastly, hockey stick materials exhibit similar properties to the rigid "bent-core" systems and their behavior is therefore transferable.

If we examine the possibilities of the molecules of bent-core materials packing together, there are a number of varieties: some have nematic arrangements, others lamellar, and some even columnar, as shown in Figure 8.2. Molecular models that illustrate the formation of lamellar phases composed of bent-core molecules are shown in Figure 8.3 for monolayer, bilayer, and interdigitated bilayer structures. Thus it seems that bent-core materials are effectively at the crossroads of molecular shape, possessing both attributes of rods and discs. Although the molecules are not rod-like, they can form layers, and in more complex packing arrangements they form disks. In addition, the molecules may possess dipoles or polarizable segments pointing along the arms or across the center of the molecular architecture, thereby providing for the possibility of symmetry breaking.

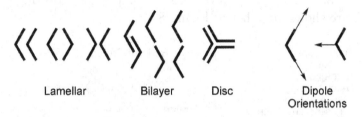

|Lamellar|Bilayer|Disc|Dipole Orientations|

Figure 8.2 Various packing arrangements for bent-shaped molecules.

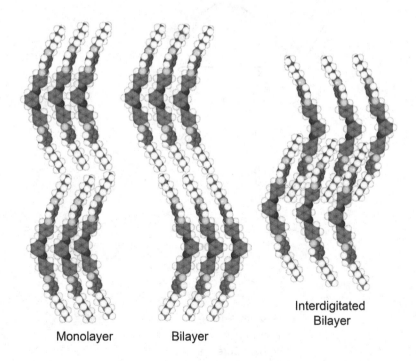

Interdigitated Bilayer

Monolayer   Bilayer

Figure 8.3 Mono-, bi-, and interdigitated layer packing arrangements for bent-shaped molecules.

The first mesophase exhibited by a bent-core system was given the classification letter of B (for banana); subsequently other examples were found, and in all there are eight, B1 to B8, phases. The numbers are not systematic apart from the time sequence for each discovery of a new phase. For some B classes there is a division into sub-phases, which do not use the letter B but instead, smectic or columnar; for example, B2 is subdivided into four smectic phases labeled Sm. A generic chemical structure for materials that exhibit "bent-core" phases is shown in Figure 8.4, and examples of various compounds, covering most B phases, are shown together in Figure 8.5.

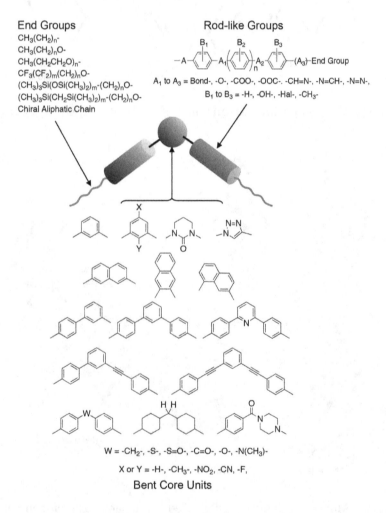

Figure 8.4 A generic template of molecules exhibiting B phases.

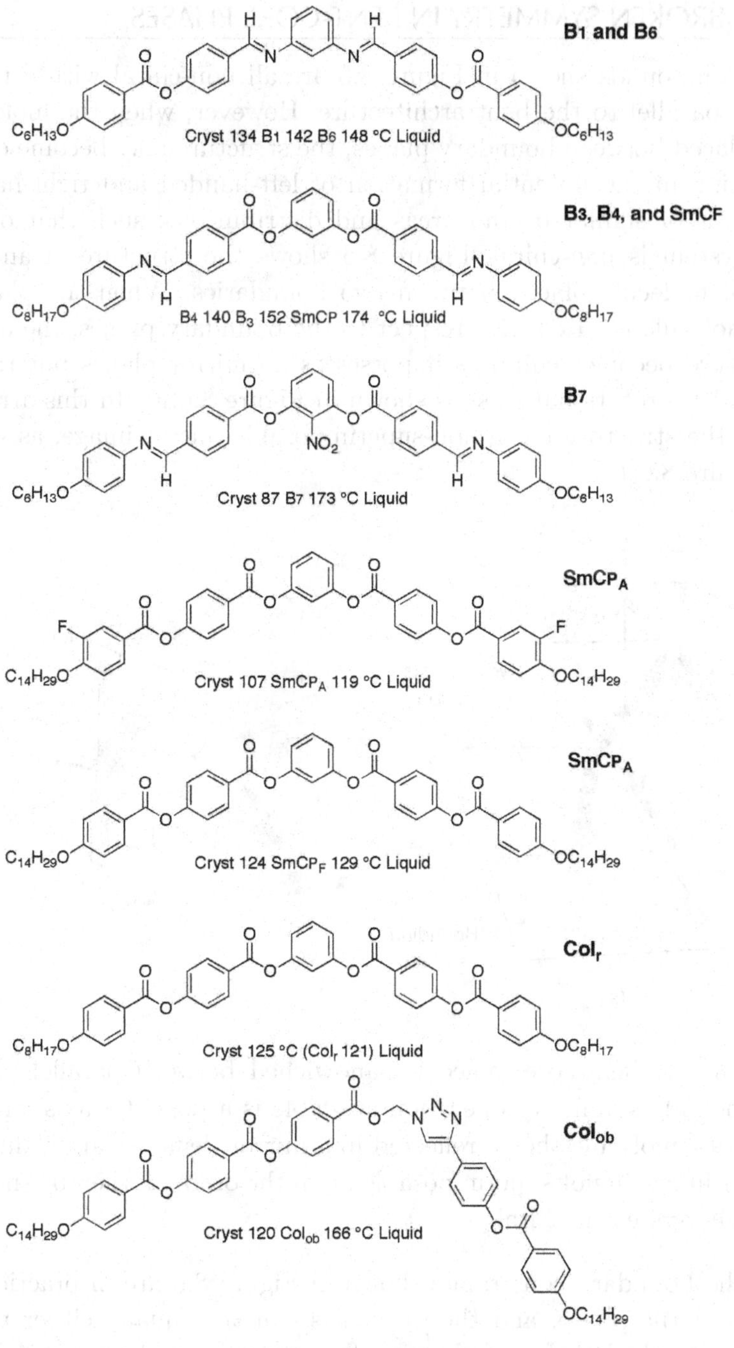

Figure 8.5 Bent-core compounds that exhibit mesomorphic behavior.

8.2   BROKEN SYMMETRY IN BENT-CORE PHASES

The compounds shown in Figure 8.5 are all non-chiral with a mirror plane parallel to the bent architecture. However, when the molecules are placed between boundary planes, the structure may become chiral, resulting in the potential formation of left-handed and right-handed domains of similar overall areas and distributions, such that overall the system is non-chiral. Figure 8.6 shows the structure of an individual molecule placed between two boundaries. When the plane of the molecule is tilted with respect to the boundary planes, the overall structure becomes chiral as it possesses no mirror planes but retains a polar C_2 rotational axis, as shown in Figure 8.6(a). In this arrangement the structure has a non-superimposable mirror image, as shown in Figure 8.6(b).

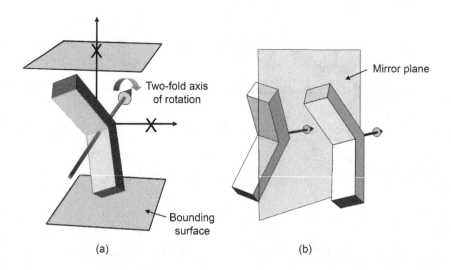

Figure 8.6 A bent-core molecule sandwiched between parallel planes: (a) the only symmetry operation available is a polar C_2 axis; and (b) the same molecule shown reflected in a mirror demonstrating that the mirror image in not superimposable upon the original, thereby showing that the system is chiral.

The boundary constraints shown in Figure 8.6 are in practice the layers of the phase, and the molecules can tilt equally either to the right or to the left. Hence the overall structure is no longer chiral, and could be referred to as being racemic. However, in terms of chemistry, the definition of a racemic mixture, or racemate, is one that has equal amounts of left- and right-handed enantiomers of a chiral molecule, and as the molecules are achiral in bent-core systems, the mixture probably should not be termed a racemate as the molecules are not enantiomers. Also, there are as many molecules tilted to the left as tilted to the right, and so there is no possibility of having associated degrees of enantiopurity. Nevertheless, there are domains of left and right tilted molecules that have sometimes been referred to as creating a conglomerate. Again in terms of chemistry, a conglomerate is defined when the molecules of a substance have a much greater affinity for the same enantiomer than for the opposite one, and a mechanical mixture of enantiomerically pure crystals results. In bent-core phases, some are referred to as being conglomerates, which suggests that they have soft-crystal structures.

If we take a much deeper examination of the structure in Figure 8.6(a), when the molecule possesses a dipole along the C_2 axis that bisects the bent structure, then the overall structure is polar as shown in Figure 8.7(a), and has the potential to be ferroelectric. Thus under the application of an electric field, it is possible to move the dipole and rotate the molecule about a cone as shown in Figure 8.7(b). For a phase composed of two differently tilted domains, the molecules rotate in opposite directions to give light and dark domains, which invert their appearance when the direction of the field is reversed. In the bulk system, ferroelectric responses are only observed when the molecules are tilted in layers and the layers are hard and not soft as in conventional smectic phases, which possess sinusoidal density variations perpendicular to the layers, as described in Chapter 5. Thus the layered phases formed by bent-core molecules are more aptly referred to as being lamellar. However, in terms of optical devices, the presence of left- and right-tilted light and dark domains means that the contrast is never great, and the ability to align the domains is poor.

In addition to ferroelectric properties in lamellar bent-core systems, it is also possible to exhibit antiferroelectric behavior, and ferroelectric properties for columnar structures.

In the following, each section describes a B phase, along with the various mesophases that are supported according to the classification scheme.

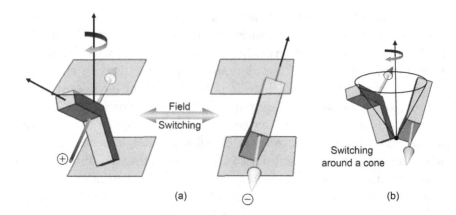

Figure 8.7 (a) The ferroelectric response to an applied electric field for two orientations of a bent-core molecule. (b) The rotation of a molecule around a cone when the electric field is reversed.

8.3 THE BIAXIAL NEMATIC PHASE

The biaxial nematic, N_b, phase was originally thought to be exhibited by bent-core materials based on derivatives of oxadiazole. Details of this phase and the architectures of the compounds that are believed to exhibit biaxiality are described in Section 5.8. A schematic picture of the structure of the biaxial nematic phase is shown in Figure 8.8. The rotations of the bent-shaped molecules about the vertical axis, shown by the director **n** in the figure, are predicted to be restricted, and hence the phase becomes biaxial, with more molecules pointing along the horizontal axis than perpendicular to it, *i.e.*, -**m** = +**m**. Although much theorized and predicted, the phase has not been realized so far, probably because of the fluidic nature of the nematic phase, in which the orientational order is quite dynamic, possibly preventing the establishment of orientational order along a transverse axis. However, for bent-core materials with smaller bend angles than for the oxadiazoles discussed in the earlier chapter, there is the possibility that biaxiality might be realized, as shown by the material in Figure 8.9. This material possesses an azo linkage that can be transformed from its bent *cis*-form to its more linear *trans*-form by illumination with light at certain wavelengths, hence changing the overall molecular shape of the material. Although it has been reported that this compound exhibits biaxiality, this is still open to debate.

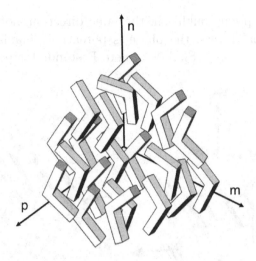

Figure 8.8 Structure of the biaxial nematic phase based on bent-core systems.

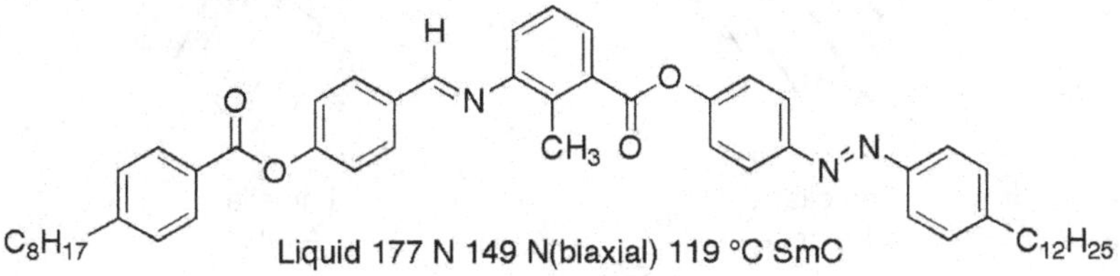

Liquid 177 N 149 N(biaxial) 119 °C SmC

Figure 8.9 An azo-linked material that has the potential to exhibit a uniaxial nematic (N_u) to biaxial nematic (N_b) phase transition as a function of temperature.

8.4 THE B2 PHASE AND LAMELLAR SUB-PHASES

Unlike Section 8.2, which highlighted the complicated symmetry issues surrounding the packing together of tilted molecules in layered structures, we now return to a simpler arrangement where the mirror planes of the molecules are on average vertical (quasi-orthogonal) with respect to the layer planes, as shown in Figure 8.10.

For bent-core systems there are three possible arrangements of the molecules in layers as shown in Figure 8.3. In a monolayer the molecules on average point in the same direction, and are positionally disordered, thus the phase has a similar structure to the calamitic smectic A phase (see Figure 8.10(a)). However, unlike the calamitic phase, as the bent aromatic units point roughly in the same direction along the C_2 axis of the individual layers, the phase is potentially biaxial, and so the phase has been labeled $SmAP_F$, where P stands for polar and F for ferroelectric.

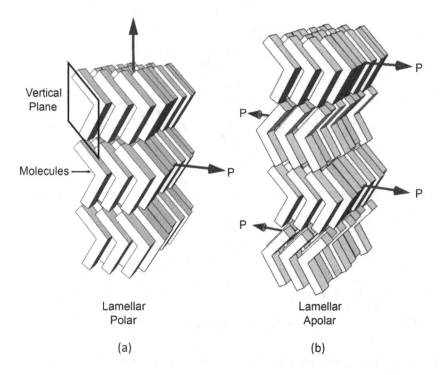

Figure 8.10 The structure of the lamellar phases of bent-core phases in which the molecular planes are orthogonal to the layer planes: (a) has a monolayer structure ($SmAP_F$) whereas (b) is a bilayer ($SmAP_A$).

Between the layers there is a problem that the terminal aliphatic chains could tend to cross over and point in opposing directions, unless their conformational structures allow the chains to point into the interfaces between the layers. The longer the chains are, the easier it is for the conformational variations to allow for interdigitation of the chains, and thereby a greater chance of observing different phase types. The interfacial organizations and the potential for biaxiality mean that the layer structure is more robust than found in calamitic phases, and phases of this type are sometimes referred to as lamellar.

A second possibility for the formation of quasi-orthogonal lamellar phases occurs when the bend directions of the molecules alternate from layer to layer, thereby giving a structure represented by the sketch in Figure 8.10(b). The repeat for this structure involves two layers, and therefore the phase is composed of bilayers. The molecules in the layers are positionally disordered, but as the bend alternates, so too does the polar axis, and for a bilayer the polar nature is averaged to zero. Thus the phase has been labeled $SmAP_A$, where A stands for antiferroelectric.

Like the monolayer variant, the interfaces between the layers are important. In the case of the bilayer structure, the arms of the molecules point into the interface at an angle, thereby meeting the arms of the molecules in the adjacent layers pointing in a similar direction, resulting in the transfer of information about molecular orientation from one layer to the next. The layers therefore are relatively well defined and the structure is of a lamellar form.

Let us now turn to layer structures for which the molecular planes are tilted with respect to the layer planes. Four possible arrangements can be defined as follows: two in which the tilts are either in the same direction from one layer to the next or they alternate in direction, and two in which the polar axes are either in the same direction or alternate from one layer to the next. The various combinations of tilt and polar directions are shown together in Figure 8.11, in which tilts in the same direction are called synclinic, whereas if they are in opposite directions they are called anticlinic. For the first structure at the top left of the figure, the molecules are shown tilted to the right and the polar axes are shown pointing out of the page. As the molecules are tilted, the phase has been called smectic C, denoted with the letter s to indicate the synclinic arrangement of the tilt, P to show the phase is polar, and F to indicate it is ferroelectric; hence the label is $SmCsP_F$. For the structure shown in the top right of the figure, the same description applies for the polarity, except the tilt is now anticlinic, and as a consequence this phase has been labeled $SmCaP_F$. For the bottom pair of structures, the direction of the polarity alternates backwards and forwards from one layer to the next, and similarly so does the tilt. Thus, the bottom left structure has synclinic layer ordering with alternating polar direction and is therefore labeled $SmCsP_A$, whereas the bottom right structure has anticlinic tilt, alternating polar direction, and a label $SmCaP_A$.

For each structure shown in Figure 8.11, there is a mirror image organization, resulting in two types of domains of opposite handedness being formed. The domains are large enough to be seen by optical microscopy at a magnification of x100. The phases so-formed exhibit

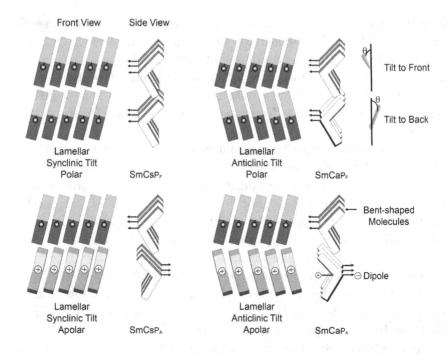

Figure 8.11 The phases formed by bent-core molecules that are tilted in layers showing head-on and side-on views.

ferroelectric or antiferroelectric properties depending on the synclinic or anticlinic arrangements of the tilt. This means that when an electric field is applied, the molecules become reoriented via rotation about a cone, as shown in Figure 8.7. The molecules in some domains rotate in one direction, whereas for the other domains the rotation is in the opposite direction. Thus the domains seen between crossed polarizers appear light or dark and invert in color when the field is reversed. The domains have the same volumes and therefore similar areas. In comparison to the ferroelectric and antiferroelectric phases of rod-like phases, there are many similarities, except that bent-core materials form domains that cannot be aligned over large areas. Monodomains are not achievable because the left and right tilted structures are equivalent and are not enantiomorphic.

8.5   THE B1 PHASE AND ITS COLUMNAR SUB-PHASES

It is possible to see how defects can be formed in the B2 phases whereby there is a step change in the layers such that the direction of the molecular bend is inverted via a half layer translation between the layers. Such changes can occur along the layers in the direction of the packing of the bent-core molecules, or perpendicular to the bend direction. These changes in structure may be random leading to a very disordered high-energy system, or they can be periodic thereby producing different phases of matter. In fact two different structures for phases can be realized, which are shown together in Figure 8.12. These two columnar phases are based on layered structures and are called the B1 and the B1rev phases.

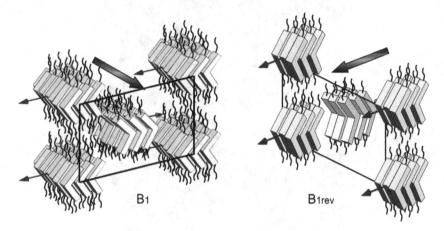

Figure 8.12   The columnar structures of the B1 and B1rev phases.

The structure of the B1 phase is shown in the left-hand part of the figure. The structure is based on the $SmAP_F$ lamellar format discussed previously, except that there is an alternation of the bend directions of the molecules occurring between the layers. This alternation is periodic resulting in the formation of columns perpendicular to the bend directions. The lattice so formed has a rectangular unit cell. For the second phase, B1rev, the structure is the same except the shift in the layers is perpendicular to the bend direction of the molecules, as shown in the right-hand segment of the figure. Again the resulting columnar lattice is rectangular. Consequently the planes of the molecules lie flat with respect to the column axes for the B1 phase and parallel for the B1rev phase.

Figure 8.13 shows the structure of the B1 phase for the material shown at the top of Figure 8.5. The structures of the space filling models of the material are minimized in the gas phase at absolute zero, and preferentially pack together to create an image of the proposed structure of the phase. It can be seen from the picture that the molecules pack together in such a way as to minimize the free volume of the system, resulting in the formation of a lattice for a soft crystal phase.

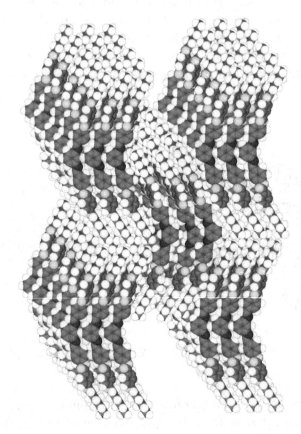

Figure 8.13 Packing of space filling models of the molecules of the bent-core material (shown in the top of Figure 8.5) in a columnar sub-phase of the B1 phase.

There is also a third variation of the B1 phase called the B1revtilt phase. This phase has the same structure as the B1rev phase except that the molecules are tilted with respect to the layer directions. All of the B1 phases, however, are not fully understood and details of their structures are not fully elucidated.

The material shown at the top of Figure 8.5 is also found to exhibit a rare example of the phase classified as 6. Unsurprisingly, the proposed structure for this phase is related to the B1 phases described above. Initial structural studies show that the phase has a periodicity smaller than half of the molecular length, indicating interdigitation and a tilting of the molecules within layers with the molecules having no periodic ordering. These results indicate that the B6 phase is similar to the B1 phases except the structure is more disordered, which might be expected as the phase appears at higher temperatures than the B1 phase in phase sequences. A sketch of the proposed structure for the B6 phase is shown in Figure 8.14.

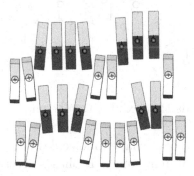

Figure 8.14 The proposed structure of the B6 phase.

8.6 SPLAY AND MODULATED STRUCTURES IN B PHASES

The B phases we have discussed so far tend to have their structures determined by directional packing of the V-shaped molecules in parallel versus antiparallel arrangements. Now we turn to the formation of B phases for which the polar directions are more relevant. Figure 8.15 shows three compounds that exhibit the B7 phase, a modulated phase that shows extraordinary spiraling defect textures in the microscope. The molecules in the figure have one structural feature in common: all possess strong dipolar groups. For the top figure, the compound has a very strong polar nitro group that bisects the arms and pulls electrons away from the apex of the central aromatic ring. The other two compounds have two polar halogens for which the lateral components effectively cancel leaving a strong component pointing towards the apex of the V-shape. Thus all three materials exhibit strong polarity along the axis that bisects the arms of the molecule.

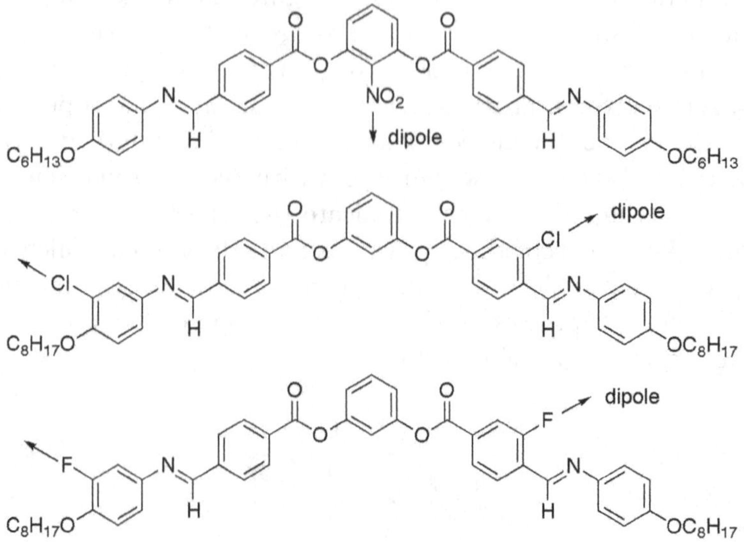

Figure 8.15 Compounds that exhibit the B7 phase.

If we think of the bent-shaped molecules packing together side by side, as in the cases of many of the B phases described earlier, the polar groups point roughly in the same direction and therefore laterally repel one another. This causes a splaying of the polar groups, as shown in Figure 8.16, and hence of the lateral packing of the molecules. The splaying of the polar groups, however, is unsustainable and as a consequence splay domains form with opposing curvature such that the overall effect of the polarity is zero, as shown in the figure. The result of this balancing of the polarity is the formation of a modulated structure that can be of a sinusoidal form with respect to the molecular ordering, as shown in Figure 8.17 for molecules with their molecular planes perpendicular to axis of the modulation.

Many of the materials that exhibit modulated B phases also form structures in which the molecular planes are tilted with respect to the layer planes. So an aspect of their modulated structures is an induced left- or right-handed spiraling of the overall structure, which can be seen from the helical fibers observed by microscopy. This arrangement is shown in Figure 8.18, where in the left half of the figure the polarity splays in a downward direction as the molecules tilt backward, and for

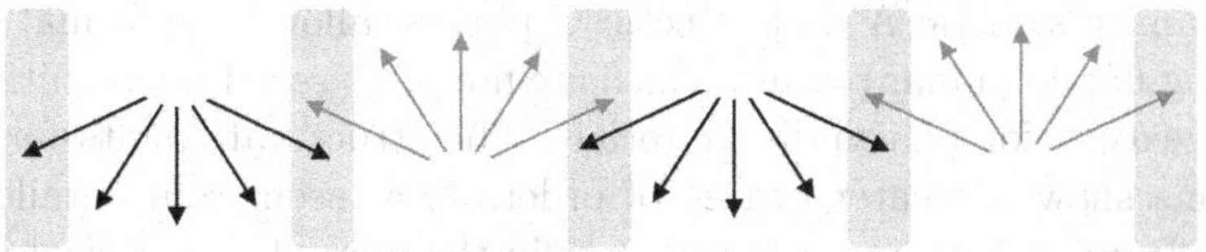

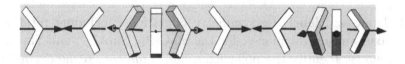

Figure 8.16 Splay in the polarity associated with the side-by-side packing of bent-core molecules.

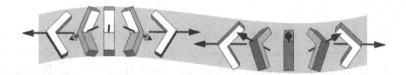

Figure 8.17 Splay in the polarity results in the molecules rotating sinusoidally in a B phase to give a modulated structure.

the right half of the figure it splays upward as the molecules tilt forward. The gray background shows the modulation of the structure coming in and out of the page. This type of structure is thought to occur most often in the B7 phase along with other variants in the structure.

Figure 8.18 The modulated structure of the B7 phase.

In this chapter an overview of the current thinking on some of the structures formed by "bent-core" materials is presented; however, the subject is still somewhat in its infancy and many of the structures and properties of B phases require further elucidation. There is the temptation to use the structures of calamitic and discotic mesophases as templates for this new family of soft matter phases, but the breaking of symmetries caused by the V-shaped structure of the molecules means that this is not always possible. Moreover, the anticlinic versus synclinic ordering of the molecules coupled with the direction of the polarity means that the layer ordering of the phases is potentially firmer than that of conventional smectics. Furthermore, the ability to form bilayer ordering and modulated structures, templated on the structures of calamitc smectic A and smectic C phases, allows the formation of rectangular columnar phases, which are not often seen for calamitic systems, except for phasmidic materials. The structural results for such systems show a greater degree of ordering, as seen from detailed X-ray diffraction patterns, than present in the more fluid classical liquid crystals phases.

EXERCISES

8.1 Draw and label the symmetry operations, including rotational axes and mirror planes, for the $SmCsP_F$, $SmAP_A$, and $SmCaP_A$ phases. Identify the polar axes, and comment on the overall polarization of the system.

8.2 For the electrical field switching of a material that exhibits a $SmCsP_F$ phase, using an arbitrary position for one molecule in a drawing, identify the location of the layers of the phase, the tilt angle of the molecule, the positions of the electrodes and the polarizers in order to observe light and dark switching on the application of a DC electric field. What tilt angle of the molecules generates the highest contrast between the light and dark domains?

8.3 The following three terms, racemic modification, racemate, and racemic mixture, often cause confusion. Compare the definitions of the three in terms of crystals and the structure of the B2 phases. What is the difference between an enantiomorph and an enantiomer?

CHAPTER 9

Lyotropic Liquid Crystals – Wet 'n' Dry

9.1 AMPHIPHILES AND LYOTROPIC LIQUID CRYSTALS

There have always been debates over when liquid crystals were first discovered and realized. Through the work of Friedrich Reinitzer, the foundation of liquid crystal science is traditionally thought to have started in 1888. Reinitzer is recognized as a botanist, although in modern terms he probably would be a biochemist. However, there are other indications from the life sciences that liquid crystallinity was actually observed much earlier. For example, in 1854 pathologist Rudolf Virchow observed a birefringent texture by optical polarizing microscopy when myelin (from the sleeves that protect nerve cells) was mixed with water. He called the resulting structures that he had found myelins. Lehmann repeated experiments performed by Quincke on myelin tubes and then unequivocally identified them as: "the myelin forms are nothing but flowing crystals".[1] However, the connection between biological systems and liquid crystals was not realized until much later. To quote J. Needham in 1950,[2] "The aspect of molecular patterns which seem to have been most underestimated in the consideration of biological phenomenon is that found in liquid crystals."

Today lyotropic liquid crystals extend well beyond aqueous biological systems: from amphiphiles and amphitropes to inverted struc-

---

[1] Dunmur, David and Sluckin, Tim, *Soap, Science, & Flat-Screen TVs: A History of Liquid Crystals*, Oxford University Press, 2011.

[2] Brown, Glenn H. and Wolken, Jerome J., *Liquid Crystals and Biological Structures*, Academic Press, 1979.

tures in lipophilic phases, and to membranes, Langmuir-Blodget films, and colloidal and chromonic liquid crystals. Their applications are widespread from soaps and detergents, to pastes, lubricants and to the processing of polymers as in the case of Kevlar® (see Chapter 10). There are even unusual accompanying, and sometimes unwanted, side effects such as the slime that forms in a soap dish, which is a liquid crystal phase generated by the dissolution of soap in water. However, interest in their biological properties in relation to medicinal applications has only begun in more recent times, partly with the realization that the B-form of DNA, as examined by Rosalind Franklin, is liquid crystalline, and drugs such as Intal®, used for the treatment of asthma, in their aqueous form are also liquid crystalline. It is not surprising, therefore, that liquid crystallinity has also been tenuously linked with various diseases such as atherosclerosis. Some of these associations are shown together in Figure 9.1, and accordingly, life, as we shall see, is itself critically dependent upon lyotropic liquid crystal phases.

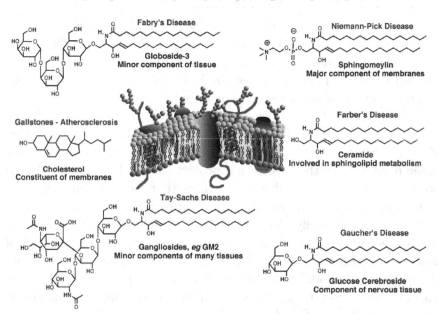

Figure 9.1 A biological membrane and diseases related to various lipids found in the membrane.

Just as the liquid crystal phases of a thermotropic material are generated by changes in temperature, lyotropic liquid crystal phases are formed on the solvation of amphiphilic molecules of a material in a solvent as a function of concentration. Equally, as there are many different types of structural modifications for thermotropic liquid crystals, there are several different forms for lyotropic liquid crystal phases. Each of the types has a different extent of the molecular ordering within the solvent matrix, and the concentration of the material in the solvent dictates the type of mesophase formed. Furthermore, like thermotropic liquid crystals, it is also possible to change the type of lyotropic phase exhibited at each concentration via a change in temperature. Some materials also have unique forms of phase behavior in that they can exhibit thermotropic mesophases without a solvent, and lyotropic phases with a solvent. This dichotomy is described as being amphitropic.

The types of molecular architectures that support the formation of lyotropic liquid crystal phases are usually amphiphilic. An amphiphile (a descriptor derived from the Greek for *amphis*, αμφις, meaning both, and *philia*, φιλια, meaning love or friendship) has a molecular architecture that is essentially dichotomous, being composed of two parts that dislike each other but are tied together so they cannot escape from one another. When amphiphiles are mixed with a hydrophilic solvent, one part of the molecule is solvated whereas the other is not, and the reverse is the case when the solvent is hydrophobic. Typically, lyotropic liquid crystals are composed of materials that have a segment that is water-loving, and another that is fat-loving. Amphiphiles are typically used as surfactants, which are compounds that lower surface tension, or interfacial tension, between two liquids or between liquids and solids. Surfactants thus have properties and applications similar to those of lyotropic liquid crystals, *i.e.*, they can act as detergents, emulsifiers, and dispersants. Surfactants often have constituent molecules, which have molecular architectures composed of a polar group joined to a non-polar unit. For example, soaps such as sodium stearate have a polar head group made up of a carboxylate salt and a non-polar unit that is a long fatty hydrocarbon chain. Synthetic detergents such as alkyl sulfates and aromatic sulfonates, shown in Figure 9.2, have analogous structures and properties to sodium stearate. Materials such as these are known as anionic surfactants because they have negatively charged head groups that are polar and aliphatic tails that are non-polar.

Cationic surfactants, being positively charged, have opposing molecular architectures to anionic surfactants, and not surprisingly exhibit different properties. Usually they do not have detergent or foaming properties, but they do have practical uses in fabric softeners, and toiletries and cosmetics, for example hair conditioners and specialized

shampoos. Cationic surfactants have also been shown to exhibit lyotropic liquid crystal phases, some typical examples are shown in Figure 9.3.

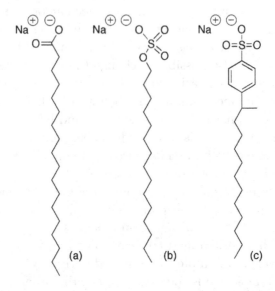

Figure 9.2 Examples of anionic surfactants, including sodium stearate,
(a), which is a typical soap, an alkyl sulfate (b), and an aromatic sul-
fonate (c).

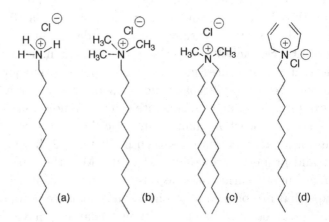

Figure 9.3 Examples of cationic surfactants: dodecyl ammonium chloride
(a), dodecyltrimethyl ammonium chloride (b), dioctadecyldimethyl
ammonium chloride (c), and a diallylamine derivative (d).

Dodecyl ammonium chloride, (a) in Figure 9.3, is a simple example of a cationic surfactant. It is composed of an amine that has been converted into the ammonium chloride salt, plus an attached long terminal chain. Accordingly, the ammonium cation constitutes the polar head group and the long terminal alkyl chain completes the amphiphilic structure in the capacity of a hydrophobic unit. Dodecyltrimethyl ammonium chloride (b) and dioctadecyldimethyl ammonium chloride (c) in the figure are more elaborate cationic surfactants. Compound (c) has two long hydrophobic alkyl chains, and related analogs are used as anti-static fabric softeners. The diallylamine derivative (d) is also an interesting example of a cationic surfactant, with two unsaturated C=C bonds in the head group and a strong electron-withdrawing group associated with the cation. It is capable of being polymerized in its liquid crystalline state to give a network polymer that has a frozen-in structure of a lyotropic phase.

Amphiphilic or surfactant materials can also be generated by non-ionic species, and for liquid crystals there are many more additional dichotomous variations and combinations that are possible, as shown below.

Hydrophilic – Hydrophobic\
Ionic – Non-ionic\
Polar – Non-polar\
Rigid – Flexible\
H-bonding – Non-H-bonding\
Fluorocarbon – Hydrocarbon\
Aromatic – Aliphatic\
π-bonded – σ-bonded\
and combinations of the above

Outside of the field, some variations may not be recognized as being amphiphilic, but in terms of liquid crystals most are associated with microphase segregation, where dissimilar parts of molecules do not mix with one another thereby causing local segregation. Examples of some non-ionic amphiphiles, which have polar-apolar dichotomies, are shown in Figure 9.4. The first material (a) has a long alkyl chain as the hydrophobic section and a hydrophilic polar head group that is composed of several ethylene glycol units. The second compound (b) is a typical case for a semi-fluorinated polar-apolar alkane; for the material shown the lengths of the hydrophilic and hydrophobic sections are essentially the same, however, the fluorinated part is less flexible than the aliphatic section. Semi-fluorinated alkanes also exhibit thermotropic liquid crystal phases as well as being lyotropic, and are hence amphitropic. Comparing these two examples, the first has a flexible structure with hydrophilic – hydrophobic and polar – apolar architectures, whereas the second has fluorocarbon – hydrocarbon, flexible – rigid, and polar – apolar architectures, which demonstrates that their structures have different structural combinations that define their amphiphilicity.

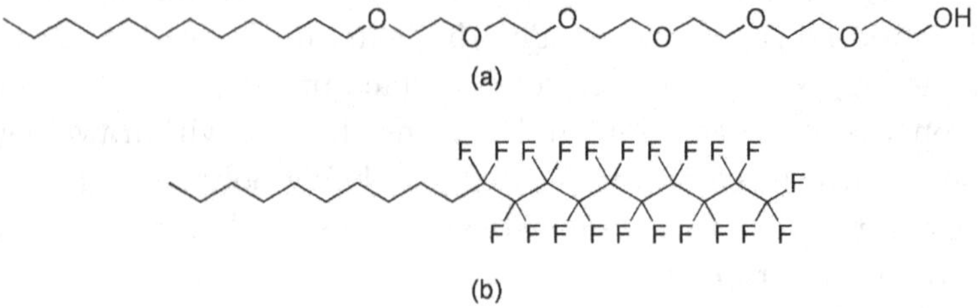

(a)

(b)

Figure 9.4 Examples of non-ionic amphiphiles.

We can make a different division in classifying liquid crystal amphiphiles based on their origins: many are synthetic, and others are natural. The ones described above in Figures 9.2 to 9.4 are synthetic, and are designed for practical applications, whereas those described in the following section, and shown in Figure 9.5, are derived from natural sources, which in many cases have been made available in larger quantities via natural products synthesis.

Major biological sources of amphiphilic materials are cell membranes, which are known as plasma membranes (PM) or cytoplasmic membranes. Such membranes separate the interior of all cells from the extracellular space of the outside environment, thereby protecting the contents of the cell from the potentially hostile environment. The importance of biological membranes to life on Earth is summed up in this quote from John Gribbin in the book *Stardust*.[3]

In January 2001, scientists from NASA's Ames Research Center and the University of California, Santa Cruz, surprised many of their colleagues and created headline news by announcing the results of experiments carried out in laboratories here on Earth which produced complex organic

---

[3]Gribbin, John, *Stardust*, Allen Lane, 2000.

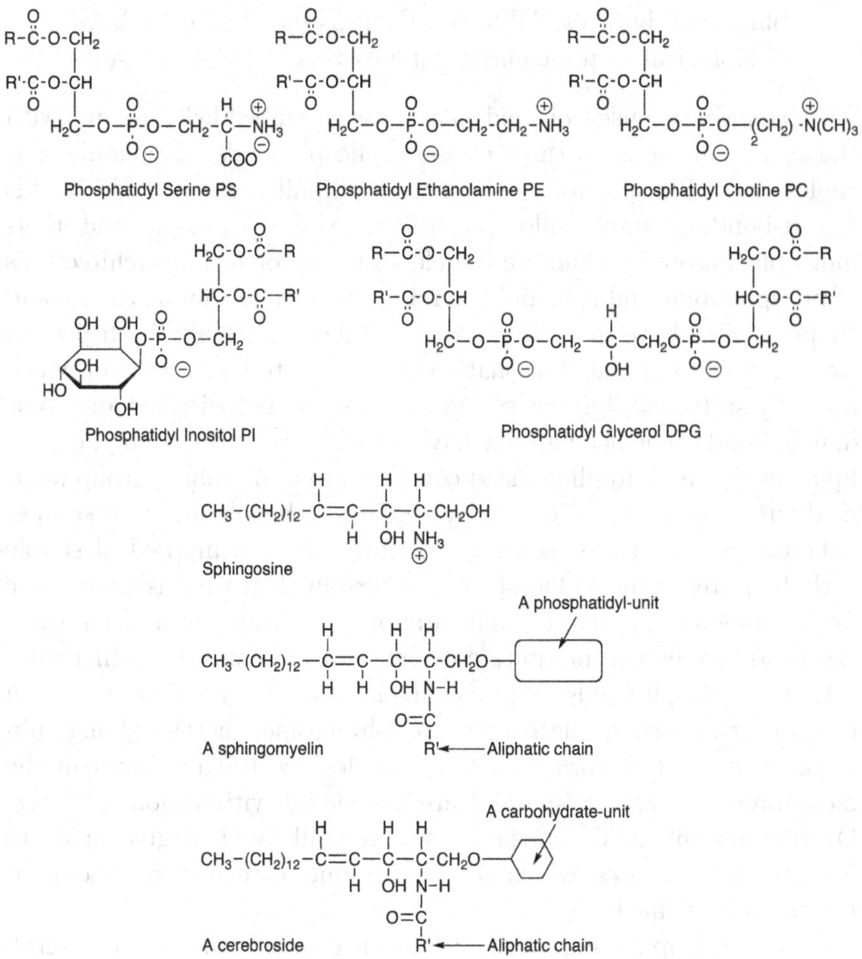

Figure 9.5 Mesogenic amphiphiles found in biological membranes.

molecules under conditions resembling those which exist in interstellar clouds of gas and dust. In these experiments, a mixture of the kind of icy material known to exist in those clouds (composed of water, methanol, ammonia and carbon monoxide frozen together) was kept in a cold vacuum and dosed with ultraviolet radiation. Chemical reactions stimulated by the radiation (typical of the kind of radiation from young stars which zaps real interstellar clouds) produced a variety of organic compounds which, when immersed in water, spontaneously created membranous structures resembling soap bubbles. All life on Earth is based on cells, bags of biological material encased in just this sort of membrane.

Membranes consist of lipid bilayers with embedded proteins, with the lipids belonging to three classes - phospholipids, glycolipids, and cholesterols. Phospholipds are ionic amphiphiles whereas glycolipids are H-bonding amphiphiles; both can exhibit lyotropic and thermotropic mesophases and are therefore amphitropic. The architectures of phospholipids and glycolipids tend to mirror one another in size and shape as they have to pack together in bilayers. Usually both possess two relatively long fatty aliphatic chains (C_{12} to C_{24}), most of which are fully saturated, but some that are unsaturated with *cis* and *trans* double bonds. The head groups have specific structures, with phospholipids having a hydrophilic head consisting of a phosphate group modified with organic moieties to give cholines, ethanolamines, or serines. Whereas the head groups for glycolipids can be composed of simple carbohydrates such as glucose or galactose in their various stereogenic forms (anomeric, pyranose, and furanose), additionally in some cases the head groups can be quite complex and large as shown in Figure 9.1. Both phospholipids and glycolipids also share another structural form, that of forming derivatives of sphingosine, thereby giving sphingomyelins and cerebrosides. Cerebrosides are usually found in the membranes of nerve tissue and are associated with various diseases. Overall, phospholipids appear to be more likely than glycolipids to form lyotropic phases, which is probably due to their greater solubilities in aqueous media.

The third lipid component of biological membranes is cholesterol, and its related derivatives. Cholesterol on its own has a rigid aliphatic fused ring structure, with only a single hydroxyl group acting as a polar head group. However, when the hydroxyl head group is derivatized with a flexible aliphatic chain, it becomes another form of an amphiphile with a rigid-flexible dichotomy. In this form such derivatives of cholesterol exhibit thermotropic liquid crystal mesophases, as shown in Figure 9.6 for cholesteryl oleylcarbonate. Apart from its role in creating steroidal-based hormones, cholesterol is also an important component of biological membranes because it helps in immobilizing the outer surface of the membrane. Consequently, cholesterol in effect reduces membrane fluidity, and makes it more difficult for small water-soluble molecules to pass through the outer surface easily. Thus, without cholesterol, cell membranes simply would not be firm enough.

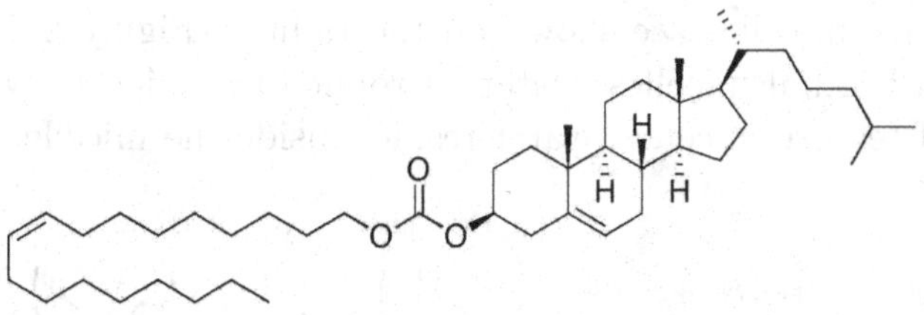

Cholesteryl oleylcarbonate COC   Cryst 26.7 N* 34 °C Liquid

COC can be mixed with cholesteryl nonanoate and cholesteryl benzoate
to give thermochromic mixtures used in strip thermometers.
It is also used in various cosmetic preparations.

Figure 9.6 Cholesteryl oleylcarbonate, an example of a thermotropic liquid crystal derivative of cholesterol.

9.2 AMPHIPHILES AND MICELLES

Micelles formed by amphiphiles come in two forms dependent on the solvent system. For water-based systems, the molecules aggregate in a way such that the non-polar chains interact together and are effectively removed from the water solvent by the surrounding polar head groups to give architectures in which the head groups are pointing into the water and the fatty chains are squeezed into the interior of the structure. Such micelles occur when the solution is relatively dilute and the solution behaves as an isotropic fluid. At low dilutions the micelles are spherical in shape and contain about 50 to 150 molecules, which are in dynamic equilibrium with single molecules. At higher concentrations of the amphiphiles, the micelles may have cylindrical, disk-like, or planar constructions. The overall forms of the micelles are also somewhat determined by the shapes of the amphiphiles and their ability to pack together (more on this later). Micelles are also stable in water provided that the concentration of surfactant is above the critical micelle concentration (CMC). The structures of spherical micelles, which are sometimes called normal micelles, are shown in the left-hand part of Figure 9.7.

Reverse micelles can also form where the non-polar chains radiate away from centrally aggregated head groups that surround the water solvent. Such reverse micelle formation usually occurs in oil-water mixtures where the amount of water is small and fills the void surrounded by the polar head groups. This phase in which the amphiphilic molecule separates the water from the oil is stable. The structures of what are called reverse micelles are shown on the right in Figure 9.7. Thus, in the case of normal micelles, water surrounds the micelles, whereas in the case of reverse micelles, water resides inside the micelle.

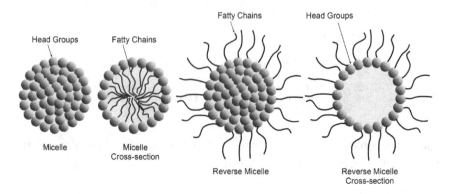

Figure 9.7 Structures of normal and reverse micelles.

One of the simplest ways to model spherical micelles is to assume the molecules have conical shapes as defined by the length of the amphiphile l_{max} and the cross-sectional area of the head group a. Then assuming self-assembly creates a spherical structure, relationships can be found between these two parameters and the molecular volume v. For the perfectly spherical micelle with radius R_{mic} shown in Figure 9.8, there are two ways to determine the number of molecules in the micelle N. Using the surface area of the micelle, one obtains $N = 4\pi R_{mic}^2/a$, and using the volume of the micelle, one obtains $N = 4\pi R_{mic}^3/(3v)$. Equating these two expressions for N yields an expression for the radius of the spherical micelle: $R_{mic} = 3v/a$. Now if the radius of the micelle is equal to the length of the amphiphile, then $v/(al_{max}) = 1/3$.

For the packing condition in which $R_{mic} \leq l_{max}$, then $v/(al_{max}) \leq 1/3$ for a spherical micelle. This condition is satisfied for sodium dodecyl sulfate with $v = 0.35$ nm^3, $a = 0.62$ nm^2, and $l_{max} = 1.7$ nm. Amphiphiles with smaller head groups ($1/3 < v/(al_{max}) < 1/2$) tend to form cylindrical structures and hence hexagonal phases, whereas those with the relationship $1/2 < v/(al_{max}) < 1$ do not form spheres or cylinders, and tend to form bilayers and lamellar phases. When $v/(al_{max}) \approx 1$, planar bilayers tend to form, and when the value is less than one, it is more favorable for vesicles to form. Lastly, when $v/(al_{max}) > 1$, such amphiphiles tend to form reverse micelles.

These relationships mirror the average shapes of the molecules,

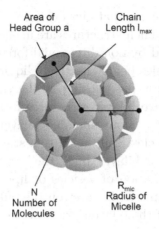

Figure 9.8 Model of a spherical micelle.

which can be rod-like, wedge-like, or conical. The curvature can be either positive or negative. When it is positive, the head groups for an ionic system are larger in cross-section than the tails, whereas the opposite is the case for small head groups. In the first case the head groups naturally are on the exterior of the micelle, and in the second the tails are on the outside of the reverse micellar structure. When the curvature is zero, no micelles form, but in this case lamellar structures are generated. The effect of curvature in relation to micellar or mesophase structure is shown in Figure 9.9. These nucleated structures form the basis for the formation of lyotropic mesophases.

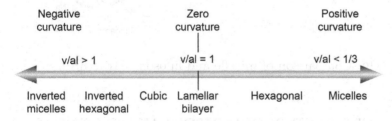

Figure 9.9 Property-structure correlations for amphiphiles as a function of curvature.

These relationships give reasonable approximations for the self-assembly and ultimately mesophase formation for amphiphiles, however, such dynamic-packing models do not actually rationalize the structures formed by molecular assemblies. There are no single crystal structures and NMR spectra that give evidence for a cone-like shape of any known amphiphile. Their structures are in effect short-lived and are only vaguely determined by non-directional forces and by solvophobic effects, with the molecules being effectively liquid like.

At zero curvature for the packing of the amphiphilic molecules, it is also possible to have bilayer structures form. However, for non-cylindrical molecules, the more unstable bilayer formation becomes, the more a tendency to elastically relax to a state of spontaneous curvature. If the layers in the bilayer differ, the bilayer becomes asymmetric resulting in the formation of vesicles or liposomes. Liposomes are usually spherical in shape, with water on the outside and water being trapped at the center of the bilayer structure, as shown in Figure 9.10. These types of self-assembled structures have been used intensively as molecular vehicles, particularly for drug delivery, where the delivery is controlled with respect to time and the drugs are protected from the environment.

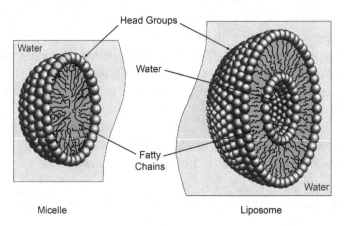

Figure 9.10 Comparison of micelle and liposome structures.

9.3 THE STRUCTURE OF LYOTROPIC LIQUID CRYSTALS

Having defined the formation of micelles and their potential shapes in the nucleation processes for the formation of lyotropic liquid crystal phases, the following section looks with more detail into the structures of the condensed mesophases, particularly as a function of curvature. Curvature in the packing of molecules is typically affected by an increase in amphiphilic concentration; possibly a much easier way to envisage this is through the swelling of head groups as they strongly interact with solvent molecules. Thus, different well-defined structures are formed, such as the "normal" continuous cubic, discontinuous cubic, and hexagonal phases, and the "reverse" continuous and discontinuous cubic, and hexagonal phases. For these phases of matter, which are effectively soft-crystals, a nomenclature classification has evolved using various capital letters and subscripts. For example, L stands for the lamellar phase, V (sometimes Q is used) for the bicontinuous cubic phase, I for the discontinuous cubic phase, and H for the hexagonal phase. Furthermore, subscripts I (1) and II (2) represent normal and reversed phases, respectively. Space groups are also used in describing the various mesophase structures, thereby placing lyotropic mesophases nearer to being solids rather than the comparative low molar mass nematic liquid crystals that are nearer to being liquid in their properties.

9.4   THE LAMELLAR LYOTROPIC LIQUID CRYSTAL PHASE

The lamellar lyotropic liquid crystal phase (termed L_α) has a relatively planar structure that is composed of lipid bilayers that are separated by water. The polar head groups of the amphiphilic molecules associate with, and are in direct contact with water, whereas the hydrophobic tails are separated from water. Thus, the amphiphilic nature of the molecules means that the self-assembly is a bilayer in nature with two layers being made up of intertwining non-polar chains from oppositely directed molecules. Consequently, the interfaces of the layers where the polar head groups meet is effectively separated by a band of water in a form of a microphase segregated system. A sketch of the structure of the lamellar phase, for a water (a polar protic solvent) based system, is shown in Figure 9.11. The bilayer thickness is usually 10-30% less than twice the length of an "all-trans" non-polar chain and the water layer thickness is between 1 and 10 nm if the water content is between 10 and 50% by weight. X-ray diffraction, obtained via small-angle X-ray scattering (SAXS), shows Bragg peaks with relative ratios of 1:2:3:4, and the study of the L_α phase by polarized transmitted light microscopy shows that the lamellar phase can exhibit "focal-conic, streaky, or mosaic-like defect textures." The focal-conic defect texture is indicative for the L_α phase, which, although possessing a layered structure, does not mean that its structure is the same as that of a smectic A phase.

Usually, lamellar lyotropic liquid crystal phases only exist down to

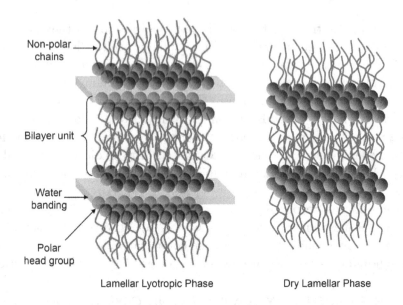

Non-polar chains

Bilayer unit

Water banding

Polar head group

Lamellar Lyotropic Phase   Dry Lamellar Phase

Figure 9.11 Structure of the lamellar (L_α) lyotropic liquid crystal phase (left), and the equivalent "dry" (thermotropic) mesophase (right).

50% surfactant (amphiphile), and below 50% the lamellar phase gives way to hexagonal lyotropic liquid crystal phases or to an isotropic micellar solution. However, in some cases the lamellar phase is even exhibited in extremely dilute solutions. Thus, rheology shows that L_α phases are less viscous than hexagonal lyotropic liquid crystal phases despite the fact that they contain less water. This is because the parallel layers are able to slide over one another with relative ease during shearing, which is quite easy to visualize from the figure. Furthermore, the lamellar phase extensively exists in organisms, and is the basic building block of cell membranes, as shown in Figure 9.1 and for the materials shown in Figure 9.5.

9.5 THE HEXAGONAL LYOTROPIC LIQUID CRYSTAL PHASES

The hexagonal lyotropic phase is one of the more common lyotropic liquid crystal phases, and as the name implies, the phase consists of a packing of cylindrical molecular aggregates into hexagonal arrangements. There are two types of hexagonal phases, the normal hexagonal phase (H_I) and the reverse hexagonal phase (H_{II}), as shown together in Figure 9.12. The normal hexagonal phase consists of micellar cylinders of indefinite length packed in a hexagonal arrangement with the head groups of the molecules pointing outwards on the surfaces of the cylinders, and the aliphatic chains located towards the central axes of the cylinders. The diameter of the micellar cylinders is typically 10 to 30% less than twice the length of the "all-trans" non-polar chains. The spacing between cylinders varies enormously between 1 and 5 nm depending upon the relative amounts of water that surround the cylinders and the nature of the surfactant. The reverse hexagonal phase is basically the same as the hexagonal phase except that the micellar cylinders are reversed with the non-polar chains radiating outwards from the cylinders and the head groups pointing inwards. Water that is present is contained within densely packed cylindrical reverse micelles, which have a typical diameter of 1 to 2 nm. It is thought that the water inside the cylindrical micelles is not in contact with any of the water outside the cylinders. The remaining space is therefore occupied by the non-polar chains, which overlap to leave the cylinders much closer together than in the normal hexagonal phase. X-ray diffraction gives Bragg peak ratios of $1{:}\sqrt{3}{:}2{:}\sqrt{7}$ using small angle diffraction. When solvated by oils, the oil molecules are situated towards the outside of the cylinders mixing with the lipid chains.

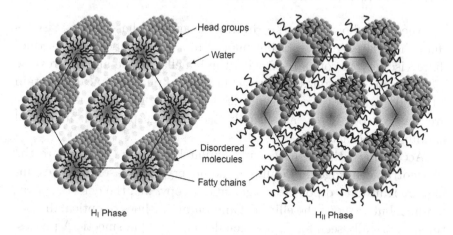

H$_I$ Phase   H$_{II}$ Phase

Figure 9.12 Structure of the hexagonal lyotropic liquid crystal phase. Left is the "normal" H$_I$ hexagonal phase, and right is the "reverse" H$_{II}$ phase.

Materials such as galactocerebrosides (Figure 9.13) that are found in nerve tissue, appear to exhibit both forms of hexagonal phase. When dry the galactose head group is still surrounded by moisture from the air and thereby forms a reverse hexagonal phase. But when wet, the molecules flip over so that the head groups are now positioned pointing outwards into the water, thereby forming a normal hexagonal phase.

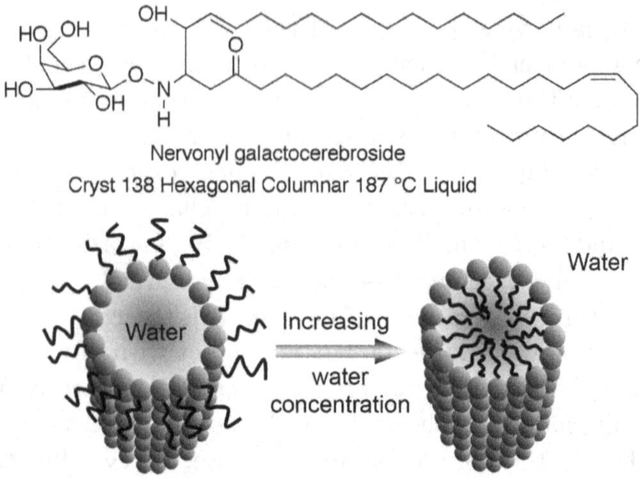

Nervonyl galactocerebroside
Cryst 138 Hexagonal Columnar 187 °C Liquid

Figure 9.13 Nervonyl galactocerebroside from bovine brain exhibits both normal and reverse hexagonal phases.

Rheology shows that the reverse hexagonal phase is less viscous than the related reverse bicontinuous cubic phase, and is of intermediate viscosity in comparison with the lamellar phase and the reverse bicontinuous cubic phase. Thus, as hexagonal phases typically contain 30 to 60% water by weight, and despite the high water content, they are very viscous. They are usually used for the practical, industrial handling of surfactants.

Accordingly, these phases give similar birefringent textures to the thermotropic hexagonal columnar liquid crystal phases when examined by transmitted polarized light microscopy, *i.e.*, the defects appear fan-like, but without the elliptical and parabolic lines of optical discontinuity normally seen for focal-conic defects found in smectic A phases.

9.6 THE DISCONTINUOUS CUBIC LYOTROPIC LIQUID CRYSTAL PHASES

Cubic lyotropic liquid crystal phases are not as common as the lamellar or hexagonal phases. However, cubic lyotropic phases do occur in different regions of phase diagrams. Accordingly, there is probably a range of different cubic lyotropic liquid crystal phases, the exact structure of which relates to their position within phase diagrams. Structurally, the cubic lyotropic liquid crystal phases are not as well-characterized as the lamellar or hexagonal phases, however, two types of cubic lyotropic liquid crystal phases have been established and each can be generated in the "normal" manner (water continuous) or in the "reverse" manner (non-polar chain continuous), which makes for a total of four different phase families, labeled I_I and I_{II}, for the discontinuous phases and V_I and V_{II} for the continuous phases. For simplicity of structure, the ones we encounter in this section are the discontinuous cubic phases, which are sometimes called the "micellar" phases. The discontinuous cubic phase consists of a cubic arrangement of molecular aggregates, where the aggregates are similar to micelles (I_I phase) or reversed micelles (I_{II} phase). The structures of the "normal" and "reverse" cubic (I) phases are illustrated in Figure 9.14. In the polarizing microscope they appear black, demonstrating their isotropic structures. Mechanical shearing of specimens in the microscope shows that the cubic phase is highly viscous in comparison to other lyotropic phases.

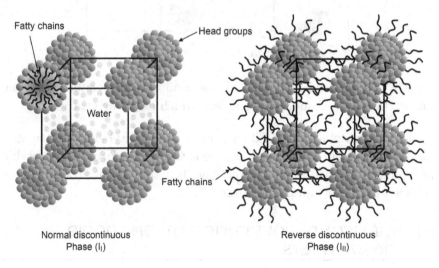

Normal discontinuous
Phase (I_I)

Reverse discontinuous
Phase (I_{II})

Figure 9.14 The simple structures of the discontinuous cubic lyotropic liquid crystal phase.

Although some reports suggest that the molecular aggregates are spherical, and others that they are cylindrical or ellipsoidal, the reverse discontinuous cubic phase is still thought to possess micelles arranged in a cubic lattice. There are potentially two main micellar cubic phases with the following symmetries, Fd3m and Fm3m. The Fd3m cubic phase, shown in Figure 9.15, has a closed structure composed of micelles of possibly two different sizes that are organized in a double diamond network. X-ray measurements give the relationship for the Bragg reflections as $\sqrt{3}:\sqrt{8}:\sqrt{11}:\sqrt{16}:\sqrt{19}$. Conversely, the Fm3m reverse cubic phase has been reported to be made up of a compact packing of large monodispersed micelles in a face-centered cubic (fcc) lattice.

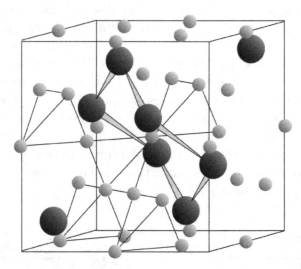

Figure 9.15 Proposed structure of the Fd3m reverse cubic lyotropic phase. Two sets of micelles are shown which are of arbitrary size.

Of the two phases, the Fd3m variant is actively being investigated for applications in drug release, whereas the Fm3m form is more stable in equilibrium with excess water, and has the potential for being a vehicle for greater uptake of drugs.

9.7 THE BICONTINUOUS CUBIC LYOTROPIC LIQUID CRYSTAL PHASES

The second family of cubic lyotropic liquid crystal phases is found to lie between the hexagonal and lamellar phases. These cubic phases also can be divided into "normal" (V_I) and "reverse" (V_{II}), and in each case they are not made up of small aggregates but composed of large, continuous channel networks that are either water continuous (V_I) or non-polar chain continuous (V_{II}). The phases can be subdivided into three further variants with the following lattices: a primitive lattice Im3, a double-diamond lattice, Pn3m, and a gyroid lattice Ia3d. These lattices are based upon relatively disordered associations of microsegregated molecules that pack together in curved arrangements. To minimize the free volume, a minimal surface is required that minimizes the local area, which is equivalent to zero mean curvature. Such surfaces can be modeled mathematically, which was achieved over 100 years ago by Schwartz using differential geometry to develop a theory of infinite periodic minimal surfaces. Thus for example, the primitive lattice is associated with a Schwartz P surface.

To explain why these fascinating and structurally unusual phases form, we can investigate what happens at a molecular level when water is added to a monolayer made up of disordered amphiphilies as shown in Figure 9.16. Let us assume that the layer is flat with as many molecules pointing up as down, as shown in the top of the figure. The addition of water swells the head groups (the reverse would be the case for the addition of oil). The swelling head groups force the layer to become distorted, which raises the free energy of the system. Rearrangement of the molecules to give wave-like curvature of the layer in effect produces better packing of the molecules and at the same time minimizes the overall curvature, as shown in the lower part of the figure. We can think of the rippling of the layer being summations of positive and negative curvature, which overall result in zero curvature. In reality this picture is more complicated for the L_α phase because it has a two-dimensional bilayer structure. If we now take a 2-D layer, the curvature could occur in different directions within the 2-D space, and for curvature directions at right angles to one another, the minimal surface is that of a saddle shape as shown in Figure 9.17. In three-dimensional systems, based on the theory of infinite periodic minimal surfaces, the three lattices listed above possess three-dimensional periodic ordering. In these structures the lipid molecules are located with either the head groups or fatty chains positioned on the minimal surfaces, with the two variations giving the "normal" or "reverse" cubic phases. X-ray diffraction can be used to distinguish between the three cubic phases with Bragg peaks as follows: Im3m $\sqrt{2}$:$\sqrt{4}$:$\sqrt{6}$:$\sqrt{8}$, Pn3m $\sqrt{2}$:$\sqrt{3}$:$\sqrt{4}$:$\sqrt{6}$:$\sqrt{8}$:$\sqrt{9}$, and Ia3d $\sqrt{6}$:$\sqrt{8}$:$\sqrt{14}$:$\sqrt{16}$:$\sqrt{20}$:$\sqrt{22}$. Figure 9.18 shows the two structures for the bicontinuous cubic phases that have primitive lattice structures based on the Im3m lattice. The structure shows that continuous pathways permeate the structure, a description which causes the structures to be called plumber's nightmares.

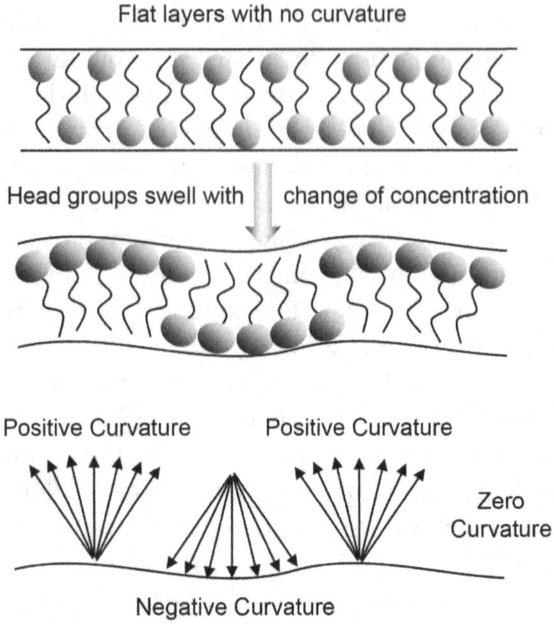

Flat layers with no curvature

Head groups swell with change of concentration

Positive Curvature   Positive Curvature

Zero
Curvature

Negative Curvature

Figure 9.16 Introduction of curvature into a monolayer by the swelling of head groups.

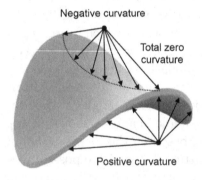

Negative curvature

Total zero
curvature

Positive curvature

Figure 9.17 Curvature in two directions leading to a saddle shaped surface.

The cubic lattice arrangements of the molecular aggregates in the bicontinuous phase means that the phases are optically isotropic. So when examined by optical polarizing microscopy, the cubic phases are optically extinct unlike the birefringent textures that are generated by the lamellar and hexagonal phases. The isotropic nature of the cubic

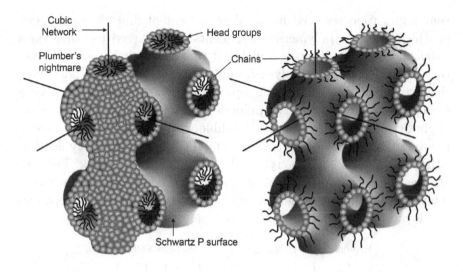

Figure 9.18 The normal and reverse structures of the bicontinuous cubic phase that have a primitive structure.

phases often makes them difficult to detect by microscopy unless a phase plate is inserted into the microscope. Like discontinuous cubic phases, bicontinuous phases are also extremely viscous, even more so than the hexagonal phases. The high viscosity results from the lack of shear planes within the structure that would allow for a sliding movement.

Minimal surfaces, minimization of the free volume, and the effects of curvature on self-assembly and the formation of quasi-soft crystal phases, extends to systems other than amphiphiles, and therefore could be considered collectively as a "universal behavior." For example, the curvature in packing affects large molecular systems such as diblock copolymers, and extends down to relatively small molecular materials that exhibit smectic cubic phases.

For polymeric materials, as with molecular materials, mixing has been thought of as a way forward to create materials with desired physical properties, particularly for applications. However, this is not usually the case for polymers. Miscible polymer blends are usually sought with the objective of creating a material with a single-phase structure and one glass transition temperature. Unfortunately, segregation often occurs, and consequently immiscible polymer blends, sometimes called "heterogeneous polymer blends" result. Most polymers fall into this group and have two glass transition temperatures. Attempts to prevent segregation resulted in the development of diblock copolymers, in which different polymeric components are tethered together via a covalent bond, resulting in a material that has a dichotomous structure for which the two polymeric units cannot separate. Figure 9.19 shows two examples of diblock copolymers, where material (A) has a structure similar to a giant amphiphile with a polar head group and an aliphatic tail, and material (B) is a diblock copolymer that has a relatively rigid aromatic block (polystyrene [PS]) and a relatively flexible, polar, poly(methyl methacrylate [PMMA]) aliphatic block. Polymer (A) has the potential to be solvated in water and thereby is able to exhibit lyotropic phases, whereas polymer (B) is unlikely to be solvated, and is more likely to exhibit thermotropic phases.

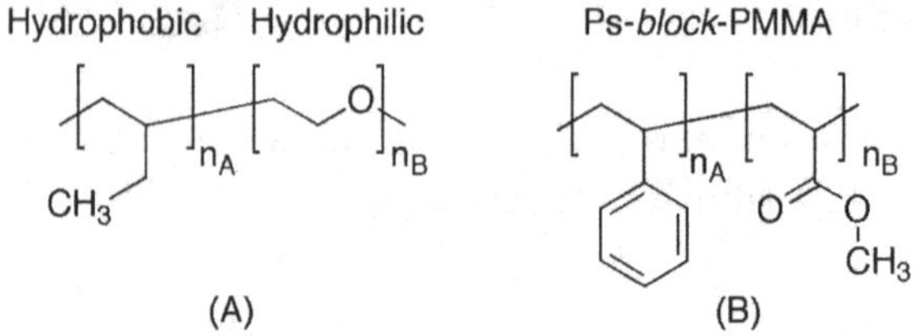

Figure 9.19 Examples of diblock copolymers.

Consider a diblock copolymer made up of segments A and B. When the two polymer chains interact, they have the possibility of A–A, B–B, and A–B interactions. As with amphiphiles, demixing occurs when the self-interactions are stronger than the opposing interactions, and thus macroscopic phase separation takes place on the nanometer length scale. The entropy for mixing depends on the degree of polymerization, $N = N_A + N_B$ (see next chapter), which drops with the increasing value of N. Therefore, the longer the polymer chain, the more likely it is that segregation occurs due to entropic gain. Thus $N > 1000$ is potentially required for segregation to occur.

For a symmetric block copolymer, when the Flory-Huggins "interaction parameter", which describes enthalpic contributions, reaches a value close to that at which the two blocks would phase separate if they were not bonded together, then both theory and experiment show that fluctuating A-rich and B-rich domains begin to appear. Under these conditions, the A and B blocks begin to stretch out so that their respective radii of gyration exceed their equilibrium values. At this point, an order-disorder transition becomes a possibility, and segregation occurs thereby creating an interface boundary between the two blocks as shown in Figure 9.20.

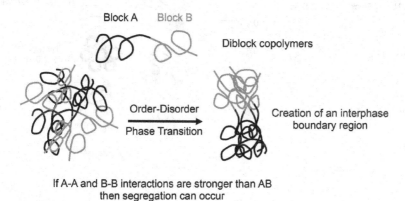

Figure 9.20 Macroscopic phase separation taking place on the nanometer scale length.

For a diblock copolymer in which the two units have roughly the same length and have equivalent interactions, the first stage in the segregation is the formation of a lamellar structure consisting of alternating A and B blocks of the two polymer units as shown in Figure 9.21. The proportion of one segment and hence the ratio of the two segments can be defined via the term for the A segment as $f_A = N_A/(N_A + N_B)$, where N is the degree of polymerization. When the two segments have similar lengths, f_A equals approximately 0.5. At this point the system curvature is approximately zero, hence relatively flat layers are formed. By varying the value of f_A, positive or negative curvature can be introduced into the structure resulting in the formation of continuous cubic phases (usually found in their gyroid lattice forms), hexagonal phases, and discontinuous phases, as shown in the figure. The approximate ranges for f_A that support certain phase types are also given in the figure. Obviously, the different sizes of the block copolymer units contribute to differing rotational volumes in a similar way to the swelling of the head groups via solvation for low molar mass amphiphilic systems, for which phase formation is driven by curvature induced packing constraints, minimization of the free volume, and minimal surfaces.

Conventional liquid-crystalline materials also have the possibility of forming phases based on the above properties. For example, small molecules such as 4'-alkoxy-3'-nitrobiphenyl-4-carboxylic acid, shown

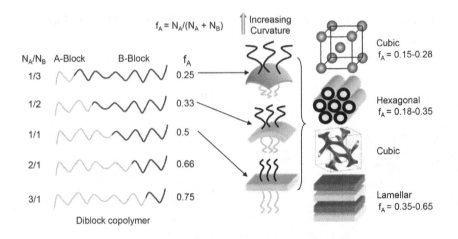

Figure 9.21 Formation of liquid crystal phases of diblock copolymers by microphase segregation.

in Figure 9.22, have been found to exhibit the rare cubic D phase. The D phase possesses a bicontinuous cubic structure, which is sandwiched between two smectic phases; the phase behavior is thermally driven and not via the addition of solvent. The phase transitions show extensive hysteresis, and under various cooling rates a hexagonal phase can replace the cubic phase. This very unusual behavior is due to the liquid crystal SmC and SmA phases having properties closer to the liquid state, whereas the D and hexagonal phases are quasi-crystalline and therefore show supercooling in the crystallization processes, which because of the complexity of the phases structures, are kinetically driven. Similar behavior for thermotropic phases also occurs for cubic blue phases, for which the phase transitions are driven by the formation of helical structures due to the materials being chiral.

9.8 PHASE TRANSFORMATIONS AND PHASE DIAGRAMS

To understand how lyotropic phases form, we need first to examine the nucleation process whereby amphiphilic molecules coalesce to generate micelles, which then in turn form self-assembled structures and hence mesophases. Two concepts and definitions are applied in this process. First is the critical micelle concentration (CMC), which is defined as the concentration of "amphiphilic molecules" for which micelles form and all additional amphiphilic molecules that are added to the system

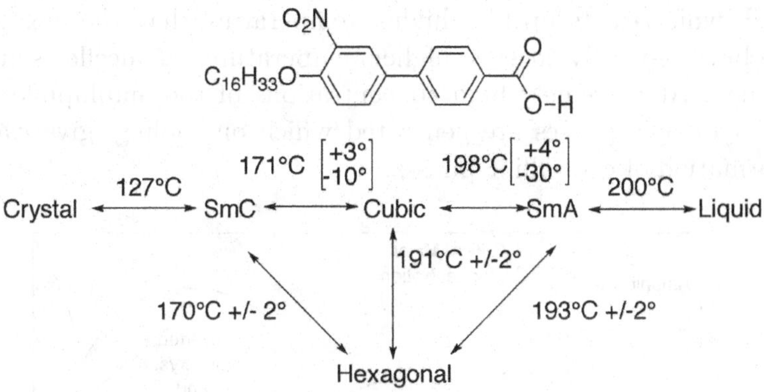

Figure 9.22 Phase behavior and transition temperatures of 4'-alkoxy-3'-nitrobiphenyl-4-carboxylic acid.

form or add to the micelles (see discussion surrounding Figure 4.17 in Chapter 4). Second is the Krafft point, which is the minimum temperature (T_K) at which a surfactant can form micelles. At this point the surfactant solubility equals the critical micelle concentration, and above the Krafft point lyotropic liquid crystal phases are generated, whereas below micelles become insoluble. The temperature for the formation of lyotropic phases should usually be increased to about 10°C above the Krafft point to ensure that the phases are stable without interfering coexistence regions. As the temperature is decreased, lyotropic phases persist until the melting point of the neat surfactant is reached. Temperature is important in the generation of lyotropic liquid crystal phases because the amphiphilic molecules need to move relative to each other. At lower temperatures, the molecules become more rigid and crystallization results.

The best way to illustrate the behavior of an amphiphilic material in water is to use a phase diagram. Phase diagrams for lyotropic phases are constructed with amphiphile concentration along the horizontal axis and temperature along the vertical axis. A typical phase diagram for a soap and water is shown in Figure 9.23, which clearly shows the critical micelle concentration (below which micelles do not form) and the Krafft point at each temperature (below which the crystal is insoluble in water). Above the Krafft point, lyotropic liquid crystal phases are generated. At relatively low concentrations, the hexagonal phase is generated up to certain temperatures when it gives way to a micellar solution. At relatively high concentrations, the lamellar phase is formed, which exists up to a higher temperature than the hexagonal phase but eventually, at even higher temperatures, a micellar solution is formed. At extremely high concentrations of the amphiphile, "reverse" lyotropic phases are generated which on cooling, give way to the formation of crystalline phases.

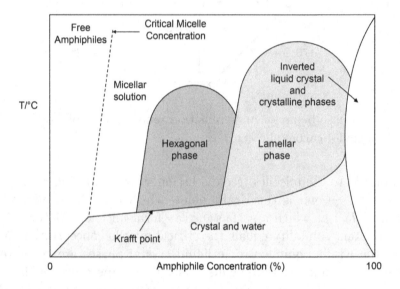

Figure 9.23 Phase diagram for typical soap in water mixtures.

As with calamitic and discotic liquid crystals, there have been attempts to rationalize the phase behavior and transitions for lyotropic phases. The first thing to note is that the lyotropic phases also exhibit nematic phases, and that the arrangements for the more ordered lamellar/cubic/columnar phases resemble those of thermotropic liquid crystals, and therefore it is tempting to carry over observations for thermotropics to lyotropics. Moreover, there are other phase variants in each individual class of lyotropic phases. However, there is a major difference, and that is the effect of curvature on phase sequences and structures, which provide for another aspect to phase classification, as shown in Figure 9.24. In the figure, lyotropic phases are shown linked to the thermotropic behavior of amphiphiles, but extending this relationship to calamitics may be a little bit more difficult because of the number of phases and extensive degree of polymorphism in small molecular systems. However, it is interesting to note the change in dimensional ordering across the phase diagram as shown in the figure. It is clear that the cubic phases are three-dimensionally ordered, but from the molecular point of view there is no positional order, and the gross packing arrangements of the molecules require minimization of the space available because of the need to reduce the free volume. Thus the periodicity effectively arises from the defects in an elastic system, and as a consequence the cubic phase is a quasi-crystal. For the hexagonal and lamellar phases, similar arguments can be made, with the hexagonal phase being a two-dimensional crystal, and the lamellar one-dimensional. Transitions for the cubic phase sandwiched in a conventional thermotropic phase sequence demonstrates that the lyotropic phases have a higher degree of crystal ordering than low molar mass thermotropic materials (note the hysteresis of transition temperatures in the formation of cubic and hexagonal phases for 4'-alkoxy-3'-nitrobiphenyl-4-carboxylic acid). Such transitions show that it is probably unlikely that for lyotropic phases there are reversible transitions between the classes of mesophase, and that the transitions have a reasonable degree of kinetic behavior associated with them because of the time required for rearrangements of the molecules.

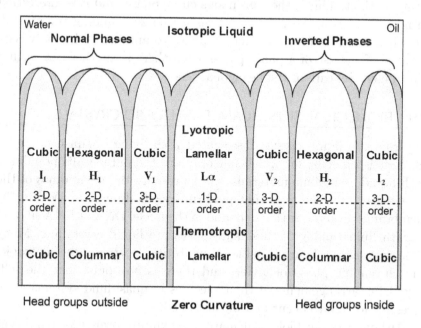

Figure 9.24 The universal phase diagram for lyotropic phases as a function of temperature.

The connectivity between thermotropic and lyotropic phases is demonstrated for octyl β-D-glucopyranoside in Figure 9.25. Octyl β-D-glucopyranoside is a non-ionic detergent that is used in the isolation of proteins, *etc.*, and it also exhibits thermotropic and lyotropic mesophases. On heating, it melts via three crystal forms to give a lamellar thermotropic phase at around 67°C, which then persists to over 100°C. On the addition of water to the crystal phase at room temperature, the glycolipid dissolves via the formation of a variety of lyotropic phases, the sequence for which is determined by the swelling of the head groups and the associated increase in curvature. The sequence is beautifully demonstrated in Figure 9.26, in which a crystal is shown in the center of the field of view of a polarizing microscope fitted with a phase plate above the objective. The sample is at room temperature and water is run under the cover-slip using capillary action to surround the crystal, which then slowly dissolves. The picture shows the crystal surrounded by rings of lyotropic phases. Outside of the crystal, the first ring is a weakly birefringent texture of the lamellar phase; the second ring shows no texture as the phase is cubic and optically extinct. This is the continuous cubic phase, and it is interesting to note there is a black line, which is a ring surrounding the crystal. This is a transition from one cubic phase to another and is only seen because of the use of a phase plate. Then there is another ring with a birefringent texture of the hexagonal phase.

9.9 BIOLOGICAL SIGNIFICANCE OF LIQUID CRYSTALS

Returning to liquid crystals associated with living organisms, the focus first is on cells, for which not only are the outer walls populated with liquid-crystalline materials, but so too are the membranes of the organelles of the interiors that perform specific functions. Within the membranes proteins are located, and in the nuclei DNA or RNA sit, and as with lipids, many of these substances are liquid crystalline. Many biological materials are capable of exhibiting lyotropic, thermotropic, and chromonic phase behavior, and there is no doubt that the self-organizing arrangements of the molecules in quasi-fluid systems are a necessity for life to occur and be sustained.

Although many biological materials exhibit a range of particular mesomorphic behavior in one family of materials, the templating for other forms of liquid crystalline behavior is not necessarily exact. For example, a family of glycolipids may exhibit thermotropic behavior

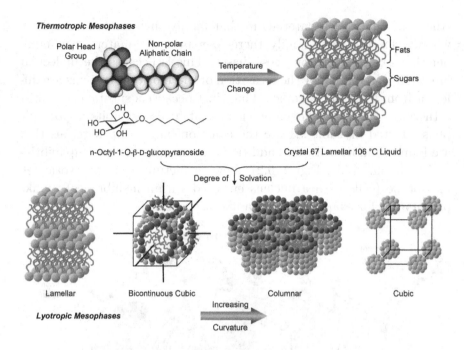

Figure 9.25 Sketch of the melting and solvation processes for octyl 1-O-
β-D-glucopyranoside.

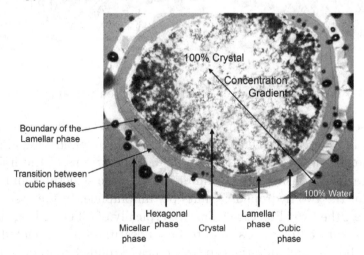

Figure 9.26 Photomicrograph of the solvation processes for octyl 1-O-β-
D-glucopyranoside observed at room temperature in a polarizing transmitted light microscope fitted with a phase plate (x100).

(there are numerous materials in that family that are mesomorphic), whereas for the same family there may be fewer materials in number that exhibit lyotropic mesophases. This is because the molecular interactions required for the formation of thermotropic systems are different from those of lyotropics. Thus, just because a substance exhibits a thermotropic phase does not mean that it also exhibits a lyotropic phase. Listed in Table 9.1 are the common families of materials that are found in most cells, and additional comments on their capabilities to be mesomorphic. Figure 9.27 shows a picture of a eukaryotic cell membrane (cells with a nucleus enclosed within membranes), unlike prokaryotes for example, bacteria and archaea.

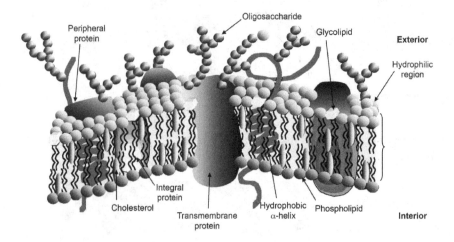

Figure 9.27   A sketch of the cell membrane of a eukaryote.

The families of lipids that are found in eukaryotic cells are shown together in Figure 9.5, whereas an example of a lipid found in an archaeal cell membrane is shown in Figure 9.28. Comparisons of the lipids of archaea versus eukaryotes lie almost solely in the structures and properties of the two hydrophobic units per amphiphile. In the case of eukaryotes, the two aliphatic chains are typically of differing lengths and either saturated or possess one to four unsaturations associated with double bonds, and are attached to the head groups by esters. The differences in the chain lengths lowers melting points much like the use of eutectic mixtures in nematic formulations. For archaea, the chains have no unsaturations, but instead have lateral methyl substitution introduced by the incorporation of saturated isoprenic units. The aliphatic units in this case are attached to the head groups by ether units, as

Table 9.1 Tendency for biological materials to support mesomorphic behavior.

Lipids Phospholipids Glyocolipids Inositols Bolaphiles Cholesterols Steroids	Phospholipids exhibit both thermotropic as well as lyotropic phases. They are usually unstable to heat but readily form lyotropic phases and membranes, therefore favoring lyo- over thermo-tropics. Glycolipids and inositols, which are found on the outer leaflets of membranes, are relatively stable, but decompose near clearing points. They tend to favor thermotropic phases over lyotropic phases. Bolaphiles are materials that have two head groups separated by a methylene chain. They readily form mesophases with the same pattern above except where one head group is a phosphate and the other is a sugar, in which case the ionic head group dominates the behavior. Cholesterol is not liquid crystalline, but cholesterols substituted with alkyl chains exhibit thermotropic behavior.
Proteins/Peptides Peripheral Transmembrane Integral	Generally proteins do not exhibit thermotropic liquid crystal phase behavior, but there are some that are still mesomorphic, for example spider silk. Oligomeric polypeptides have been found to exhibit chiral nematic phases, and for those with short chain lengths, some exhibit mesophases where there is microphase segregation. Also, amyloid fibrils, which are associated with plaques found in brain tissue of Alzheimer patients, are thought to be liquid crystalline.
Biopolymers Polysaccharides DNA RNA	Biopolymers refer to materials such as cellulose. In the presence of water, cellulose has been shown to be mesomorphic and to have properties such as the selective reflection of light. For the more complex polysaccharides found on the exterior surfaces of cells, very few studies have been performed to prove liquid crystallinity. On the other hand polynucleotides, such as DNA and RNA, have been found to exhibit lyotropic mesophase behavior, and it has been suggested that their behavior is chromonic.

shown in Figure 9.28, for which some are trans membrane and thereby make gigantic molecular loops. The differences between the two sets of cells are due to the chemical properties desired by the organisms.

Archaea bacteria exist under extreme conditions, for example at the bottom of oceans, near volcanoes, in the Dead Sea, *etc.*, and therefore chemical stability is important, hence the replacement of ester units with ethers and the removal of unsaturations in the aliphatic chains. The lipid chains still do the same job in eukaryotes as in archaea.

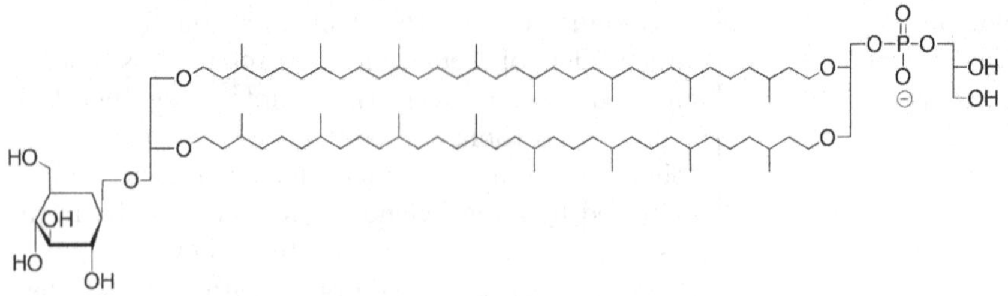

Figure 9.28 An example of an amphiphile found in the cell membrane of an archaea bacteria.

Most amphiphiles that are found in biological membranes possess the simple architecture of a head group with two aliphatic chains attached. When dry and heated to examine their thermotropic mesophases, the mesophase that usually dominates is the "reverse" hexagonal phase in which the head groups point inwards in the columns with the fatty chains to the exterior. In aqueous media the structure is often lamellar at lower concentrations of water, thereby stabilizing the formation of bilayer structures. As the water concentration increases, hexagonal phases can be found as shown for nervonyl galactocerebroside in Figure 9.13. Consider now what happens when only one aliphatic chain is present. (2S,3R,4E)-2-amino-3-hydroxy-4-octadecen-1-yl β-D-galactopyranoside, or psychosine as it is known, is the parent of cerebrosides (see Figure 9.29). On heating it does not form a hexagonal phase and instead forms a lamellar phase, and in water it is likely to support formation of hexagonal phases at lower concentrations simply because the head group swells to give a larger cross-sectional area, whereas the single aliphatic chain does not change and the fatty part of the molecular architecture always has a smaller cross-section in comparison to the cerebroside. Hence the curvature is larger.

Psychosine is a highly cytotoxic lipid that accumulates in the nervous system in the absence of the enzyme galactosylceramidase. Psychosine and its relatives are bioactive compounds that appear as intermediates of sphingolipid metabolism and cell signaling, and related

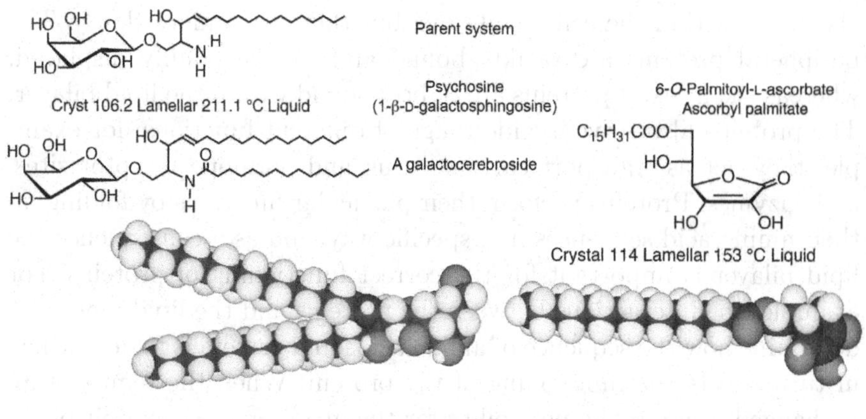

Figure 9.29 Amphiphiles psychosine and ascorbyl palmitate, which exhibit lamellar mesophases.

diseases. For example, Krabbe disease results from a mutation in galactocerebrosidase, an enzyme used in the degradation of galactocerebroside to ceramide and of psychosine to sphingosine.

It is interesting to note that galactocerebroside is a major component of myelin. In the nervous system, axons of nerve cells may be myelinated, or unmyelinated. This is the provision of an insulating layer, called a myelin sheath. And so we go full circle back to the discovery of lyotropic mesophases by Rudolf Virchow in 1854.

There are numerous other biologically active amphiphiles that have only one fatty chain and one head group; some are natural and some are synthetic. For example, 6-O-palmitoyl-L-ascorbic or ascorbyl palmitate (marketed as vitamin C ester) is an ester formed by the reaction between ascorbic acid and palmitic acid (see Figure 9.29). It is produced to create a fat-soluble form of vitamin C, and as a source for vitamin C as ascorbyl palmitate breaks down into its components before being digested. It is also known as an antioxidant food additive (E number E304), and like psychosine, it exhibits a lamellar thermotropic liquid crystal phase.

We now look at the thermal properties associated with liquid crystals and their affects on biological membranes. In the liquid crystal form, the head groups of the amphiphiles do not have any periodic ordering and the hydrocarbon chains are not rigid but liquid-like. The liquidity of the structure allows the movement of molecules of the amphiphiles in the cell membrane. Of course, the associated proteins also move within the cell membrane, but they do so more slowly. The peripheral proteins are weakly bound and can be readily displaced, whereas the integral proteins are tightly bound within the lipid bilayer. The proteins also serve a wide range of different functions; for example, they act as transport carriers, drug and hormone receptor sites, and enzymes. Proteins perform their particular functions by folding up their amino acid sequences in a specific way, and as a consequence the lipid bilayer is important for the correct functioning of proteins. For example, the interactions between the proteins and the lipid molecules determine how the sequence of amino acids in a protein is folded, which in turn affects the functioning of the protein. When the temperature is changed, there is the possibility for the membranes to exhibit phase transitions, just like a thermotropic liquid-crystalline compound. In this case the phase transition is called a gel point and the structural change at this phase transition is illustrated in Figure 9.30. On cooling the environment of the cell membrane to just below the normal ambient temperature, the head groups become arranged in a relatively ordered hexagonal manner and the hydrocarbon chains become elongated and stiffer. The temperature of the liquid crystal phase to gel phase transition depends upon the environment and the organism concerned. For example, homeothermic animals control their own body temperatures and so the cells are not exposed to marked changes in temperature. Accordingly, the lipid membranes of such animals are composed of a high proportion of saturated fatty acids, which give rise to a relatively high gel phase transition. On the other hand, poikilothermic organisms (*e.g.*, fish) are subject to relatively large ranges of temperature. Consequently, the cell membranes tend to include lipids with a relatively high proportion of unsaturated fatty acids to provide a relatively low gel phase transition temperature.

Lowering the environmental temperature below the gel transition changes the functioning of the cell membrane and can lead to the death of the organism. Similarly, organisms that are used to high-pressure environments (deep under water) have a gel transition temperature to suit this environment. However, when such organisms are brought to the relatively low-pressure environment on land, their gel transition temperatures are a long way below the environmental temperature, which can also cause death due to cell membrane rupture. Consequently, in addition to requiring a gel transition temperature below ambient temperature, organisms must also have the gel transition temperature close to ambient temperature and not too far below it.

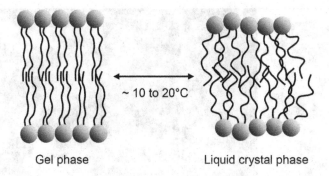

Gel phase   Liquid crystal phase

Figure 9.30 The change of structure at the gel to liquid crystal transition.

In addition to lipids and proteins, there are other liquid crystalline polymeric systems to consider. In fact in terms of lyotropic polymers, the majority of lyotropic polymers are actually found in biological systems, for example, cellulose and some of its derivatives exhibit lyotropic phases. Probably the most well-known biopolymer is DNA, however, it is generally not known that there are two forms, A and B, and that it was the B form that ultimately revealed the structure of DNA (see Figure 9.31(a)). It was Rosalind Franklin who produced the now famous X-ray diffraction pattern of a cross of diffraction spots that gave the clue that a helical structure is present. The experiment was achieved by pulling a fiber of the B form from a solution of DNA (see Figure 9.31(b)) under controlled humidity, and then subjecting it to fiber diffraction. Resulting structural studies showed that DNA is made up of pairs of templating bases. There are four types; adenine (A), thymine (T), guanine (G), and cytosine (C). The base pairs span two polymer chains that are made up of phosphate and sugar groups that form the double helix. In recent years there have been many studies of the liquid crystal properties of DNA (and RNA) fibers, including natural polymeric chains and synthetic constructions that are prepared by machine synthesizers. The studies include those by polarized light microscopy, which have revealed defect textures characteristic of liquid crystals, as shown by the texture of the chromonic M phase shown in Figure 9.31(c).

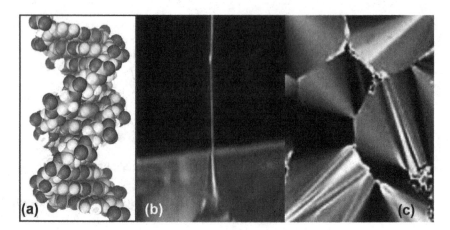

Figure 9.31 Structural model of DNA (a), fiber of DNA being pulled from a solution (b), and texture of the columnar chromonic phase of the B form of DNA.

9.10 CHROMONIC LIQUID CRYSTALS

Certain dyes, drugs, and short strands of nucleic acids form assemblies that result in liquid crystal phases in a slightly different manner from the amphiphilic molecules discussed so far. Rather than having a polar head group and a non-polar tail, these molecules are more disk- or plank-like. The inner portion of the molecule usually contains aromatic ring structures while the outer portion usually contains polar groups. When such a molecule is mixed with a polar solvent, typically water, the flat molecules tend to stack on top of each other, forming weak bonds between the π-electrons on the aromatic rings, while presenting the polar groups to the solvent. If the stacks of molecules form long enough assemblies and are concentrated enough, liquid crystal phases form. These are called lyotropic chromonic liquid crystal phases or just chromonic liquid crystal phases.

The same basic explanation reveals why short oligomers of nucleic acids form chromonic liquid crystals. Nucleic acids are soluble in water because the polar phosphate and sugar groups shield the non-polar base pairs from the water. Short oligomers, however, have a high proportion of end base pairs, and these are not shielded from the water. But just like chromonic molecules, the ends of the oligomers can form weak bonds and lower the free energy. The result is a dynamic distribution of longer nucleic acid rods and the formation of liquid crystal phases.

Examples of chromonic liquid crystal molecules are shown in Figure 9.32. Disodium cromoglycate is an asthma drug that has been around for many years. Because it absorbs light in the ultraviolet rather than the visible, it is the most studied of all the chromonic liquid crystals. Sunset Yellow FCF is a common food dye that goes by the trade name Yellow 6 in the U.S., but is banned from use in some countries. As might be expected, fairly high concentrations are necessary for liquid crystal phases to form, about 12 wt% in the case of disodium cromoglycate, about 30 wt% in the case of Sunset Yellow FCF, and about 7 wt% for the perylene dye shown in Figure 9.32.

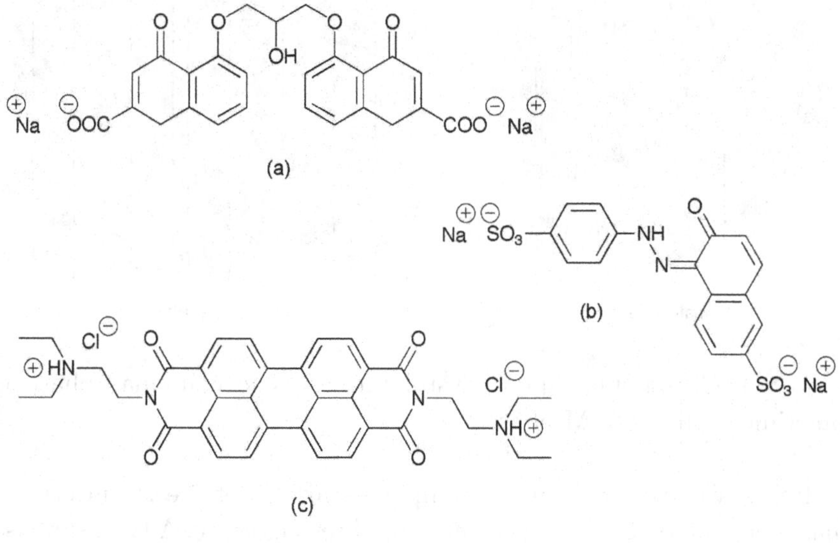

Figure 9.32 Typical chromonic liquid crystal molecules. (a) disodium cromoglycate, (b) Sunset Yellow FCF, (c) bis-(N,N-diethylaminoethyl)perylene-3,4,9,10-tetracarboxylic diimide dihydrochloride.

The most common chromonic liquid crystal phases are the nematic phase and columnar phase (also called the M phase). In the nematic phase, the stacks of molecules that are long enough to act as rods, tend to orient along a preferred direction just as the molecules in a thermotropic nematic liquid crystal tend to orient. At higher concentration, a second phase usually appears in which the assemblies arrange themselves in a hexagonal lattice with the preferred direction perpendicular to the plane of the lattice. Both of these phases are depicted in Figure 9.33. Thus while the assemblies in the nematic phase possess only orientational order, the assemblies in the columnar phase possess both orientational and positional order. These two phases have much in common with the nematic and columnar phases of discotic liquid crystals, which is not surprising given that the basic molecular shape of both is disk-like. But keep in mind the difference: the molecules of a chromonic liquid crystal are in a solvent while discotic liquid crystals contain no solvent.

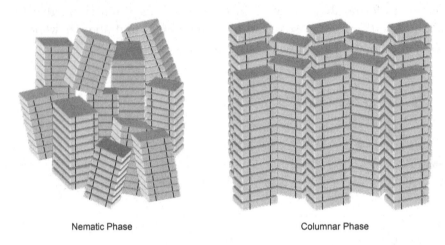

Nematic Phase   Columnar Phase

Figure 9.33 Chromonic liquid crystal phases. The columnar phase is sometimes called the M phase.

Being a lyotropic liquid crystal, the stability of the liquid crystal phases depend on both concentration and temperature. A typical phase diagram is shown in Figure 9.34, where it is clear that it differs from the phase diagrams of most lyotropic liquid crystals. Note the existence of broad two-phase regions between the phases. These no doubt stem from the large distribution in assembly size. Some interesting phase sequences are possible. For example, for a fixed concentration at low temperature (the arrow in the figure), the sample might show a mixture of columnar and nematic phases. Upon heating, the columnar phase disappears and the entire sample is nematic. Upon further heating, the nematic and isotropic phase coexist. With continued heating, the nematic regions transition to columnar regions, until finally the columnar regions melt and the entire sample is isotropic.

The most fundamental difference between chromonic liquid crystals and most other lyotropic liquid crystals is that the molecules in the stacks of chromonic liquid crystals are bound together quite weakly,

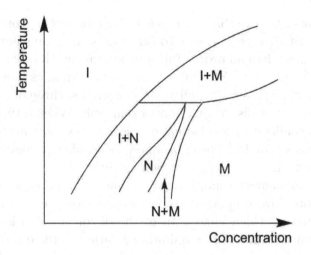

Figure 9.34 Typical phase diagram for a chromonic liquid crystal: isotropic (I), nematic (N), columnar (M). The phases are separated by two-phase regions.

typically by between 5 and 10 k_BT. This means that the molecules are much more dynamic, entering and exiting assemblies at a rapid rate. Also, at equilibrium the distribution of assembly length is very large (the width of the length distribution is on the order of the average length). The most simple model for this is discussed in Section 4.8, in which the assumption is made that the change in free energy for a molecule to join or leave an assembly is independent of the length of the assembly. This isodesmic assembly (sometimes called living polymerization) results in a decreasing exponential dependence on the concentration of assemblies with increasing assembly length. Unlike most lyotropic liquid crystals that possess a critical micelle concentration or CMC, assembly formation in chromonic liquid crystals occurs at all concentrations as shown in Figure 4.19.

The nematic phase of chromonic liquid crystals is macroscopically identical to the nematic phase of thermotropic liquid crystals. Both phases possess an orientational order parameter, elastic constants, and viscosities that decrease with increasing temperature in similar ways. If the molecule forming a chromonic or thermotropic nematic is chiral, or if a chiral dopant is added to either system, then the chiral nematic phase forms in place of the nematic phase. Solid surfaces and fluid interfaces also tend to orient the director in both nematic phases. But there is one difference that turns out to have important consequences, and that is that the twist elastic constant for a chromonic liquid crystal is usually more than an order of magnitude smaller than the splay and bend elastic constants. So while thermotropic nematics tend to adopt director configurations with all three distortions, chromonic nematics tend to favor twist distortion almost exclusively. When a thermotropic nematic is confined to a spherical droplet or a cylindrical capillary, it often adopts a radial (droplet) or an escaped radial (cylinder) director configuration. Both of these contain no twist distortion. But under the same confinement conditions, a chromonic liquid crystal adopts a twisted radial (droplet) or a twisted escaped radial (cylinder) director configuration in which almost all of the distortion is twist. When a colloidal sphere is placed in a uniform nematic and the director prefers to be parallel to the surface of the sphere, the director distorts in regions on opposite sides of the sphere. In a thermotropic nematic, these regions contain all three types of distortion. In a chromonic nematic, the distortion in these regions is nearly all twist.

Finally, it should be pointed out that there are some chromonic liquid crystals that possess more complicated assemblies than simple stacks of molecules. Not very much is known about these materials and they are difficult to work with, probably because equilibrium is not easily achieved. The characteristic that unites these compounds is their ability to form liquid crystal phases at very low concentration, often below 1 wt% at room temperature. The conjecture is that the assemblies contain significant amounts of water, perhaps in the interior of a thin cylindrical shell of chromonic molecules. Measurements seem to point to structures with diameters much larger than the size of a molecule, but more evidence is needed to form a firm conclusion.

EXERCISES

9.1 Explain the following experimental observations. A crystal of a glycolipid is viewed at a magnification of x100 in a polarizing light microscope fitted with a temperature-controlled stage. On heating the crystal melts at 50°C to give a birefringent texture of a liquid crystal phase. Further heating produces a transition to the isotropic liquid. Just before this point, water is run into the specimen. The sample becomes an isotropic liquid, but on further heating a new phase is formed, which is stable up to a temperature of 120°C before becoming liquid again.

9.2 Consider two related glycolipids, one has a single aliphatic chain attached to the head group, whereas the other has two aliphatic chains attached. For both glycolipids the radius of the head group is 0.6 nm, and the aliphatic chain length is 2.0 nm for a nonyl aliphatic chain. Determine for each glycolipid what type of self-organized structure is formed on passing from the micellar solution to the lyotropic state. Note the volume of a methylene group is 0.0239 nm^3 and a methyl group is 0.0478 nm^3.

9.3 Compounds **1** to **3** self-assemble to give condensed phases.

(a) Predict, with the aid of drawings, the structures of the thermotropic phases formed by compounds **1** to **3** when they self-assemble on cooling from the liquid phase.

(b) Predict the condensed phase formed when **1** and **3** are mixed together in equimolar proportions. Sketch the structure of the phase.

9.4 Two glycolipids **A** and **B** are composed of single sugar head groups and relatively long aliphatic chains of the same length. Glycolipid **A** has one head group and two aliphatic chains, whereas glycolipid **B** has two head groups and one chain. Both materials form self-assembled phases with similar temperature ranges. Predict the self-assembled phase formed by the following mixtures as a function of temperature and as a function of addition of water.

A:B 50:50 %
A:B 20:80 %
A:B 80:20 %
A:B 40:60 %
A:B 60:40 %

9.5 Sketch the typical phase diagram for a chromonic liquid crystal shown in Figure 9.34. On it, draw lines at constant concentration that represent increasing the temperature and observing the following phase sequences: (a) nematic – nematic/isotropic coexistence – isotropic; (b) nematic – nematic/isotropic coexistence – columnar/isotropic coexistence – isotropic, (c) columnar – nematic/columnar coexistence – isotropic/columnar coexistence – isotropic, and (d) columnar – isotropic/columnar coexistence - isotropic.

9.6 For chiral chromonic liquid crystals, the inverse pitch is proportional to the chiral dopant concentration as long as the chiral dopant concentration is low. If the pitch of a chiral chromonic liquid crystal is 50 μm with 5 wt% chiral dopant, what is the pitch with 12 wt% chiral dopant?

CHAPTER 10

Polymers, Oligomers, and Dendrimers – Big and Beautiful

10.1   MACROMOLECULAR MATERIALS

Hermann Staudinger (1181-1965) was the father of polymer science but his research was considered to be controversial. At the time he was deeply involved in developing polymers, his contemporaries commented, "Dear colleague. There are no organic molecules with a molecular mass over 5000. Purify your products and they will crystallize and reveal themselves as low molecular weight substances."[1]

Previous chapters have discussed how "molecular" liquid crystals are built from combinations of atoms in various arrangements and functional groups. In this chapter we look at supermolecular systems, which are composed of building blocks that are in effect molecular moieties. When the molecular entities are essentially the same in structure, polymers, oligomers, and/or dendrimers result. On the other hand, if the entities are different, the resulting supermolecular architecture is unique and therefore structurally discrete, resulting in the formation of a supermolecular material. Common biological macromolecules, such as peptides, have unique primary structures, whereas synthetic equivalents such as poly(analine) do not, but instead are composed of polymer chains of differing length, and therefore have an associated polydisper-

---

[1]Staudinger, H., *The Foundation of Polymer Science by Hermann Staudinger*, American Chemical Society, 1999.

sity γ. Figure 10.1 shows a diagrammatic comparison of a dendrimer, which is a star-shaped polymer (top), and a supermolecular material that has a unique structure (bottom). The dendrimer possesses 12 dendrons (arms) as shown, thus it is multi-legged (polypedal) and an oligomer. The feet can be varied in number, and therefore the material can be composed of a variety of oligomers with differing values of n. The properties of this material therefore depend on the distribution of the oligomers in a sample. Conversely, for the supermolecule there is a single architecture; therefore the properties of the material are unique and reproducible, *e.g.*, the material has a distinctive melting point.

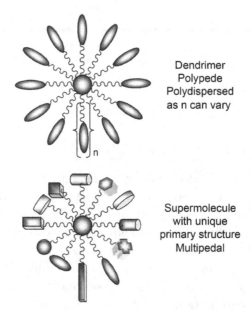

Figure 10.1 Comparison between the structure of a dendrimer (top) and that of a supermolecule (bottom). The structure of the dendrimer is not unique and can vary as a function of n, whereas the structure of the supermolecule is unique.

Having defined the general structures of supermolecules and macromolecules, the finer details of macromolecular materials are now described, again using a dendritic system for simplicity. An example of an actual dendrimer is shown in Figure 10.2. The material is an oligomer with four ($n = 4$) dendrons and twelve mesogenic moieties attached. The center of the dendrimer is based on pentaerythritol, to which the mesogenic building blocks are laterally attached via linking chains; these combined units make up the individual dendrons. In this example mesogenic units are laterally appended, which usually favors the formation of nematic phases; in this case this oligomer melts at 50.9°C to give a nematic phase that clears to the liquid at 70.1°C.

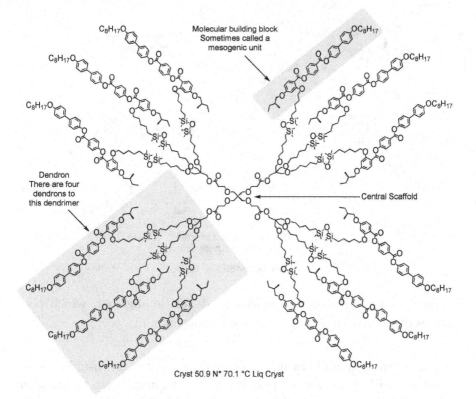

Cryst 50.9 N* 70.1 °C Liq Cryst

Figure 10.2 Example of a nematogenic dendritic oligomer, which possesses four dendrons and twelve mesogenic units. The material is an oligomer of pentaerythritol, which is located at the center. Although shown with a circular structure, in reality the molecular shape is likely to be rod-like.

In the example the number of repeat units is four, however, it is also possible to have other oligomers where the repeat number is different. This is shown for the oligomer in Figure 10.3, which has three repeat units. Therefore, the oligomer has three dendrons and nine mesogenic units, and melts at 49.7°C to a nematic phase and clears to the liquid at 69.8°C.

The examples in Figures 10.2 and 10.3 are featured because they

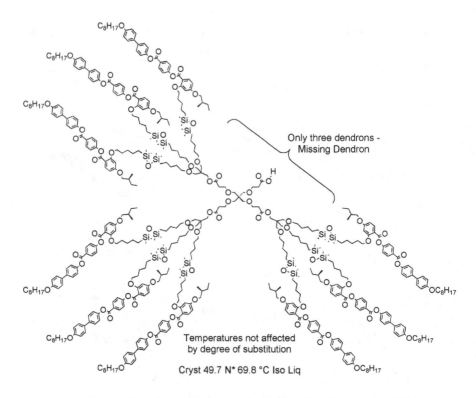

Figure 10.3 Example of a nematogenic dendritic oligomer, which possesses only three dendrons and nine mesogenic units.

are real examples and not mixtures. Synthesis, however, usually yields mixtures with an associated polydispersity, but in this case the number of possibilities for n is small and the individual oligomers are separable. The two oligomers shown have very similar melting and clearing points, and within experimental error they could be considered to be the same. But for more complex mixtures, this is not likely to be so, particularly for linear polymers as described later in this chapter.

For supermolecular materials that have unique chemical structures, this is not the case, and materials that have identical chemical formulae do not exhibit the same transition temperatures or behavior. For example, Figure 10.4 shows two "Janus" supermolecules, which have identical chemical formulae and similar structures. The only difference between the two is the orientation of the central unit, shown in gray (one is inverted with respect to the other). The figure shows that compound (a) has very different liquid crystal properties from compound (b).

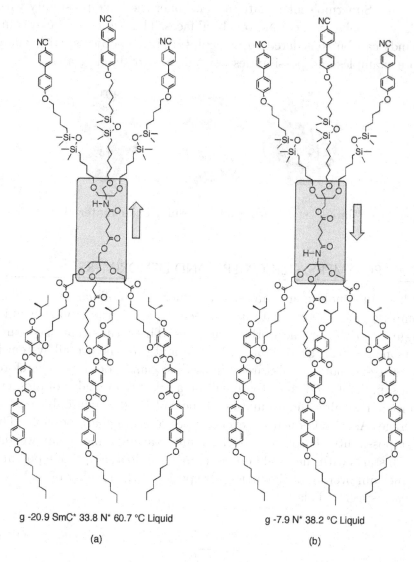

g -20.9 SmC* 33.8 N* 60.7 °C Liquid

(a)

g -7.9 N* 38.2 °C Liquid

(b)

Figure 10.4 Supermolecular materials that possess the same mesogenic units in the same orientations, but with inverted central core units.

The supermolecular materials above were described as being "Janus". Janus is a Roman god that is identified with doors, gates, and all beginnings, and is depicted with two opposite faces, as shown in Figure 10.5. Pierre de Gennes in his Nobel Prize Lecture of 1991 described Janus colloidal grains as having two sides: one polar and the other apolar, thereby having certain features in common with surfactants. Supermolecular materials also offer the same possibility with polar-non-polar, and chiral-non-chiral faces. Thus supermolecular compounds can be considered to be sculptured nano-objects, with many more examples and possibilities beyond those of Janus grains.

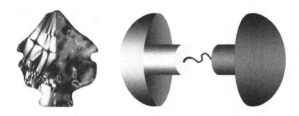

Figure 10.5 "Janus" structuring for supermolecular materials.

10.2 POLYMERS, OLIGOMERS, AND DENDRIMERS

Unlike supermolecular materials, polymers are modular in that they comprise one or more continually repeating structural monomer units to give "string" or "network" structures. This means that the chains of monomers can be variable in length, and thus the overall system is not discrete, but is a mixture of individual giant molecules. Therefore, in a sample of a polymer there will be a mixture of chains of different molecular weights and evaluation of the molecular weight only gives an average. As a consequence, several types of average are possible. The first possibility to consider is the number-average molecular weight, which starts with the total number of moles N in a sample of a polymer as the sum over all of the molecular species of the number of moles N_i of each species. This can be written

$$N = \sum_{i=1}^{\infty} N_i. \qquad (10.1)$$

The total weight of a sample can then be expressed as the sum of the weights of each molecular species w_i, which for each species i is equal to N_i times its molecular weight M_i, which can be written as

$$w = \sum_{i=1}^{\infty} w_i = \sum_{i=1}^{\infty} N_i M_i. \qquad (10.2)$$

The number average molecular weight $\overline{M}_n$ is the weight of the sample per mole, and can be written as follows:

$$\overline{M}_n = \frac{w}{N} = \frac{\sum\limits_{i=1}^{\infty} N_i M_i}{\sum\limits_{i=1}^{\infty} N_i}. \tag{10.3}$$

Another way to express the molecular weight is as the weight average. Typically for polymers, the molecular weight is determined via light scattering experiments, or by gel permeation chromatography (GPC). For light scattering studies, the intensity of the scattering is proportional to the square of the particle mass. Thus the molecular weight average $\overline{M}_w$ is given by

$$\overline{M}_w = \frac{\sum\limits_{i=1}^{\infty} N_i M_i^2}{\sum\limits_{i=1}^{\infty} N_i M_i}. \tag{10.4}$$

The relationship between the number average and the weight average as a function of the molecular weight and the percentage in the polymer are shown together in Figure 10.6. The closeness of the system to being monodisperse can be assessed through the polydispersity γ, which is given by the molecular weight average divided by the molecular number average. The closer the value γ is to unity, the closer the sample is to having polymer chains of similar length.

$$\gamma = \frac{\overline{M}_w}{\overline{M}_n} \tag{10.5}$$

Thus a polymer mixture can be defined by the distribution of masses of the individual chains or by the distribution of their lengths and the polydispersity. For liquid crystal polymers, the phase transitions tend to occur over a wider temperature range when the polydispersity is large (>4), and over a shorter range when the polydispersity is nearer to one. Moreover, for polydispersities nearer to one, the more likely it is for more phases to be observed because of the sharper nature of the phase transitions as the polymer mix acts as a single giant molecule. Furthermore, for polymers with higher degrees of polymerization DP, the differences in the chain lengths are relatively small and the polydispersity then appears to be nearer to unity.

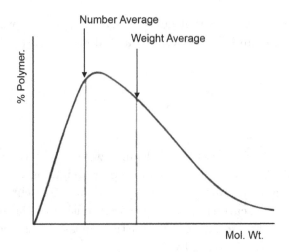

Figure 10.6 Comparison between the number average $\overline{M}_n$ and the weight average $\overline{M}_w$ as a function of molecular weight.

The properties of polymers, such as strength, also vary in a similar way. For short chain lengths (oligomers), properties can change greatly as the degree of polymerization is varied. Eventually for long chains, the properties level out, as shown in Figure 10.7. This is the main reason for creating and producing polymers. Production can be relatively reproducible yielding unique materials, and often extreme properties such as strength, and in the case of liquid crystals, processing in an oriented state, can add further to properties like yield strength as is the case with Kevlar® and Vectran®.

The varieties of structures that polymers can have are expressed in the schematic shown in Figure 10.8 for homopolymers and copolymers. In the figure the monomer units are colored gray and black; they can be rod-like, disk-like, or spherical, thereby mimicking material design of low molar mass liquid crystals. For homopolymers, the monomers can be strung together in a linear fashion, or in a network, or strung out from a central point, as shown in the top part of the figure. For copolymers, the monomers can be randomly joined together to give a statistical arrangement, or be joined together sequentially. For both homopolymers and copolymers, lateral grafts can be appended from the polymer backbones, and the grafts can be either mesogenic or not. There are also a number of arrangements in copolymers for which the monomers form blocks. The blocks themselves can be sequential, for ex-

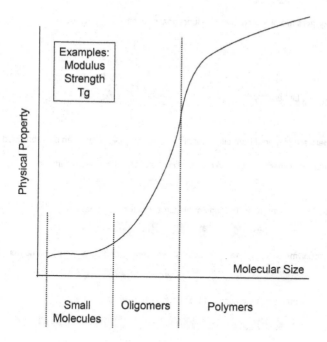

Figure 10.7 Variation of physical properties as a function of increasing polymer chain length.

ample diblocks (a segment of the polymer composed of one monomer A attached to a segment of monomer B). Triblock polymers can be composed of ABA or BAB arrangements of monomers, and longer polymers can be composed of alternating blocks. For some of these variations, such as diblock copolymers, the overall structure is similar to a supermolecule, which would have polydispersity near to one. In all of these examples, the monomer units may have flexible connectors leading to a flexible system, and in some cases the monomers are tied together in a fixed arrangement.

The simplest form for a liquid crystal polymer is based on monomers directly connected to one another where the linkage is relatively rigid. For a copolymer for which the monomers are rod-like and the linkages are directional, the polymer itself is rod-like (calamitic). Such polymer systems can exhibit thermotropic or lyotropic properties. Vectran® is an example of a thermotropic rod-like copolymer produced via the condensation of 4-hydroxybenzoic acid and 6-hydroxynaphthalene-2-carboxylic acid (see Figure 10.9). The monomers are bifunctional (OH and COOH) and can react with one

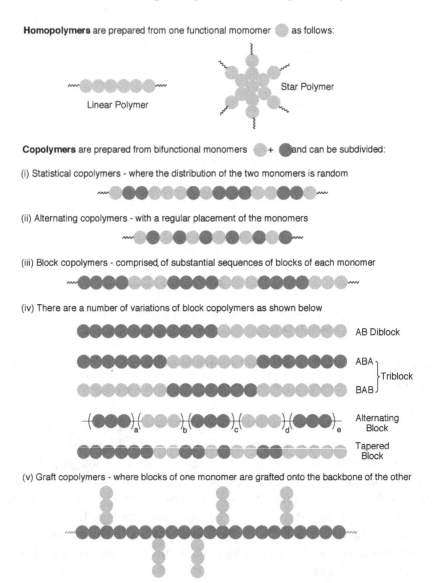

Homopolymers are prepared from one functional momomer ● as follows:

Linear Polymer

Star Polymer

Copolymers are prepared from bifunctional monomers ● + ● and can be subdivided:

(i) Statistical copolymers - where the distribution of the two monomers is random

(ii) Alternating copolymers - with a regular placement of the monomers

(iii) Block copolymers - comprised of substantial sequences of blocks of each monomer

(iv) There are a number of variations of block copolymers as shown below

AB Diblock

ABA
BAB } Triblock

Alternating Block

Tapered Block

(v) Graft copolymers - where blocks of one monomer are grafted onto the backbone of the other

Figure 10.8 General structures of homo- and co-polymers.

another or can self-condense; thus the monomer units are statistically arranged along the polymer backbone. Neither of the monomer units is liquid crystalline, but the polymer product is.

Vectran® was invented at the Celanese Corporation and is now manufactured by Kuraray; it is a high-performance multifilament yarn

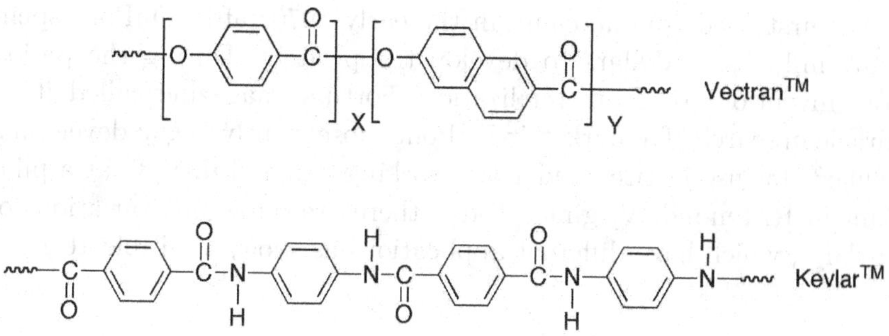

Figure 10.9 Chemical structure of Vectran® and Kevlar®.

spun from a liquid crystal polymer (LCP). It is commercially available as a melt-spun LCP fiber, and exhibits exceptional strength and rigidity. Pound for pound, Vectran® fiber is five times stronger than steel and ten times stronger than aluminum. It is resistant to chemical and organic solvents, and to acids and bases, and durable to abrasion and flex fatigue.

Similarly, Kevlar® is a high yield strength liquid crystal polymer that is processed in a mesomorphic state, but this time in the lyotropic phase. Lumps of the polymer are dissolved in strong sulphuric acid, and the polymer is spun from the liquid matrix by forcing the solution through a fine pore filter. As the solution flows through the holes in the filter, the mesophase is forced into being orientationally ordered giving parallel alignment to the fibers.

Kevlar® is poly(para-phenyleneterephthalamide), and is also known as a para-aramid. The aramid system gives Kevlar® thermal stability, whereas the para-substitution gives it high strength and modulus. Kevlar® is an alternating copolymer in which the monomers, terephthalic acid (terephthaloyl chloride) and para-phenylene diamine, are not mesogenic and only react with each other to give amide linkages, which are the same as the linkages found in nylons and proteins. Like these materials, the polymer chains interact through hydrogen bonds, which aid in the orientation of the chains. Thus the rod-like form of the para-aramid chains, the lateral H-bonding, and the extrusion process make Kevlar® fibers anisotropic such that they are stronger and stiffer in the axial direction than in the transverse direction.

Kevlar® was developed by chemist Stephanie Kwolek at Dupont when she devised a liquid crystal solution that could be cold-spun. Like Vectran®, Kevlar® is five times stronger than the same weight of steel.

It was first used commercially in the early 1970s after DuPont spent many millions of dollars to develop the product. During the period from invention to commercialization, Fortune magazine called it "a miracle in search of a market."[2] DuPont subsequently began developing Kevlar® for use in tires under the working name "Fiber B" at a pilot plant in Richmond, Virginia. Today there are numerous variations of Kevlar®, which have different applications as shown in Table 10.1.

Table 10.1 Variations of Kevlar®.

Product	Description
Kevlar K29	used in cables, replacement for asbestos, brake linings, and body/vehicle armor
Kevlar K49	high modulus polymer used in cable and rope products
Kevlar K100	a colored version of Kevlar
Kevlar K119	higher-elongation polymer, flexible and relatively more fatigue resistant
Kevlar K129	used in ballistics

Although Vectran® has a similar structure to that of Kevlar®, its applications are somewhat different. For example, Vectran® fibers are used as reinforcing fibers for ropes, cables and sailcloth and in advanced composites, bicycle tires, and electronics. It is used as one of the layers in NASA's spacesuits, and in the fabric for the airbag used in Pathfinder landings on Mars. It has also been used in tethers for military surveillance blimps, and in accurately machined parts for applications in the fuel injectors of sports cars.

These two polymers are made up of linear chains, and as such are termed thermoplastics. However, it is also possible to have polymers in which a number of chains originate from a single point, and these are usually called "star" polymers as shown in Figure 10.8. For homopolymer systems, the polymer chains attached to the center can be the same or of different lengths. The chains can also be bifurcated in a regular way if the chains are composed of monomeric units that have more than two reaction sites. The first set of monomeric units attached to the central core is called Generation 1, the second set is called Generation 2, and so on, as shown in Figure 10.10. As with linear polymers, dendrimers do not necessarily possess identical star-shaped molecules; they can have mixtures of differing sized molecules or molecules with

---

[2]L. Smith, Fortune, December 1, 1980, p. 92.

differing chain lengths. Copolymers can also have analogous structures to those of linear block copolymer systems. The lowest member for block copolymers is equivalent to a tri-copolymer in which there are three arms attached to a central point. These materials are described as Miktoarm Star polymers, where the term Miktoarm is derived from the Greek word μικτός meaning mixed.

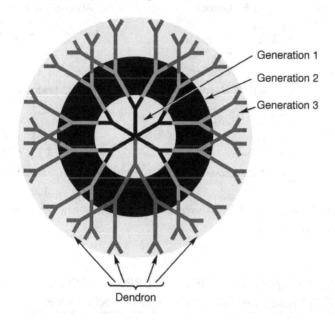

Figure 10.10 General structure of a dendrimer showing the various generations. One set of monomers coming from the origin is colored gray, and, as described earlier, is sometimes referred to as a dendron.

Some examples of Miktoarm Star polymers are shown in comparison to linear block copolymers in Figure 10.11. In the schematic there are three differing segments labeled A, B, and C. Each segment is made up of repeat units similar to those found in an oligomer. If there are three segments in a linear polymer for which two are the same, then there are two possible structures: an ABA-triblock polymer or a BAB-triblock polymer, whereas for the Miktoarm Star polymer the possibilities are an A_2B-3-Miktoarm and an AB_2-3-Miktoarm, as shown. Other examples among very many are also shown in the figure.

One interesting aspect of the Miktoarm systems is that they have more contorted shapes, such as wedge-like or conical, in comparison to those formed by linear block polymers. This means that they can easily pack into more complex structures, such as columnar or cubic, as well as into lamellar or nematic-like architectures. These structures are more usually soft-crystals as opposed to conventional liquid crystals.

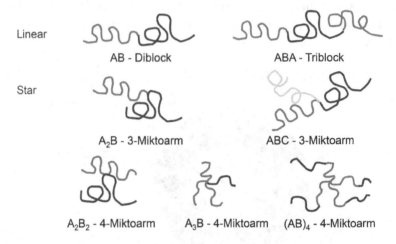

Figure 10.11 Examples of Miktoarm Star polymers and a comparison to linear block polymers.

10.3 MESOGENIC MOIETIES INCORPORATED IN POLYMERS, OLIGOMERS, AND DENDRIMERS

There is another set of liquid crystal polymers that are based on the architectures described above, but which are composed of conventional polymers that have incorporated mesogenic units. If we take a polymer such as Vectran®, which has non-mesogenic monomers, and replace them with bifunctional mesogenic units, then the chances are high that the resulting polymer is also liquid crystalline. These polymers are often called main chain liquid crystal polymers (MCLCPs), and can incorporate mesogenic units that are spherical, disk-like, or rod-like in shape. The cores of the mesogenic units are usually tied together with flexible hydrocarbon chains, as shown in Figure 10.12. A discotic MCLCP and a calamitic MCLCP are shown together in Figure 10.13. The discotic polymer exhibits a columnar phase, whereas the calamitic polymer shows a calamitic phase that is classified as nematic.

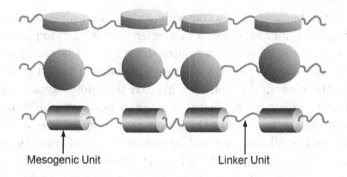

Mesogenic Unit   Linker Unit

Figure 10.12 Molecular architectures for the design of main chain liquid crystal polymers (MCLCPs).

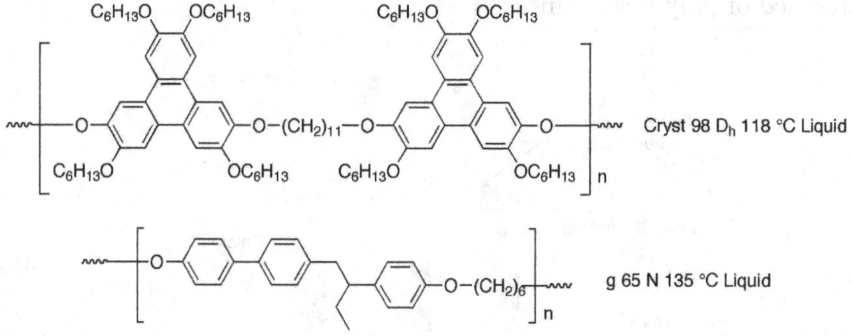

Cryst 98 D_h 118 °C Liquid

g 65 N 135 °C Liquid

Figure 10.13 Examples of two main chain liquid crystal polymers: (top) incorporating disk-like mesogens, and (bottom) incorporating rod-like mesogens.

10.4 MESOGENIC MOIETIES ATTACHED TO POLYMER BACK-BONES

Other than including mesogens in the main chains of polymers, there is another way that mesogenic units can be incorporated into polymer architectures, and that is via the use of graft structures as shown in the bottom of Figure 10.8. In these systems the mesogenic units are appended to a polymer backbone, as shown in Figure 10.14 for calamitic, spherulitic, and discotic shaped mesogenic units. These types of mesomorphic polymers are called side-chain liquid crystal polymers (SCLCPs). In a true graft polymer, the backbone is already present, as in a polysiloxane, and the mesogenic units are attached via functionalization of the backbone. Alternatively, side chain polymers can be created by the polymerization of a monomer that is carrying a mesogenic unit, *e.g.*, mesomorphic polyacrylates. Examples of laterally appended calamitic and discotic mesogenic units are shown together in Figure 10.15. In these cases the units are grafted to a polysiloxane backbone of about 40 repeat units, where the overall polymer polydispersity is dependent on the backbone. Both polymers form glasses at low temperature, and exhibit associated second-order glass transitions. They differ in that the calamitic SCLCP crystallizes from the glass on heating, and then melts to form a smectic A phase. For the discotic SCLCP, the glass transforms directly into a discotic phase without crystallization, which is probably due to the higher disordering of the mesogens caused by the inability of the disk-like moieties to pack together in the presence of polymer chains.

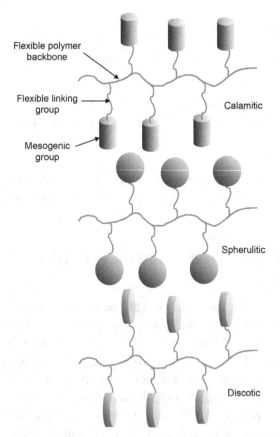

Figure 10.14 Structure variations for side chain liquid crystal polymers.

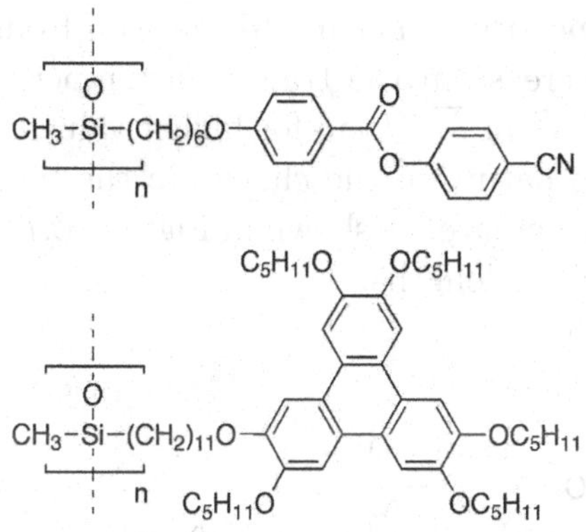

Figure 10.15 Calamitic (top) and discotic (bottom) side chain polymers (SCLPs) with mesogenic units grafted on to a polysiloxane backbone.

With these types of thermoplastic polymers incorporating mesogenic units and linking groups, there is always an issue of flexibility. The polymer backbone may be flexible, semi-rigid, or rigid; the mesogenic moieties similarly may be relatively rigid (which they are usually selected to be); and the linking group may be short or long in length. For an aliphatic linking group, the longer it is the more likely it is that the mesogenic units are decoupled from the backbone, and thereby such polymers might exhibit behavior similar to those of their low molar mass analogs; hence the interest in them for high technology applications. Unlike low molar mass materials, there are numerous possibilities for the design of materials with potentially novel properties. For instance, copolymers can be designed to have properties in between those of the two monomers, but in a polymer structure, the monomers do not have the ability to segregate.

In the following, the various possibilities for designing SCLCPs are described with examples. Although there appears to be many possibilities and advantages for design, there are also many drawbacks with respect to processing and organizing the bulk polymer.

First, we start with the variation in number average molecular weight $\overline{M}_n$ and the effect it has on phase transition temperatures. Figure 10.16 shows the effect of increasing the value of $\overline{M}_n$ for a typical side chain, nematogenic, polyacrylate. As the average polymer chain length is increased, the clearing point from the nematic to the isotropic liquid also increases, similarly so does the glass transition temperature. As expected, the transition temperatures are dependent on $\overline{M}_n$, and *inter alia* $\overline{M}_w$, and the polydispersity γ. It can also be seen from the results in the figure that the increase in the transition temperatures occurs more rapidly at lower values of $\overline{M}_n$ than for higher values. This is consistent with the predictive nature of the change in the physical properties as a function of molecular size, as shown in Figure 10.7, with saturation occurring at longer chain lengths.

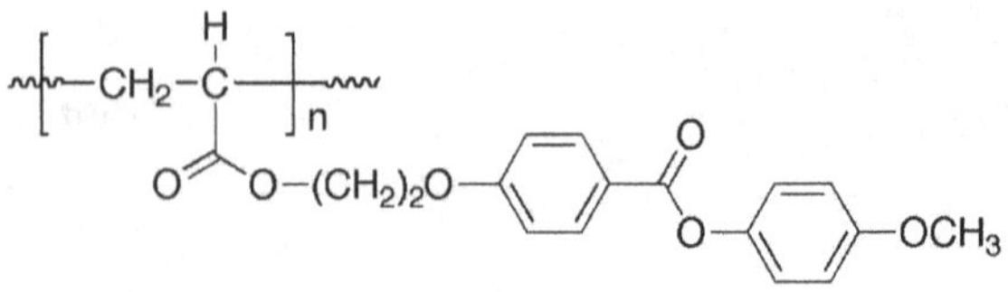

$\overline{M}n$	n	Transition Temperatures (°C)
4500	13	Glass 53 N 100 Liquid
14000	41	Glass 59 N 114 Liquid
3900	114	Glass 62 N 116 Liquid

Figure 10.16 The effect number average molecular weight has on transition temperatures for a typical side chain liquid crystal polyacrylate.

Secondly, the nature of the parent polymer is important to how liquid crystal phases are formed and stabilized. The system needs to be relatively flexible and malleable such that the attached mesogenic units can interact and form local packing arrangements that support certain mesophases. For example, in the formation of layered mesophases, the rod-like mesogenic moieties need to pack roughly parallel to one another; this is easiest to achieve if the polymer backbone can flex and bend rather than being too firm or that the mesogenic units are too far apart. Obviously some of these properties support certain applications whereas others do not. Figure 10.17 shows three different polymers. The top left structure is a substituted polysiloxane, in which the side chain mesogenic units are attached at every silicon atom making it a densely substituted system. However, the Si-O bond is relatively long and the siloxane backbone is therefore flexible, thus siloxanes support mesophase formation. In comparison, although the SCLC polyacrylate shown in Figure 10.16 has a similar repeat of a mesogenic unit at every second atom along the backbone, the acrylate backbone is stiffer and the transition temperatures are lower. Conversely, the polyoxetane shown at the top right of Figure 10.17 has a substitution at every fourth atom along the backbone, and because of its flexibility this material has better mesogenic properties.

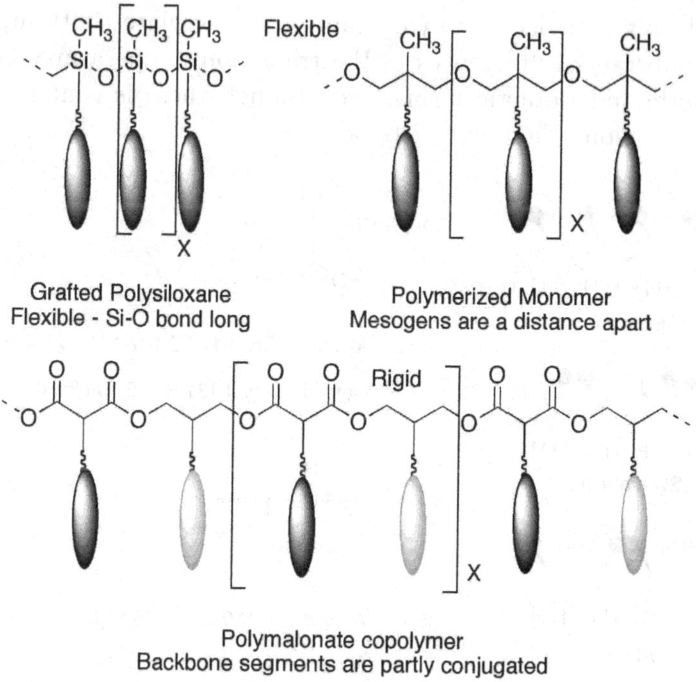
CH3
Si
O
X
Flexible
CH3
O
X
Grafted Polysiloxane
Flexible - Si-O bond long
Polymerized Monomer
Mesogens are a distance apart
O
Rigid
X
Polymalonate copolymer
Backbone segments are partly conjugated

Figure 10.17 Comparison of the polymer backbones of SCLCPs and the spacing between the mesogenic units.

The polymer in the lower part of the figure is a copolymer based on a condensation between a 2-substituted malonoylchloride and a 2-substituted 1,3-propanediol, which gives a polymalonate. Although the mesogenic units are well spaced apart along the backbone, the backbone itself is much more rigid than that of polyoxetane. Thus, the malonate has weaker liquid crystal properties.

Thirdly, in addition to the nature of the polymer backbone, its chemical structure is also important. Often the attachment of the mesogenic group involves substitution at a position where there is a stereochemical choice; there are three possible choices as shown in Figure 10.18. In an isotactic liquid crystal polymer, the mesogenic units are located on the same side of the polymer backbone, whereas for a syndiotactic analog the substituents alternate along the polymer backbone, and for an atactic arrangement the substituents are randomly placed. Although there is little information on the effects of tacticity on mesophase formation, results show that there can be substantial differences between the various forms. The figure shows that for a substituted polymethylacrylate (top) and a polyacrylate (bottom), there can be substantial differences in the transition temperatures between the atactic and isotactic forms, even though there is conformational disordering along the polymer backbone.

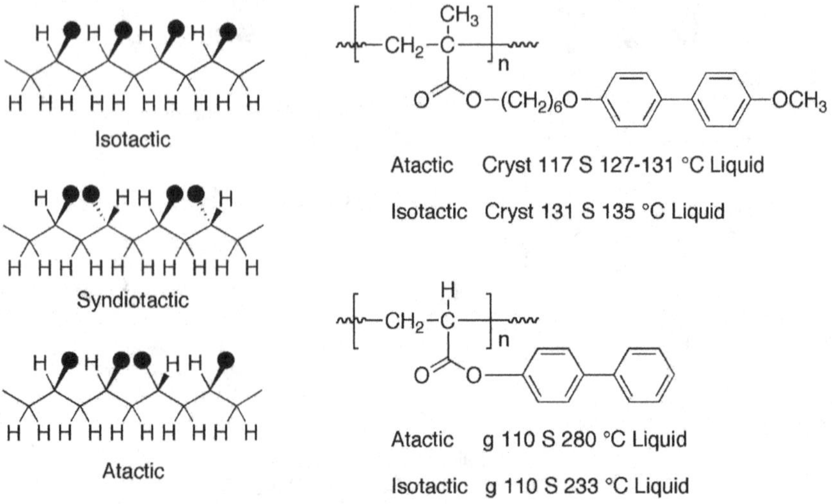

Isotactic

Syndiotactic

Atactic

Atactic   Cryst 117 S 127-131 °C Liquid

Isotactic Cryst 131 S 135 °C Liquid

Atactic   g 110 S 280 °C Liquid

Isotactic g 110 S 233 °C Liquid

Figure 10.18 Effect of tacticity in polymers on the transition temperatures of SCLCPs.

Having examined the properties of the polymer backbone, we need to examine the effects of the structure of the linking group of the mesogenic units on mesophase formation. A simple prediction is that as the linking group is extended, the more likely it is that the mesogenic units, particularly the cores, become decoupled from the polymer backbone. When this occurs, it is expected that there is a change in polymorphic behavior as the possible number of interactions of the cores changes. This effect is shown clearly in Figure 10.19 as the linking group is increased in length for a substituted polymethylacrylate. When the linker is short, nematic phases seem to be preferred, but for longer linking groups, smectic phases are favored due to decoupling of the mesogenic cores. This is similar behavior to that observed for low molar mass materials. There is one contradiction to this observation, which is for mesogenic groups possessing terminal nitriles. In this situation the aromatic nitriles of the mesogenic groups cannot form antiparallel packing arrangements because the backbones prevent this from occurring, and as a consequence, such SCLCPs are more likely to form smectic phases.

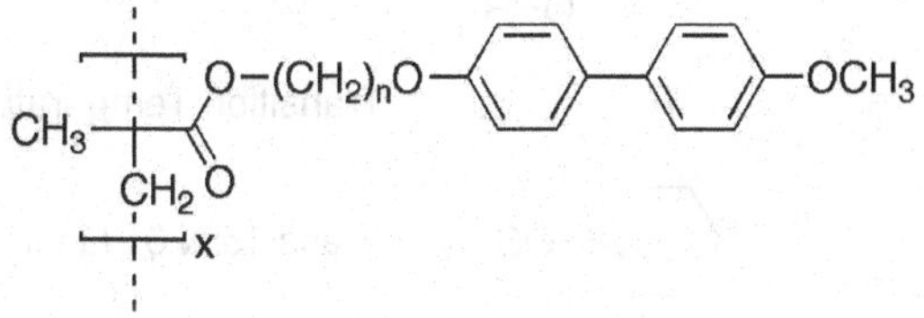

Spacer Length (n)	Transition Temperature (°C)
0	S 255 Liquid
2	g 120 N Liquid
6	C 119 S 136 Liquid
11	g 54 SmC 87 SmA 142 Liquid

Figure 10.19 Effect of the length of the aliphatic spacer on mesophase formation and transition temperatures in a family of polymethyl acrylates.

The effect of the structure of the mesogenic unit on the formation and properties of SCLCPs is similar in context to the design principles for low molar mass materials, as described in previous chapters. For rod-like side groups, the larger the aspect ratio is, the higher are the transition temperatures. The results for a study in which the mesogenic core is systematically extended in length by the addition of phenyl rings, and the polymer backbone is constant and based on a flexible polysiloxane, are shown in Figure 10.20. In this study the phase types exhibited are the same, with a glass transition followed by a nematic phase with the clearing point substantially increasing, and saturating as the mesogenic length reaches four rings.

The details of the polarities and polarizabilities of mesogenic groups can be assessed as described in previous chapters, with many of the interactions between mesogenic groups being extrapolated from low molar mass materials. As mentioned previously, there is one exception to this method, and that is for materials with a terminal nitrile group. Normally low molar mass materials with relatively short terminal aliphatic chains and nitriles, tend to exhibit nematic phases. However, for SCLCPs it is more often the case that smectic A phases

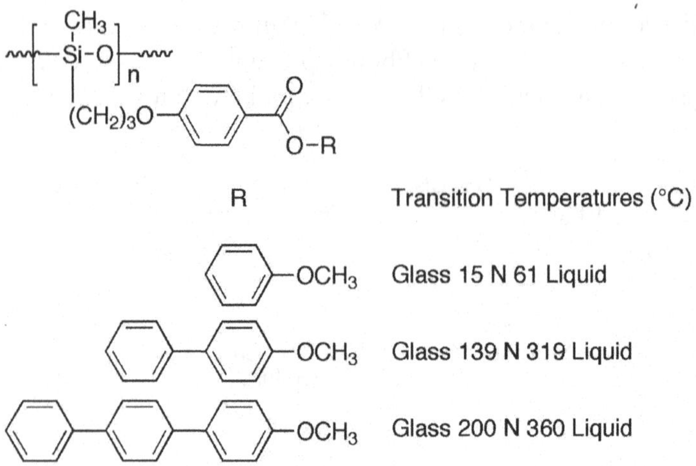

R	Transition Temperatures (°C)
—OCH$_3$	Glass 15 N 61 Liquid
—OCH$_3$	Glass 139 N 319 Liquid
—OCH$_3$	Glass 200 N 360 Liquid

Figure 10.20 Effect of the length of the mesogenic unit on the properties of substituted polysiloxanes.

are found, as shown in Figure 10.21 for the results derived from substituted polysiloxanes. The probable reason for this is that the polymer backbone interferes with the antiparallel packing that is required for nematic formation. This is the case for rod-like mesogens that are appended terminally to the polymer backbone; however, for laterally appended mesogens, nematics are preferred, as described in the following section.

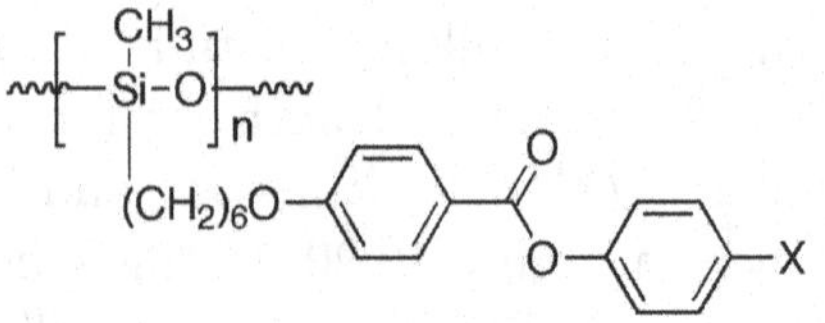

X	Transition Temperatures (°C)
CN	Glass 14 Cryst 37 SmA 165 Liquid
CH$_3$O	Glass 0 Cryst 23 N 97 Liquid

Figure 10.21 Effect of the polarity of the mesogenic unit on the phase formation of substituted polysiloxanes.

Of the mesogenic units that can be appended to polymer backbones, only the rod-like mesogen has a choice of the orientation of the attachment. For example, for a side chain liquid crystal polymer possessing a disk-like mesogenic group, it doesn't matter where the attachment of the linker is on the disk-edge as the orientation of the disk is the same relative to the polymer backbone, as shown in Figure 10.22. However, this is not the case for rod-like mesogenic units because it is possible to attach the core of the mesogen either terminally or laterally with respect to the backbone, as shown in Figure 10.23. Generally, terminal attachment of the mesogenic unit gives it a preference for smectic phase formation, whereas lateral attachment supports nematic phase formation. This is probably because terminal attachment allows the mesogenic units to pack into layers where the polymer backbone can insert either between the layers or perpendicular to them. Conversely, for laterally appended mesogenic units, the polymer backbone disrupts the packing into layers thereby allowing only for the orientational ordering of nematic phases.

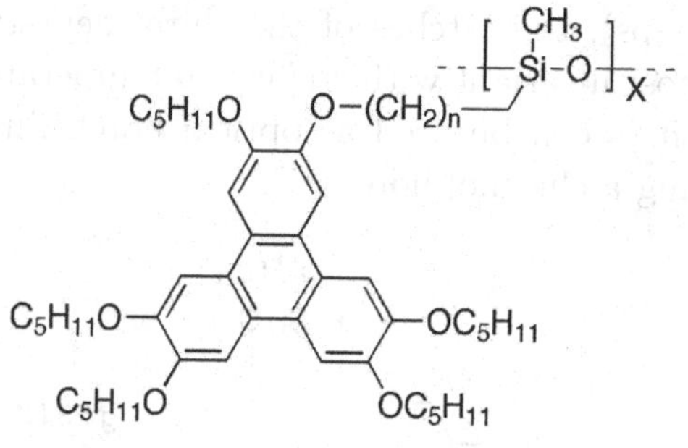

Figure 10.22 An SCLCP that possesses an attached disk-like mesogen.

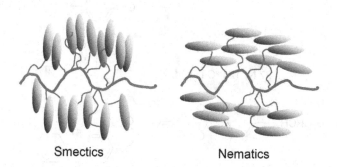

Smectics   Nematics

Figure 10.23 Possible orientations of attachment of rod-like mesogenic units to a polymer backbone.

This argument is supported by the introduction of chirality into the system via the incorporation of a stereogenic center into the structure of the mesogenic unit, as shown for the two polymers in Figure 10.24. In both cases chiral nematic phases are formed, which means that the spiraling orientational order is not affected by the presence of the polymer backbone. For the two polymers shown, it is important to note that the point of attachment of the linking chain to the aromatic core of the mesogenic group can be at a central position of the core (lower), or to one corner (upper). However, the clearing points of the two polymers are effectively the same, given that the lengths of the polysiloxane chains are the same, and the glass transitions are within four degrees of one another. These comparisons suggest that in each case the mesogenic groups are essentially decoupled from the polymer backbone and are behaving in the same way in the polymer matrix. Moreover, unlike the strong variation in the pitch as a function of temperature for the chiral nematic phases of the two mesogens (minus their linking groups), the pitches of the chiral nematic phases of the SCLCPs are almost invariant with respect to temperature. This means that these polymers can be used as optical notch filters or pigments without possessing a chromophore.

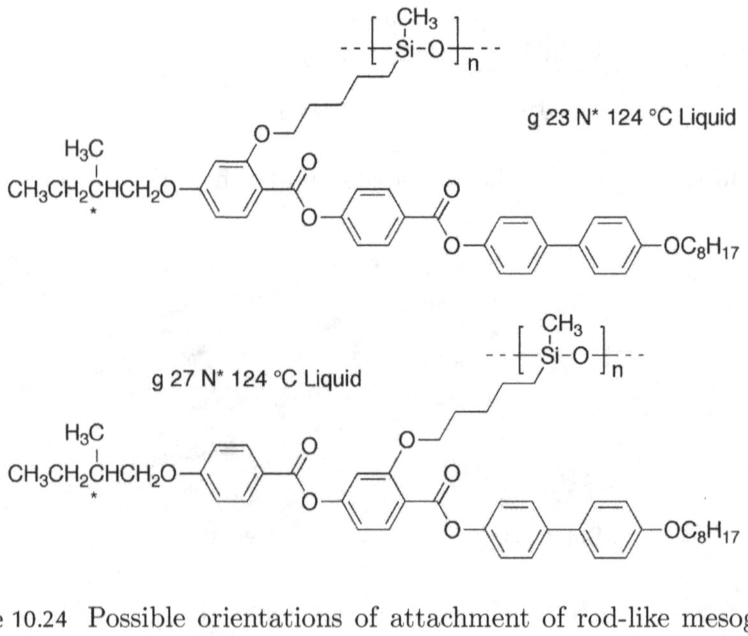

Figure 10.24 Possible orientations of attachment of rod-like mesogenic units to a polymer backbone.

10.5 MESOGENIC MOIETIES ATTACHED TO DENDRIMERS

In principle the property-structure correlations that are found for the design of SCLCPs, as described in the previous section, apply also in the design of dendrimers substituted with mesogenic units, for which the mesogens are located at the ends of the dendrimer's arms. However, there is a difference when it comes to the overall shape of a liquid crystalline mesogen terminated dendrimer (LCMTD), as shown in Figure 10.25. The gross shape of an LCMTD is determined to a large extent by the density of the substitution at the central point of the structure. For a low density of substitution where there are few arms, there is a tendency for the arms to clump together through self-interactions of the mesogenic groups. In doing so the structure becomes rod-like as shown in the figure. At higher densities, there can be too many mesogenic units and eventually the LCMTD becomes disk-like, and at higher densities eventually the structure becomes spherical. Across this range of shape, rod-like LCMTDs support the formation of calamitic phases, disk-like support columnar phases, and spherulitic support cubic phases. Of course there is also the issue of packing dendrimers together, and distortions can occur through conformational change and inter-dendrimer interactions as they bump into one another as shown in Figure 10.26.

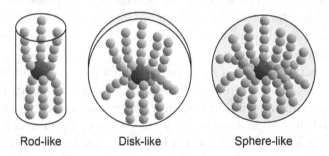

Rod-like   Disk-like   Sphere-like

Figure 10.25 The effect of the density of arm packing on the formation of mesophases by liquid crystalline mesogen terminated dendrimers (LCMTD).

A good example of how the shape distortion can occur for an LCMTD is shown in Figure 10.27. The dendrimer shown possesses a silicon central point of substitution, it has 36 arms, based on 12 dendrons, and decorating the exterior are 36 cyanobiphenyl mesogenic units. As drawn, the LCMTD appears disk-like, and could even be

Figure 10.26 Packing of dendrimers together causing shape distortions.

spherulitic. However, on heating, it forms a lamellar smectic A phase from a glass. The smectic A phase is the only liquid-crystalline phase that is observed, thereby indicating that the rod-like conformational form is the one most preferred for this material.

For this particular dendrimer, it appears that the center is soft and therefore collapses inwards to allow the mesogenic units to parallel pack. In doing so the mesogenic groups cannot easily form overlapping arrangements as found for low molar mass equivalents. With this possibility suppressed, the mesogenic units simply form monolayers as found in a lamellar phase. Thus the process of mesophase formation is dominated by the interactions of the mesogens and not the dendrimer scaffold, and as with the SCLCPs described above, a lamellar phase is favored over a nematic.

The central point of a dendrimer scaffold can be expanded in size via a judicious choice on what to build upon as a scaffold. In Figure 10.28, the dendrimer is built upon a central octasilsesquioxane cubic cage, which has eight locations for substitution, upon which bifurcation can be used to produce dendrons. In this context the material could be considered as a dendrimer or a supermolecule; either way the end product is effectively discrete with a single structure and identity. The lateral chiral 2-methylbutyl substitution in the mesogenic units means that packing of the mesogens is not straight-forward and causes disordering, so this material exhibits a chiral nematic phase from 162 to 165°C, but again smectic phases dominate and a chiral smectic C* phase exists from the melting point of 133°C to the transition to the chiral nematic at 162°C. As the smectic C* phase is chiral, the material exhibits ferroelectric properties.

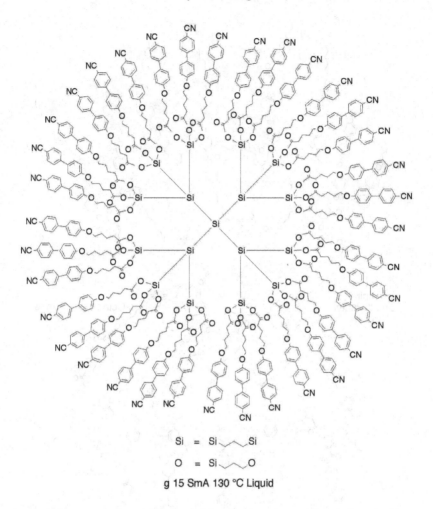

Si = Si $\diagdown\diagup$ Si

O = Si $\diagdown\diagup$ O

g 15 SmA 130 °C Liquid

Figure 10.27 A dendrimer possessing cyanobiphenyl mesogens terminating the dendritic scaffold.

If we now just switch over the position of the 2-methylbutyl group with the point of attachment of the mesogen to the location of the bifurcation, the mesogen laterally appends to the central scaffold, as shown in Figure 10.29. Nothing else has changed in the structure except that mesogenic units become more difficult to pack together. This causes the overall shape of the dendrimer to become bulkier and no longer rod-like. Due to the lateral attachment, the material exhibits a chiral nematic phase, but again only over a short temperature range. However, instead of a smectic phase being formed, columnar phases are

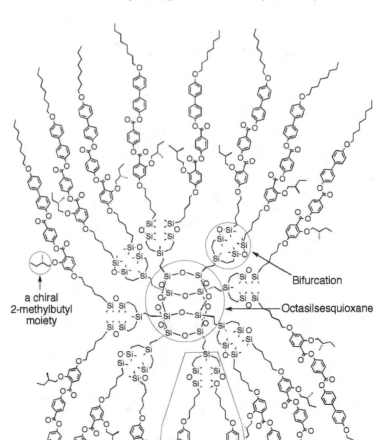

a chiral
2-methylbutyl
moiety

Bifurcation

Octasilsesquioxane

Dendron

Cryst 133 SmC* 162 N* 165 °C Liq

Figure 10.28 A chiral nematic and ferroelectric dendritic supermolecule.

stabilized indicating the overall molecular structure is probably more disk- or wedge-like than rod-like. Thus even with the same chemical constitution, nano-molecular engineering can be used to control the structures of the mesophases formed and their physical properties.

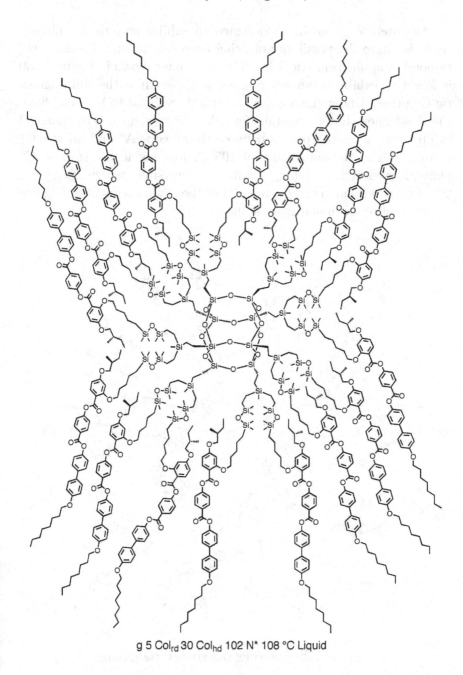

g 5 Col$_{rd}$ 30 Col$_{hd}$ 102 N* 108 °C Liquid

Figure 10.29 A dendrimer/supermolecule with laterally appended mesogens and chiral terminal chains.

As noted for dendrimers designed to exhibit smectic C* phases, they also have the possibility of being ferroelectric, and therefore will respond to applied electric fields. The dendrimer shown in Figure 10.30 is found to exhibit a number of smectic phases plus the chiral smectic C* phase. Under the application of a directional DC electric field, the dendrimer shows two-state ferroelectric switching. The material exhibits two unknown phases between the smectic A* and smectic C* phases, and at a temperature of 105°C, just inside the smectic C* phase, ferroelectric switching produces a reversible switching angle of 26°. This result again demonstrates that the dendrimer is behaving like a collection of low molar mass liquid crystals.

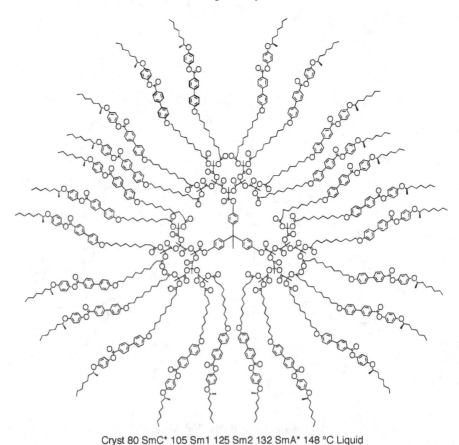

Cryst 80 SmC* 105 Sm1 125 Sm2 132 SmA* 148 °C Liquid

Figure 10.30 A ferroelectric dendrimer that exhibits two-state switching in a DC electric field.

Having examined the effects of lateral versus terminal attachment of mesogenic units on mesophase formation, it is also interesting to investigate the effect of the number of generations on the incidence of mesophases and their associated transition temperatures. Starting with the lateral mesogenic unit (top right shown in Figure 10.31), a single mesogenic unit is attached to a carbosilazine scaffold unit, and then a second unit is attached to a similar scaffold, and then a third until a scaffold is complete. This can be considered as generation 0, as shown in the figure. Thus as the number of mesogens attached to the central scaffold becomes complete, the melting points transform to glass transitions, and over the same sequence of structural changes, the clearing points steadily increase. Expanding the dendritic core to accommodate the additional generations 1 and 2, crystal phases are not found, but the temperatures for the glass transitions are very similar and so too are the clearing points. This result demonstrates that as the numbers of generations increase, the proportion of the mesogenic units to scaffold does not greatly change, and so the transition temperatures and glass transition do not vary greatly either. In addition, mesophase formation does not appear to be affected. For instance, for the comparative system that has a mesogenic unit with four phenyl moieties, an additional smectic C phase is found. However, again the clearing points and transition temperatures to the smectic C phase do not vary greatly.

Given the large number of possible mesogenic units combined with the different structural alternatives for polymers, it is not surprising that an enormous number of SCLCPs have been prepared and evaluated. Combinations of mesogens, *e.g.*, disk-like and rod-like, in main chain polymers (Figure 10.32) have been prepared and evaluated. Combinations of main chain and side chain polymer systems, such as structures (b) and (c) in the figure, are also known, and classified as combined liquid crystal polymers (CLCP). Even these classes can be subdivided as the side chain mesogenic units can be attached, via a spacer unit, to a mesogenic main chain either at the linking unit or at the mesogenic unit. Although the syntheses and mesogenic behaviors of many liquid crystalline polymers have been published, in the vast majority of cases the molecular weights and polydispersities have not been determined.

Further variations on MCLCPs, SCLCPs, and CLCPs are also possible through cross-linking of the polymer chains. There are two structural variants, one in which the cross-linking is weak or low density, and

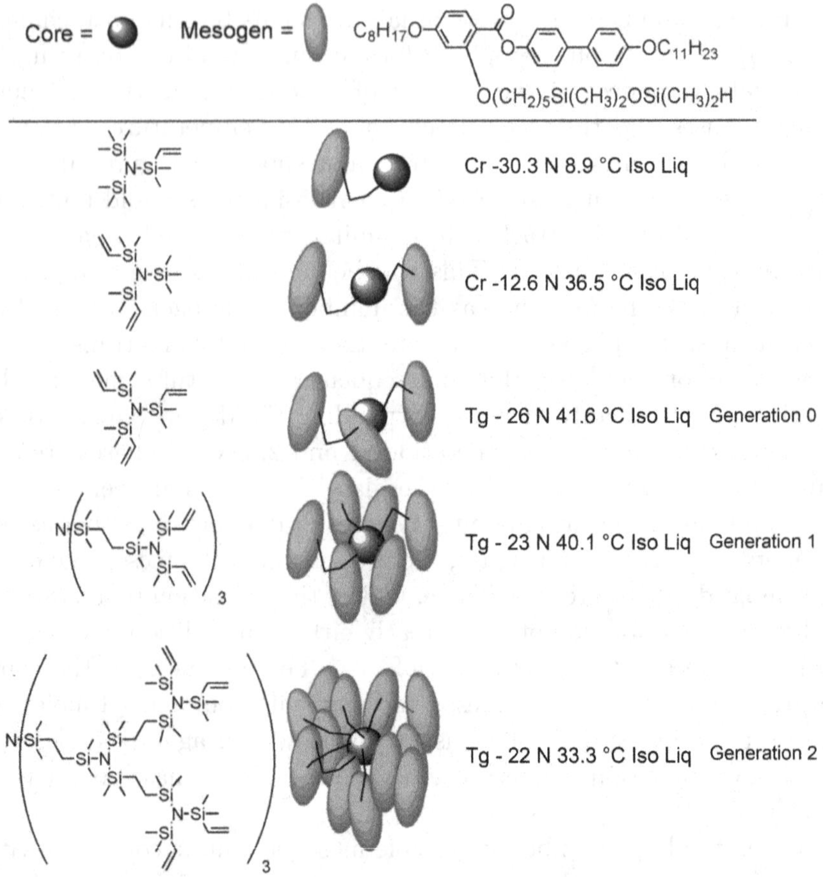

Figure 10.31 Effect of the number of generations on the transition temperatures of dendritic materials that possess laterally appended mesogenic units.

another in which the cross-linking is dense. Low density cross-linking leads to elastomeric materials, which are employed in mechanical devices, such as artificial muscle, whereas high density cross-linking leads to harder and potentially brittle materials, which are found in optical films. The low-density linking materials are often prepared via grafting, whereas the high-density materials are produced via the UV cross-linking of bifunctional low molar mass liquid crystals.

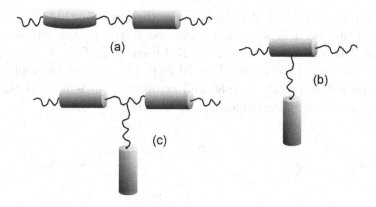

(a)

(b)

(c)

Figure 10.32 Variations in designs of SCLCPs.

10.6 PHOTOPOLYMERIZATION IN ORGANIZED MEDIA

The processing of main-chain polymers in organized media to give polymer blends (*e.g.*, Kevlar® and Vectran®) that are stronger than steel have been described earlier in Section 10.2. The processing techniques use mechanical methods, such as injection molding, fiber spinning, *etc.* In the following, processing through reaction chemistry in organized media using reactive mesogens is described. One of the driving forces for the development of this technique is the commercial production of anisotropic electro-optic films, which are used in a wide variety of devices such as LCDs and OLEDs.

Reactive mesogens, known as RMs, are usually monomeric calamitic liquid crystals that have structures in which the terminal end groups are designed to be mutually reactive with one another through photopolymerization under ultraviolet (UV) light in the presence of a suitable initiator. Most RMs have rod-like architectures, such that they can be used to create anisotropic networks. Anisotropic components of thermoset networks are also associated with high technology properties: for example, rheology, toughness, brittleness, solvent resistance, processing characteristics (fabrication times and shaping of materials), and cost/property profiles determine their individual fields of application. Composite networks can also be produced, which are often used in large-scale applications for which high mechanical strength in two dimensions is required (*e.g.*, airplane wings).

Polymerization can also be achieved in a variety of mesophases, for example, nematic, smectic A, and smectic C phases. As a consequence, very strong and highly stable composite thermosetting 2D/3D networks can be produced. The structures of some typical RMs are shown in Figure 10.33. Compound (a) is an RM used in the formation of optical films, compound (b) is an RM used in producing gels for applications of ferroelectric liquid crystals, and compound (c) is an RM that has only one reactive group and is used to soften networks.

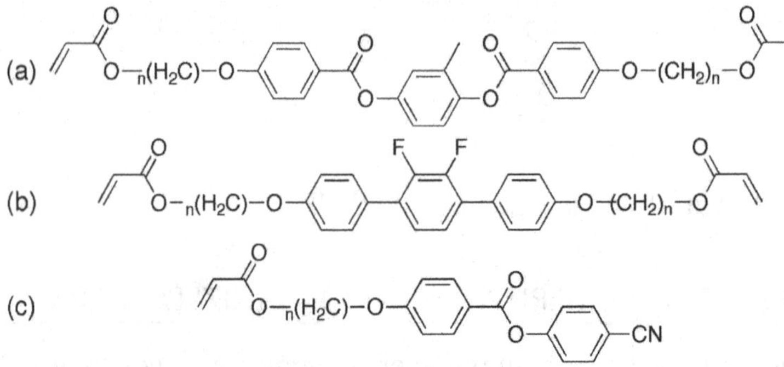

Figure 10.33 Examples of reactive mesogens.

Photopolymerization in the liquid crystal phase to make commercial optical retardation films is done using roll-to-roll technology. The plastic substrate used in this procedure is flexible and is usually coated with either a photoaligning material or a surface-treated polymer in order to orient the RM mixtures, which are usually deposited in solution so that when the solvent evaporates the RMs become oriented on the surface. In addition, the surface can be patterned to give additional orientational structures to RMs before polymerization. The RMs are not usually single materials but are complex mixtures designed to give desired optical properties to the resulting film, *i.e.*, they may be calamitic, discotic, or chiral. When exposed to unpolarized UV light, the RMs polymerize to give a rigid film, as shown in Figure 10.34. These types of films are found on most displays (LCs and OLEDs) that are manufactured across the world.

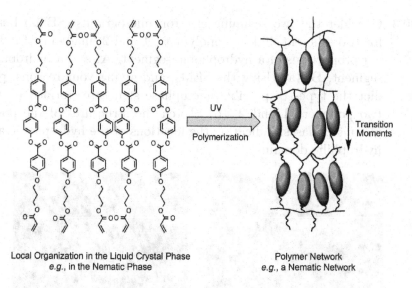

Local Organization in the Liquid Crystal Phase
e.g., in the Nematic Phase

Polymer Network
e.g., a Nematic Network

Figure 10.34 Polymerization of reactive mesogens under the exposure of UV light to give a solid oriented network.

EXERCISES

10.1 A polymer sample is subjected to fractionation using gel permeation chromatography (GPC). The fractionation gives three fractions with molar masses of 20,000, 70,000, and 200,000 in a mole ratio of 1:3:1. Calculate the number average molecular mass $\overline{M}_n$, the weight average molecular mass $\overline{M}_w$, and the polydispersity, $\gamma = \overline{M}_w / \overline{M}_n$.

10.2 (a) What is meant by the glass transition temperature T_g of a polymer? (b) What three molecular factors strongly affect the glass transition temperature? (c) Put the following polymers in increasing order of glass transition temperature.

A

B

C

D

E

10.3 Consider the two scanning electron micrographs (SEMs) below for two different diblock copolymers, **1** and **2**, made up of differing proportions of a hydrophobic segment, **A**, and a hydrophilic segment, **B**. (a) Using the SEMs, and giving your reasons, predict the types of possible microphase structures formed by the two polymers **1** and **2**. (b) Sketch the structures of the possible microphases, indicating the locations of the hydrophobic and hydrophilic domains.

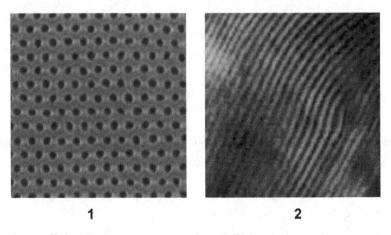

1   **2**

10.4 Sketch and compare the interchain interactions of Vectran® and Kevlar®. Using the comparisons, discuss the order parameters of the two materials while in their mesomorphic states.

10.5 Label the Miktoarm structures of the following block co-polymers.

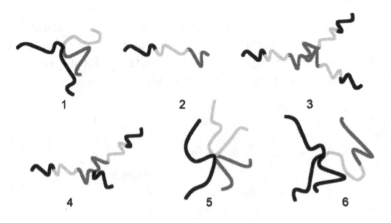

CHAPTER 11

Liquid Crystal Science – Techniques and Instruments

11.1 INTRODUCTION

Every laboratory that performs research on liquid crystals is different, especially because often a laboratory contains very specialized equipment that is not duplicated in many other laboratories across the globe. Nevertheless, there are some capabilities of liquid crystal laboratories that are always present, making them look very much the same in some respects. The reason for this is that certain fundamental measurements must be made on liquid crystals in order to understand them to the point of uncovering new information. These fundamental measurements utilize common techniques, and these techniques are the subject of this chapter. Some of the techniques probe the liquid crystal material itself (polarized optical microscopy, differential scanning calorimetry, refractometry/conoscopy, and X-ray diffraction). Other techniques investigate the physical properties of the materials (nuclear magnetic resonance, electron and atomic force scanning microscopy, and electric/magnetic, optical, and mechanical measurements). Still some techniques are required in order to develop applications of liquid crystals (alignment procedures, device construction, measuring the optical/electrical response).

Some of the techniques have been described in earlier chapters. These are only slightly extended in this chapter to show the range of their capability. Other techniques are not discussed previously, and these are described from the beginning.

11.2 POLARIZED OPTICAL MICROSCOPY

Perhaps the most universal technique used in liquid crystal laboratories is polarizing optical microscopy (POM). Such a microscope is described in Section 2.1, including a schematic diagram of a microscope and an image of a liquid crystal in the midst of the nematic-isotropic transition. Because the liquid crystal sample is in a temperature-controlled stage between crossed polarizers, one of these microscopes makes it simple to measure at what temperature the phase transitions take place. One reason for this is that the various liquid crystal phases appear differently in the microscope, mainly because the liquid crystal is usually not perfectly aligned so there are defects present, and sometimes these defects separate different domains within the sample. Fortunately for liquid crystal researchers, liquid crystal phases tend to have their own set of defects that appear differently in the microscope. These defects can be points in the bulk or on the glass surface at which the orientational order is undefined. Both of these are shown schematically in Figure 3.16. Disclinations, which are lines along which the director is undefined, are discussed in Section 3.8. Finally, dislocations are lines along which the positional order is undefined.

As mentioned earlier and explained at length in Chapter 13, a liquid crystal viewed between crossed polarizers appears dark anywhere the director is oriented parallel to the transmission axis of one of the polarizers. This gives defects a specific appearance. For example, the left-most point defect or disclination shown in Figure 3.13 indicates that the director is horizontal or vertical (assumed to be the transmission axes of the polarizers) just to the left and right of the defect.

Thus the liquid crystal appears bright around the defect except for two "brushes" that extend horizontally to the left and right of the defect. On the other hand, the director is either horizontal or vertical in the other two images of Figure 3.13 both left and right of the defect and up and down from the defect. These types of defects or disclinations have four brushes extending perpendicular to each other from the center of the defect. In practice, Figure 11.1 is a POM image of a nematic phase with the director more or less restricted to the plane of the image. It is called a *schlieren* texture and both two- and four-brush defects or disclinations are visible. If you arbitrarily assign left/right or up/down to one brush, then the director orientation for all other brushes in the image is determined. A 4-brush defect is shown in the top right of the photomicrograph, and a 2-brush defect in the center left. The director fields for both defects are shown in the diagrams to the left of the texture. Rotation of the sample between the crossed polarizers allows for the identification of the defects as + or − and the number of brushes divided by 4 classifies the defect strength as $s = \pm 1/2$ or ± 1. A clockwise rotation of the sample resulting in a clockwise rotation of the brushes is positive, whereas if the brushes rotate in the opposite direction, it is negative.

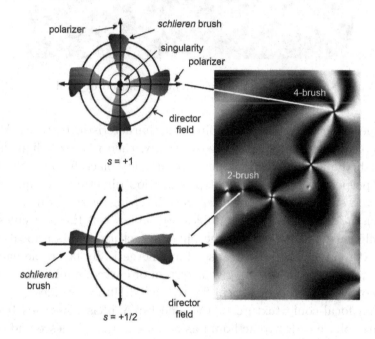

Figure 11.1 *Schlieren* texture of a nematic liquid crystal.

A *schlieren* texture can also appear in a smectic C liquid crystal when the layers are more or less perpendicular to the light path. Since the director is tilted relative to the light path, there is a component of the director in the plane of the image. This component can point in any direction just like the director of a nematic can point anywhere in the plane. However, the lamellar structure of the smectic C phase precludes the structural situations that lead to the two-brush *schlieren* texture. Accordingly, only four-brush *schlieren* defects are seen for the smectic C phase, as shown in Figure 11.2.

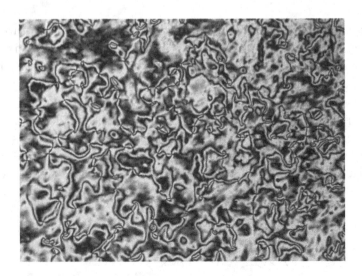

Figure 11.2 *Schlieren* texture of a smectic C liquid crystal.

The smectic A phase exhibits two characteristic textures. When the molecules are aligned homeotropically, then polarized light is extinguished and the texture appears completely black because all directions perpendicular to the light propagation direction are equivalent. However, such black regions often have birefringent rings due to disclinations at the edges of air bubbles. Second, when the molecules are not aligned homeotropically, the phase structure adopts a focal-conic fan texture, which arises because of an energetically favorable packing of the lamellar structure to give a system of curved, equidistant layers that correspond to geometrical figures called Dupin cyclides.

The focal-conic texture of the smectic A phase develops from a nematic phase or isotropic liquid as *bâtonnets* that coalesce and eventually generate the focal-conic texture. The dark lines that are visible arise from optical discontinuities generated from the structural arrangement of the molecules within the phase. The discontinuities are seen in the form of ellipses and hyperbolae (or parts of these), where the two lines are related to one another as a confocal pair. The "plan" and "side" views of a focal-conic domain are shown together in Figure 11.3. The molecular packing arrangements of the layered smectic A structure are such that the layers pack in concentric spheres, like layers in an onion. Two sets of these concentric spheres appear to merge into each other as shown side on and top down in the figure.

The ellipses and hyperbolae of the focal-conic texture appear as

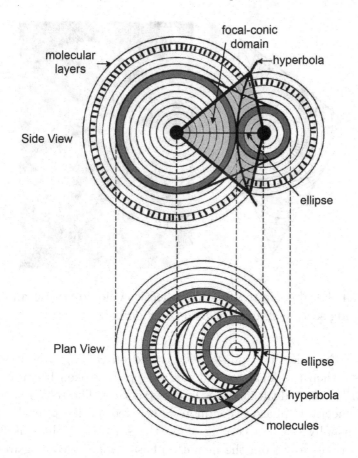

Figure 11.3 "Side" cross-section through a set of Dupin cyclides in a focal-conic domain (top). A "plan" view of the ellipse where two sets of concentric arcs meet (bottom).

black lines of optical discontinuity because these are the areas where sharp changes in the director are found. These lines do not possess orientational order, *i.e.*, they are isotropic, and therefore they appear black when viewed using POM. Figure 11.4 shows an image of a focal-conic texture for a smectic A liquid crystal along with a section through a focal-conic domain.

The smectic C phase is basically the tilted analog of the smectic A phase and POM reflects this structural situation. Two textures are exhibited by the smectic C phase. One is the *schlieren* texture as described previously, the other is based on the focal-conic texture of the smectic A phase except that the molecules within the layers are

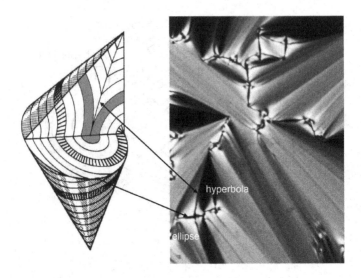

Figure 11.4 Focal-conic texture of a smectic A liquid crystal and a corresponding section through a focal-conic domain.

tilted. If the smectic C phase is produced on cooling from a smectic A phase, then the focal-conic domains appear broken because of the differently oriented tilt domains in the structure. The *schlieren* texture of the smectic C phase is commonly observed as the natural texture, and is usually seen on cooling when the smectic C phase is formed paramorphotically from the nematic phase, (*i.e.*, by cooling from another liquid crystal phase). If the smectic C phase is generated from cooling a smectic A phase, then those areas of the smectic A phase that appeared dark (homeotropic alignment) fill with the *schlieren* texture as the molecules tilt on cooling. If, as is desirable, both textures are present in one sample, then the appearance of the *schlieren* texture accompanied by the focal-conic fan texture, which appears broken in the smectic C phase, is an identifier of the phase. If the focal-conic fan texture is seen on cooling the nematic phase or isotropic liquid, the fans are not usually broken since there is no previous layer ordering. An image of a paramorphotic focal-conic texture in a smectic C phase is shown in Figure 11.5.

Other smectic mesophases are much less commonly encountered than the smectic A and smectic C phases, which is, to some extent, a reflection of the relative emphasis of research on smectic materials for devices, especially for ferroelectric displays. However, other smectic

Figure 11.5 Focal-conic texture of a smectic C liquid crystal.

phases are investigated using POM and some description of the optical textures of these phases is warranted.

Like the smectic A phase, the hexatic B phase exhibits focal-conic and homeotropic textures, but a transition between the hexatic and smectic A phase is still observable. Small changes in birefringence occur on cooling the smectic A phase into the hexatic B phase, and a rippling effect is sometimes seen in the focal-conic fan texture. In the hexatic B phase, a very smooth cone is seen with clear, blemish-free sides, and there are no lines running from the apices of the fans. On heating the hexatic B phase to the lesser ordered smectic A phase, wishbone or parabolic defects occur. Accordingly, the smectic A to hexatic B transition is more clearly observed on heating than on cooling.

The tilted analogs of the hexatic B phase are called the smectic I and smectic F phases. Both give *schlieren* textures similar to the smectic C phase and are often difficult to distinguish from the smectic C phase. However, the smectic I and F phases are more viscous and their textures appear out of focus when compared with the smectic C phase. Furthermore, the smectic I and F phases are virtually impossible to distinguish when both are not observed in the same material.

Unlike the hexatic B phase, the crystal B modification can exhibit three textures: homeotropic, mosaic (see Figure 11.6), and focal-conic. The homeotropic and mosaic textures are the natural textures observed on cooling from the isotropic liquid. The homeotropic texture is as previously described for the smectic A phase. However, the mosaic texture arises because it is much more difficult to generate curved layers in the 3D ordered crystal B phase than in the less ordered smectic A phase. In the mosaic texture, the layers are in planes and are not curved as they are in the focal-conic texture of the smectic A, C, and hexatic B phases, and the different levels of color within the mosaic areas result from the different angles the layers (and hence the molecules) make with the surface. When one of the polarizers is rotated, the birefringence changes in each mosaic domain, except where the molecules are homeotropically oriented. Even when the microscope stage is rotated, there can be color and light intensity changes for the mosaic domains due to different alignment directions of the molecules and hence the layers. The focal-conic texture of the crystal B phase is paramorphotic and only generated from the analogous texture of the smectic A or hexatic B phases.

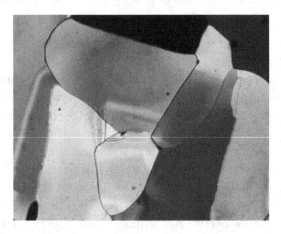

Figure 11.6 Mosaic texture of a crystal B liquid crystal.

The crystal B phase possesses an even more ordered arrangement of molecules in which the correlation of the hexagonal arrays of molecules extends three-dimensionally over large distances, and hence this phase is a soft crystal rather than a liquid crystal phase. Accordingly, the B phase does not produce a genuine focal-conic fan texture because curved layers are not really possible. However, what appears to be a focal-conic texture does form on cooling from the smectic A phase. Instead of the fans being rounded, as in the hexatic B phase, the conic sections are angled, which reflects the crystal nature of the crystal B phase. As the smectic A phase is cooled, the crystal B phase begins to build up in the bulk smectic A phase and a biphasic region develops, which produces a series of lines (called transition bars) that run across the backs of the fans (see Figure 11.7). In the case of the hexatic B phase, transition bars are not usually seen because both phases are liquid crystalline.

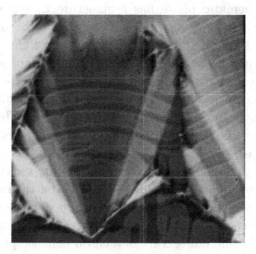

Figure 11.7   Crystal B liquid crystal showing transition bars.

The crystal E phase is more ordered than the crystal B phase, and has its board-like molecules perpendicular to the layers. The crystal E phase has an orthorhombic lattice and therefore it is optically biaxial and does not exhibit homeotropic alignment. The crystal E phase does exhibit a paramorphotic focal-conic texture that has arced lines and banding running across the domains. The arcs are a result of different orientations of the orthorhombic packing and their widths are related to the out-of-plane correlation lengths of the phase. The arcs remain a part of the texture throughout the phase range; this contrasts with the temporary nature of the transition bars described for the smectic A to crystal B transition. The crystal E phase also has a platelet texture which results from the cooling of a homeotropic texture of, for example, the smectic A or hexatic B phase. Upon cooling, the dark, homeotropic area becomes weakly birefringent and the translucent platelets generated tend to overlap. The crystal E phase also exhibits a paramorphotic mosaic texture, which is generated on cooling the mosaic texture of the crystal B phase. In the crystal E phase, the mosaic areas are crossed with parallel lines, similar to the lines in the paramorphotic focal-conic texture.

In general, the quasi-crystalline smectic mesophases (B, J, G, E, K, H) can easily be distinguished from the genuine liquid crystal phases by examining the defect textures. In the crystalline mesophases, long-range order is seen in the form of grain boundaries.

The chiral nematic phase has a structure that is similar to the conventional, achiral nematic phase except that the director gradually rotates to describe a helix. When viewed by POM, the chiral nematic phase can exhibit a few different textures depending upon the alignment and cell thickness. When the molecules of the chiral nematic phase are aligned such that the helical axis is perpendicular to the glass surfaces, then a Grandjean or planar texture is seen. This texture contains oily streaks and is often iridescent. If the helical axis is parallel to the glass surfaces, then a characteristic fingerprint texture is observed because of the periodic extinction of light as the molecular direction changes with the helical structure. Where no particular alignment is obtained, then the chiral nematic phase produces a pseudo focal-conic fan texture, which is similar to that observed for the smectic A phase. When the focal-conic fan texture of a chiral nematic phase is sheared, a Grandjean or planar texture results due to re-alignment of the helical axis. Both the Grandjean and fingerprint textures are shown in Figure 11.8.

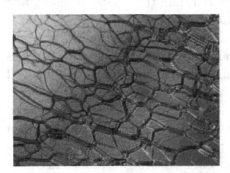 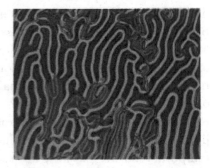

Figure 11.8 Grandjean texture with oily streaks (left) and fingerprint texture (right) of a chiral nematic liquid crystal.

The tilted, chiral smectic C phase is somewhat special because the reduced symmetry of the liquid crystal phase leads to a different structure from the achiral smectic C phase. As described in Chapters 1, 5, and 7, the chiral smectic C phase has a helical structure as a result of the spiraling tilt angle from layer to layer within the phase structure. Accordingly, this different structure is reflected in the texture observed by POM. The alternative to the *schlieren* texture is called the pseudo-homeotropic texture. In the achiral smectic C phase, the homeotropic texture is not possible because of the tilted molecules. However, in the chiral smectic C phase, the tilt angle spirals to describe a helix and on average the tilt angle throughout the phase is random. Accordingly, light shining down the averaged helical axis is extinguished. When the pitch of the helix is in the region of visible light, then colors are reflected (blue at lower temperatures and red at higher temperatures) and the pseudo-homeotropic texture becomes colored and this special case is called the petal texture. The standard focal-conic fan texture is also exhibited in a similar manner to that for the achiral smectic C phase. However, often lines are seen on the sides of the fans, which arise as a consequence of the helical structure. The lines that are seen in the focal-conic fan texture of a chiral smectic C phase are often called pitch lines. The helical structure comprises a spiral of changing tilt directions and as these helices attempt to fill space in the wedge-shaped focal-conic domains, structural distortions affecting the director occur usually at half pitch intervals leading to periodic dark lines in the texture. The distance between these lines can often be a measure of half the pitch but not always, depending upon the molecular alignment. Figure 11.9 shows a chiral smectic C texture with pitch lines visible.

There is a wide range of other chiral mesophases that are somewhat unconventional in terms of their structure and many of these are rather recent developments. Blue phases are cubic mesophases that have a double twist structure (see Chapter 7). They are observed in highly chiral materials and appear at temperatures above the chiral nematic phase. These phases give characteristic platelet-type textures that are often blue (hence the name) but they can also appear green, yellow, and red. The twist grain boundary phase is a smectic phase with a frustrated, complex helical structure. In this case the normal lamellar smectic A structure is broken into separate blocks that are twisted from each other at a gradual angle to describe a helix. A helical screw dislocation marks each separation of the smectic A blocks. This phase appears from the homeotropic texture of the normal smectic A or cholesteric phases as worm-like tapes of birefringence. This texture is called the *vermis* texture (from the Greek meaning worm-like).

Some of the smectic phases are difficult to distinguish using POM and samples between two pieces of glass. However, it is possible to dis-

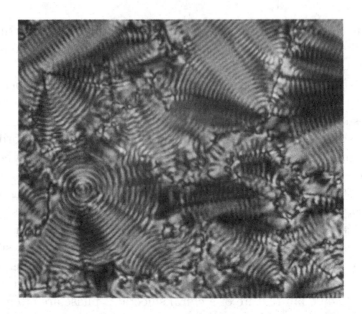

Figure 11.9 The focal-conic texture of a chiral smectic C liquid crystal.

tinguish the smectic I and F phases and the smectic J and G phases by using POM on free-standing films. A free-standing film is prepared by drawing the liquid crystal material (in the smectic A or C phase) across a small (1 or 2 mm) hole in an aluminum plate that is countersunk from beneath to give a very sharp edge. As the thin film is drawn, the smectic layers are suspended across the hole rather like a suspension bridge across a river. In the smectic A phase, the molecules are normal to these layers rather like the cables holding the road of the suspension bridge. Accordingly, this homeotropic alignment leads to optical extinction when crossed polarizers are used and this is also the case for the hexatic B and crystal B mesophases. When the molecules tilt, then a birefringent texture is observed. Free-standing films generate much clearer optical textures that are derived from the homeotropic texture of orthogonal phases. The *schlieren* texture of the smectic C phase is similar to that seen using conventional glass slides, but clearer. Cooling to the smectic I phase gives a very clear transition, which settles to the same *schlieren* texture but is much more difficult to focus. The smectic C to F phase transition is also clear to see, but the resulting *schlieren* texture in the smectic F phase is sectioned by mosaics. As the tilted smectic mesophases become more ordered, it becomes more difficult to detect a difference. In each case, cooling into the crystal J or G phase produces dendritic growth into a full mosaic texture. The dendrites of the crystal G phase are usually more rounded than the dendrites of the crystal J phase, which are often angular but not quite spiky. Proper identification is more difficult than it sounds; nevertheless distinction is possible. The crystal K and H phases are more ordered and accordingly, it is very difficult to tell one phase from the other.

The different lyotropic liquid crystal phases also show distinctive textures. If only micelles or small vesicles are present, the sample appears dark using POM with crossed polarizers. The lamellar phase has a layered structure like the smectic A phase, so one observes both focal-conic and homeotropic textures. Due to the symmetry of the cubic phase, it generally appears dark. The hexagonal phase, much like the hexagonal columnar phase in discotic liquid crystals, is dominated by bend distortion. As a result, the columns bend in arcs creating a fan texture, but with no confocal lines of optical discontinuity. If the lyotropic or columnar phase is rectangular rather than hexagonal, the fan texture appears broken due to rectangular domains with different orientations within the fans.

11.3 DIFFERENTIAL SCANNING CALORIMETRY

Differential scanning calorimetry (DSC) is an important tool in the identification and characterization of mesophases and almost all research groups working in the field of liquid crystals utilize it. As described in Chapter 2, DSC reveals the presence of phase transitions in a material by detecting the enthalpy change at each phase transition. The precise identity of the phase(s) cannot be obtained, but the level of enthalpy change involved at the phase transition does provide some indication of the types of phases involved. Accordingly, DSC is used in conjunction with polarizing optical microscopy to determine the type of mesophase that a material exhibits.

When a material melts, a change of state occurs from a solid to a liquid and this melting process requires energy (endothermic) from the surroundings. Similarly the crystallization of a liquid is an exothermic process and energy is released to the surroundings. The DSC instrument measures the energy absorbed or released by a sample as it is heated or cooled. The melting transition from a solid to a liquid is a relatively drastic phase transition in terms of the structural change and this is reflected by the relatively high energy of transition. However, DSC can be used similarly to detect the much more subtle phase transitions that are involved in liquid crystalline materials. These more subtle structural changes produce relatively small enthalpy changes.

As shown in Figure 2.8, DSC employs two furnaces, one to heat the sample under investigation and the other to heat an inert reference material (usually gold). The two furnaces are separately heated but connected by two control loops to ensure that the temperature of both remains identical through a heating or cooling cycle and the heating or cooling rate for each is constantly identical. If the sample melts, for example from the crystalline solid to the smectic A phase, then energy must be supplied to the sample to prevent an imbalance in temperature between the sample and the reference. This energy required to maintain identical temperatures is measured and recorded by the instrument as a peak on a baseline. The instrument is pre-calibrated with a sample of known enthalpy of transition, and this enables the enthalpy of transition to be recorded for the material being examined. The sample (usually 5 to 10 mg) is weighed into a small aluminum pan, which is crimp-sealed with an aluminum top. The sample pan is then placed into a holder in a large aluminum block to ensure good temperature control. The sample can be cooled using liquid nitrogen and a working temperature range of between -180 and 600°C is possible.

Although the enthalpy changes at a transition cannot identify the types of phases associated with the transitions, the magnitude of the enthalpy change is proportional to the change in structural ordering of the phases involved. Typically, a melting transition from a crystalline solid to a liquid crystal phase or the isotropic liquid phase generates an enthalpy change of around 30 to 50 kJ/mol, which at least indicates that a considerable structural change is occurring. However, liquid crystal to liquid crystal and liquid crystal to isotropic liquid transitions are characterized by much smaller enthalpy changes. For example, the smectic A to isotropic liquid transition involves an enthalpy change of 4 to 6 kJ/mol; unfortunately, other smectic and crystal smectic mesophases (*e.g.*, smectic B, smectic C, crystal B, and crystal E) can also give similar values. However, the nematic phase usually gives a smaller enthalpy change (1 to 2 kJ/mol) on transition to the isotropic liquid. Enthalpy changes at transitions between different liquid crystal phases can also be extremely small. For example, the smectic C to smectic A transition is often difficult to detect by DSC because the enthalpy change is typically less than 300 J/mol. The enthalpy for the transition from smectic A to nematic is also fairly small (1 kJ/mol) and the smectic C to nematic transition often has an even smaller enthalpy (<1 kJ/mol). However, the smectic A to smectic C transition is usually identifiable as a step on the DSC baseline as the heat capacity changes without an enthalpy change. These changes are, however, readily detected by optical polarizing microscopy.

DSC often does not reveal the presence of chiral liquid crystal phases. In the case of the blue phases and the twist grain boundary phase, the mesophase ranges are often too short to provide a distinct enthalpy peak. Of course, in some materials this situation is also true for the more usual liquid crystal phases. The transitions between the chiral smectic C(ferri) phase and chiral smectic C(anti) phase and their transitions with the chiral smectic C phase have extremely small enthalpy values, which makes detection by DSC very difficult. However, the latest DSC equipment enables the detection of such remarkably small enthalpy transitions.

As discussed in Chapter 4, in general there are two types of phase transitions, discontinuous and continuous (sometimes called first-order and second-order, respectively). In order to understand the difference, a brief mention of the thermodynamics of phase transitions is necessary. The Gibbs free energy G is defined as

$$G = H - TS, \tag{11.1}$$

where H is the enthalpy, T is the absolute temperature, and S is the entropy. The first derivative of G with respect to T at constant pressure P gives the negative of the entropy.

$$\left(\frac{\partial G}{\partial T}\right)_P = -S. \tag{11.2}$$

The second derivative is related to the heat capacity at constant pressure C_p,

$$\left(\frac{\partial^2 G}{\partial T^2}\right)_P = -\frac{C_P}{T}. \tag{11.3}$$

If the entropy at a phase transition shows a discontinuity, then a discontinuous (first-order) phase transition occurs. However, phase transitions can occur without a change in entropy and enthalpy. In such cases there is a discontinuity in the second derivative (*i.e.*, $-C_P/T$) and a continuous (second-order) phase transition occurs.

Most liquid crystal to liquid crystal transitions are discontinuous, but some, such as the smectic C to smectic A transition, are often continuous. Glass transitions in liquid crystal polymers are also continuous phase transitions. Such continuous phase transitions are revealed by a slight inflection in the baseline on the DSC scan rather than the peak that is seen for a discontinuous phase transition.

If a transition between mesophases is missed by optical microscopy, then DSC may reveal the presence of a transition at a particular temperature, or vice versa. After a DSC scan, the material should be examined very carefully by optical microscopy to provide information on phase structure, to check that transitions are not being missed by DSC, and hopefully to lead to the likely identity of the mesophases. Accordingly, optical polarizing microscopy and differential scanning calorimetry are significant, complementary tools in the identification of the types of mesophases exhibited by a material.

11.4   REFRACTOMETRY AND CONOSCOPY

As is clear from a good deal of the discussion in previous chapters, the indices of refraction, $n_\parallel$ and $n_\perp$, and the birefringence, $\Delta n = n_\parallel - n_\perp$, are important material properties. The birefringence can be measured in many ways, and these are discussed in Chapter 13. On the other hand, only certain experiments are capable of measuring the indices of refraction themselves. The most common instrument used to measure the refractive indices is a refractometer. This is an instrument that uses the fact that light passing from a higher index material to a lower index material is completely reflected if the angle of incidence is above a critical value. This angle is called the critical angle and the sine of this angle equals the ratio of the refractive index of the material forming the interface off which the reflection occurs and the medium through which the light is propagating. Imagine light of one wavelength propagating through a prism and striking the interface between the prism and the liquid crystal sample. As shown in Figure 11.10, the light is focused on the interface, meaning that it strikes the interface with a minimum angle of θ_{min} and a maximum angle of θ_{max}. The light is reflected with the same minimum and maximum angles, but between θ_{crit} and θ_{max}, 100% of the light is reflected, where θ_{crit} is given by the critical angle formula, $\sin\theta_{crit} = n_{lc}/n_p$ (n_{lc} and n_p are the indices of the liquid crystal and prism, respectively). Since n_p is known, the refractometer measures θ_{crit} and from it determines the index of refraction of the liquid crystal sample.

The refractive index of a liquid crystal is wavelength dependent. Some refractometers only supply light with a single wavelength, usu-

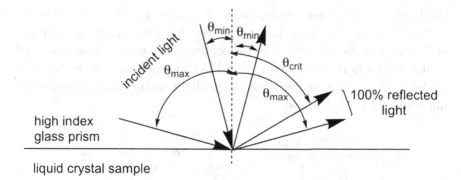

Figure 11.10 Simple refractometer. The incident light strikes the prism-liquid crystal interface at angles between θ_{min} and θ_{max}, but total internal reflection means the light is totally reflected between the angles of θ_{crit} and θ_{max}. As explained in the text, the index of refraction of the liquid crystal can be determined if θ_{crit} is measured.

ally 589 nm, the wavelength of the sodium D line. More sophisticated refractometers allow measurements at other wavelengths.

Refractometers are normally used to measure the refractive index of an isotropic fluid. In order to measure both indices of refraction for a nematic liquid crystal, two conditions must be met. First, the incident light must be polarized or the reflected light of only one polarization must be detected. Second, the liquid crystal sample has to be aligned with the director oriented either parallel or perpendicular to the light polarization. This is normally achieved by preparing the surface of the prism in the proper way.

Conoscopy is similar to refractometry, except that no reflection takes place. For example, light of a single wavelength and linear polarization coming from many directions (usually defined by a cone) strikes a sample located at the front focal plane of a lens. The lens then focuses the light coming from a specific direction to a specific point on its back focal plane through a polarizer oriented at 90° to the initial polarization of the light (see Figure 11.11). Thus the intensity at each point in the back focal plane contains information on how the light traveling through the liquid crystal along a specific direction is affected by the liquid crystal. In general, the light emerging from the liquid crystal is elliptically polarized and not extinguished by the second (crossed) polarizer. But for the right combination of direction and birefringence, the light can be of zero intensity or of maximum intensity. Thus the pattern seen at the back focal plane can be quite complex with lots of bright and dark bands. Such a conoscopic figure can be analyzed to determine both the value of the birefringence and the orientation of the director. Conoscopy is also capable of distinguishing between a uniaxial and biaxial phase.

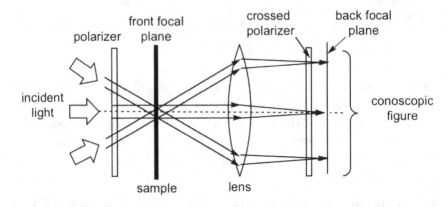

Figure 11.11   Formation of a conoscopic figure. Each point of the image on the back focal plane represents the effect the liquid crystal has on light propagating through the liquid crystal in a specific direction.

11.5   X-RAY DIFFRACTION

X-ray diffraction is a powerful technique to characterize the orientational and positional order present in a liquid crystal. For that reason, it is discussed at some length in Chapters 2 and 3. A typical X-ray experiment starts with a source of X-rays of a single wavelength, collimated and directed at a liquid crystal sample. The X-rays are diffracted by the sample and captured by an area detector. In this way, data are obtained simultaneously at many angles. The distance on the area detector from the center of the diffraction pattern is proportional to the tangent of the diffraction angle. Since the diffraction angle is usually small, the distance from the center of the pattern is inversely proportional to the repeat distance in the liquid crystal (see Section 3.6). Hence, the smaller the repeat distance, the larger the distance between the diffraction maximum and the center of the diffraction pattern. Thus an aligned nematic phase produces two peaks far from the center coming from the side-to-side molecular separation repeat distance and two peaks close to the center from the end-to-end molecular separation repeat distance. In addition, these two pairs of peaks should be oriented along perpendicular directions. This is shown schematically in Figure 11.12(b). Notice that the diffraction pattern from the isotropic phase contains peaks for approximately these same distances, since the distance between molecules is still correlated. But due to the lack of orientational order, the peaks are spread out in all directions, creating circles in the diffraction pattern. The diffraction patterns for the smectic A and C phases are also shown in Figure 11.12, with multiple peaks from the layered structures but the same single peak from the side-to-side molecular separation repeat distance.

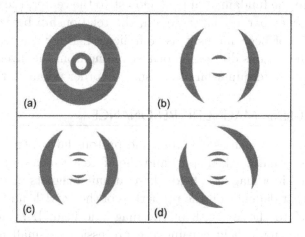

Figure 11.12 X-ray diffraction patterns for various phases: (a) isotropic, (b) aligned nematic, (c) aligned smectic A, and (d) aligned smectic C. The director is oriented vertically in (b) and (c) and at an angle to the vertical in (d). The smectic layers are horizontal in (c) and (d).

Notice that in Figure 11.12 the X-ray peaks coming from the side-to-side molecular separation repeat distance are not located along the perpendicular to the director, but are spread out in arcs. This is because the orientational order is not perfect in a liquid crystal, so the long axes of molecules are distributed around the orientation of the director. As discussed in Section 1.2, the orientational order parameter S is a measure of the width of this distribution. Therefore, it is possible to measure S from the distribution of intensity in the arcs of the X-ray diffraction pattern. The connection between S and the X-ray intensity within the arcs is not simple, but measurements of the latter do allow the former to be determined.

It is also possible to determine the smectic order parameter ψ (see Section 1.2) from the X-ray diffraction pattern. ψ is the amplitude of the sinusoidal variation in the density with a spatial repeat distance equal to L, the distance between the layers. The density variation never follows such a sinusoidal dependence perfectly, meaning there are density variations with spatial repeat distances of $L/2$, $L/3$, $L/4$, *etc.* As explained in Section 3.6, these variations give rise to the X-ray diffraction peaks at equal intervals. Hence the smaller these higher-order peaks at larger distances from the center of the diffraction pattern are compared to the fundamental peak closest to the center, the higher the positional order parameter ψ. Again, the relationship between ψ and the variation of peak intensity is complicated. But measuring the intensity of the various diffraction peaks coming from the layer structure allows for the determination of the smectic order parameter.

11.6   NUCLEAR MAGNETIC RESONANCE

Charged particles such as electrons and protons have intrinsic angular momentum (usually called spin) and therefore act as atomic magnets. When placed in a magnetic field, these atomic magnets precess with a frequency that depends on the properties of the particle and is directly proportional to the strength of the magnetic field. If in addition to the static magnetic field causing the precession, a small alternating magnetic field is applied, this additional magnetic field can affect the precession of the particles. The effect is strongest when the frequency of the alternating magnetic field is equal to the frequency of precession, a condition called resonance. So a nuclear magnetic resonance (NMR) experiment just examines the precession of the particles (each is a nucleus of an atom) when the alternating field equals the precession frequency.

Since the precession frequency is directly proportional to the static magnetic field strength, anything that alters the magnetic field at the site of the nucleus affects the resonant frequency. For example, the electrons around the nucleus produce their own magnetic field that changes the value of the magnetic field at the nucleus and shifts the resonant frequency slightly. This is called a chemical shift. Another example is if two nuclei are close to one another on the same molecule. The magnetic field of one nucleus changes the magnetic field at the second nucleus, causing the resonant frequency to change slightly. In fact, since the neighboring nucleus's magnetic field can add to or subtract from the static magnetic field, the result is two slightly different resonant frequencies. This phenomenon is called dipolar splitting and occurs for hydrogen and fluorine nuclei, among others. The resonant frequency of another type of nucleus, with the deuteron being the best example, is affected by the chemical bond the atom of that nucleus shares with the rest of the molecule.

In all three cases, the frequency shift or splitting $\Delta\nu$ depends on the angle a bond or the line between two nuclei makes with the static magnetic field. This means that $\Delta\nu$ depends on (1) the angle between the bond or line between nuclei and the long axis of the molecule, and (2) the angle between the long axis of the molecule and the static magnetic field. Since both the intramolecular motion and intermolecular motion are much faster than the precession, the resonant frequency depends only on the average angles. If the intramolecular and intermolecular motions are uncorrelated, which usually is a good assumption, then $\Delta\nu$ is given by

$$\Delta\nu = \Delta\nu_0 \left\langle \frac{3}{2}\cos^2\beta - \frac{1}{2}\right\rangle \left\langle \frac{3}{2}\cos^2\theta - \frac{1}{2}\right\rangle, \qquad (11.4)$$

where $\Delta\nu_0$ is a constant that can be measured or calculated, β is the angle between the bond or line between nuclei and the long axis of the molecule, θ is the angle between the long axis of the molecule and the static magnetic field, and the brackets denote a time average. If the director is aligned with the static magnetic field, then the average containing θ is just the order parameter S. Thus if $\Delta\nu_0$ and the average containing β are known, S can be determined from a measurement of $\Delta\nu$.

Figure 11.13 shows schematics of typical NMR spectra. Notice that there is no chemical shift or splitting in the isotropic phase because $S = 0$ and therefore $\Delta\nu = 0$. Also notice that even when the liquid crystal sample is not aligned, the NMR spectrum still allows a measurement of S because the largest value of $\Delta\nu$ occurs when the angle between the director and the magnetic field is $0°$ and the smallest values occur when the angle between the director and the magnetic field is $90°$. If multiple shifts or splittings are present, it is sometimes possible to determine both nematic order parameters S an D. Thus NMR represents a powerful technique for the measurement of order parameters and is used routinely in many laboratories investigating liquid crystals.

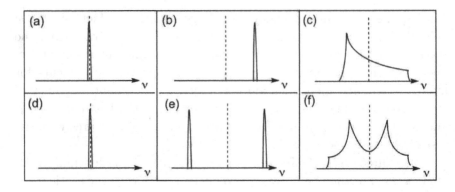

Figure 11.13 NMR spectra showing a chemical shift in (a) - (c) and a splitting in (d) - (f). Isotropic phase: (a) and (d); aligned nematic phase: (b) and (e); randomly aligned nematic phase: (c) and (f). The dotted lines represent the resonant frequency of the free nucleus.

11.7 ELECTRON MICROSCOPY

The resolution of light microscopy is limited by the wavelength of light, although very recent efforts have developed techniques that push the resolution to values shorter than the wavelength of light. In general, therefore, traditional light microscopes, including polarized optical microscopes, cannot detect features that are smaller than a quarter of a micrometer. On the other hand, the wavelength of an electron beam can be much shorter. For example, electrons accelerated through 100 KV have a wavelength of 0.004 nm. Unfortunately this is not the resolution, since other factors limit the resolution, for example, the size of the electron beam spot as it is scanned across the sample. Still, a resolution of less than a nanometer is quite standard, making electron microscopy a potentially powerful technique to image features much smaller than is possible using optical microscopy.

However, there are major challenges when using electron microscopy to investigate liquid crystals. First, the samples must be prepared in specific ways in order to perform electron microscopy, potentially causing the liquid crystal to change from its undisturbed form. For example, thin samples are required for transmission electron microscopy (TEM), which means the entire liquid crystal is constrained by nearby surfaces and not free to adopt its equilibrium structure. Second, the electrons striking the sample have enough energy to cause the liquid crystal sample to respond to the electron beam, again potentially causing the liquid crystal to adopt undesired structures. For these reasons, some of the most successful methods being used to investigate liquid crystals involve freezing the sample extremely rapidly, thereby locking the molecules in place for further investigation. In what is called freeze-fracture TEM, a thick sample of liquid crystal is frozen rapidly and fractured while it is still frozen. A replica of the surface is made by depositing a thin metal layer on it at an oblique angle to bring out the surface features, followed by a supporting carbon layer on top of that. The liquid crystal is thawed and removed, and the metal replica surface is examined by traditional electron microscopy. Another technique that is effective is cryo-TEM. A thin film of liquid crystal is rapidly frozen, and while still frozen it is examined by traditional TEM. The first technique suffers from the fact that the entire structure of the liquid crystal is not available from observing the replica surface, while the second technique has the problem of using thin samples for which the surface can have a large effect on the liquid crystal. Still, these methods have the potential to image the molecules themselves, and certainly assemblies of molecules.

11.8   ELECTRICAL AND MAGNETIC MEASUREMENTS

The electric property of interest in characterizing a liquid crystal is the relative dielectric permittivity ϵ_r (see Section 12.1). This parameter describes how large the electric polarization P is when an electric field E is applied, $P = \epsilon_0(\epsilon_r - 1)E$, where ϵ_0 is the permittivity of free space. The relative permittivity is a tensor quantity, so for a nematic liquid crystal there is a relative permittivity for an electric field applied parallel to the director and a different relative permittivity for an electric field applied perpendicular to the director. Many effects contribute to the relative permittivity of a material. A list of some of these in order of how fast they respond looks like this: molecular dipole moments re-orient, ions move, parts of molecules deform collectively, atomic dipole moments change strength, atomic bonds are altered, and electronic structures respond. Therefore, if one measures the relative permittivity when an alternating electric field is applied to a material, its value depends critically on the frequency ω of the applied electric field. If the frequency is low, many of the listed mechanisms can respond and a larger relative permittivity is measured. If the frequency is high, few of the listed mechanisms can respond and a smaller relative permittivity is measured. This is why static ($\omega = 0$) relative permittivity measurements are not the only ones performed. Rather, a dielectric spectrosopcy measurement is made, recording the relative permittivity as a function of a wide range of frequencies.

When a sinusoidally varying electric field $E(\omega)$ is applied, the electric polarization also varies sinusoidally with the same frequency. There is an amplitude to this sinusoidal variation of the electric polarization and it is also shifted in phase relative to the electric field variation. Another way of describing the amplitude and phase properties of the electric polarization is to state that there is a relative permittivity that describes the electric polarization variation that is in phase with the electric field $\epsilon'(\omega)$ and a relative permittivity that describes the electric polarization variation that is out of phase (shifted by 90°) with respect to the electric field $\epsilon''(\omega)$. Notice that the subscript r has been dropped for convenience.

When the frequency passes through the frequency at which one of the mechanisms in the list responds most strongly, $\epsilon'(\omega)$ and $\epsilon''(\omega)$ respond quite differently. $\epsilon'(\omega)$ simply decreases as the frequency gets too high for the mechanism to respond. However, $\epsilon''(\omega)$ peaks at the frequency for which the response is maximum. This process can be described as a relaxation mechanism with a relaxation time of $\tau = 1/\omega_0$, where ω_0 is the frequency at which the mechanism responds most strongly (often referred to as its resonant frequency). If $\epsilon'(0)$ is the in-phase relative permittivity far below ω_0 and $\epsilon'(\infty)$ is the relative permittivity far above ω_0, then the relaxation process yields

$$\epsilon'(\omega) = \epsilon(\infty) + \frac{\epsilon(0) - \epsilon(\infty)}{1 + \omega^2\tau^2} \quad \text{and} \quad \epsilon''(\omega) = \frac{\omega\tau[\epsilon(0) - \epsilon(\infty)]}{1 + \omega^2\tau^2}. \quad (11.5)$$

Both of these functions are graphed in Figure 11.14.

Notice that when $\epsilon''(\omega)$ is graphed versus $\epsilon'(\omega)$, the curve becomes semi-circular. This is called a Cole-Cole plot and is shown in Figure 11.15 for the same data used in Figure 11.14.

Often the data on real materials show several relaxation mechanisms. If the resonant frequency of these relaxation mechanisms are far enough apart, the plot of $\epsilon'(\omega)$ shows a series of downward steps as the frequency increases, while the plot of $\epsilon''(\omega)$ shows a series of peaks. If the resonant frequencies of different relaxation mechanisms are not far enough apart, the changes in the relative permittivity overlap. Still, knowing the functional form of each relaxation mechanism often allows $\epsilon'(\omega)$ and $\epsilon''(\omega)$ for each mechanism to be found so that when added together they agree with the data.

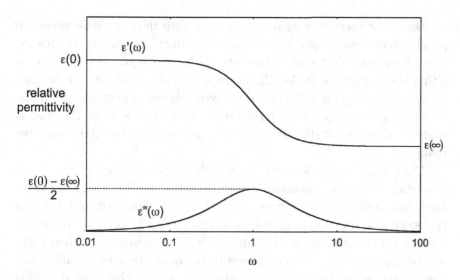

Figure 11.14 In-phase and out-of-phase relative permittivity, $\epsilon'(\omega)$ and $\epsilon''(\omega)$, respectively, for a mechanism with a relaxation time of 1. Notice that the frequency ω axis is logarithmic.

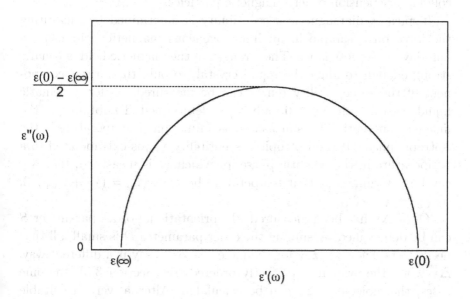

Figure 11.15 Cole-Cole plot, $\epsilon''(\omega)$ versus $\epsilon'(\omega)$, of the same data as shown in Figure 11.14.

Dielectric spectroscopy is usually done with the sample between two parallel plate electrodes. With surface treatment to the parallel plates, it is often possible to align the director of the liquid crystal so that either the parallel or perpendicular component of the relative permittivity tensor is probed. In addition, typically measurements are made at different temperatures. Armed with data of this kind, the goal is to determine both static molecular properties and molecular/collective motions.

Most liquid crystals are diamagnetic, meaning the induced magnetization is in the opposite direction to the applied magnetic field. Both magnetic susceptibilities, $\chi_{\parallel}$ and $\chi_{\perp}$ are negative, but because $|\chi_{\parallel}| < |\chi_{\perp}|$ for most calamitic liquid crystals, $\Delta\chi$ is positive. If the liquid crystal contains a metal center with an unpaired spin, then there will be a paramagnetic contribution to the magnetic susceptibility. For this contribution, the induced magnetization is in the same direction as the applied magnetic field. Ferromagnetic liquid crystals are an area of current research, either by incorporating an iron or vanadium compound into a discotic liquid crystal or forming a liquid crystal from a colloidal suspension of ferromagnetic particles.

Typically, diamagnetic susceptibility is determined by measuring the force on a sample in an inhomogeneous magnetic field using a Faraday or Gouy balance. The strength of the magnetic field is usually strong enough to align the liquid crystal, so only the largest component of the susceptibility tensor can be measured. So for a nematic liquid crystal with $\Delta\chi > 0$, only $\chi_{\parallel}$ is determined. To obtain $\chi_{\perp}$, the magnetic susceptibility is measured as a function of temperature in the isotropic phase. If the isotropic susceptibility χ_{iso} is extrapolated to a temperature in the nematic phase for which $\chi_{\parallel}$ is measured, then $\chi_{\perp}$ can be determined at that temperature because $\chi_{\mathrm{iso}} = (\chi_{\parallel} + 2\chi_{\perp})/3$, so $\chi_{\perp} = (\chi_{\parallel} - 3\chi_{\mathrm{iso}})/2$.

Once $\Delta\chi$ has been measured, the orientational order parameter S can be determined. Assuming the order parameter D is small, all that needs to be known is $\Delta\chi$ for the molecule, or to say it a different way, $\Delta\chi$ when the system is perfectly ordered (see Section 3.3). In some cases, the molecular $\Delta\chi$ can be calculated. Alternatively, a suitable theoretical function for the dependence of $\Delta\chi$ on temperature can be fit to the data in order to extrapolate a value for $\Delta\chi$ at $T = 0$, where it is assumed that $S = 1$.

11.9 OPTICAL MEASUREMENTS

Perhaps the most important optical measurement on a liquid crystal concerns the refractive indices, $n_\parallel$ and $n_\perp$, and birefringence, $\Delta n = n_\parallel - n_\perp$. One convenient way to measure the two indices of refraction is with a wedge sample (two pieces of glass making an angle α with liquid crystal in between). Such a wedge sample can be made by placing a spacer of thickness d between the two pieces of glass at one end and no spacer between the pieces of glass at the other end. If the distance between the ends of the glass is L, then $\alpha = \tan^{-1}(d/L)$. If a laser is directed at such a wedge sample, the beam will be refracted as shown in the left diagram of Figure 11.16. The deflection angle β is easily measured, so that if α is known, then by Snell's law the index of refraction of the liquid crystal is $n = \sin(\alpha + \beta)/\sin\alpha$. As shown in the right diagram of Figure 11.16, an easy method to determine α is to determine the angle γ a laser beam is reflected from an empty wedge cell. Once γ is measured, then $\alpha = \gamma/2$.

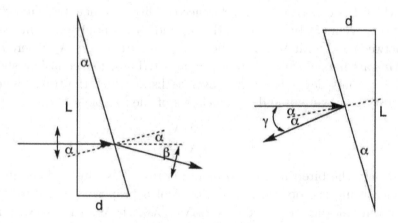

Figure 11.16 Left: measuring the index of refraction using a wedge cell. The double arrows indicate the polarization direction of the light. Right: measuring the wedge angle.

Since there are two indices of refraction for a liquid crystal, alignment layers must be applied to the glass surfaces of the wedge cell to make sure the director is oriented either parallel (to measure $n_\parallel$) or perpendicular (to measure $n_\perp$) to the light polarization direction.

Another method to measure $n_\parallel$ and $n_\perp$ relies on the interference that takes place when collimated light is incident on a cell made from two parallel pieces of glass. The light transmitted by the cell is partially made up of light that comes right through the cell and partially made up of light reflected by the front surface of the second piece of glass, then reflected by the back surface of the first piece of glass, and then transmitted by the second piece of glass. Actually, the transmitted light is made up of many components of light, each reflected a different number of times inside the cell. Since the intensity of the component decreases with each reflection, the maximum transmission occurs when the component going straight through and the component reflected back and forth once are in phase and add constructively. For this to happen, the extra path length of the reflected component $2d$ must equal a multiple of the wavelength of the light in the liquid crystal λ_0/n, where λ_0 is the vacuum wavelength of the light. If the incident light is polarized either parallel or perpendicular to the director of the liquid crystal, the proper index to use in this relation is either $n_\parallel$ or $n_\perp$. If the wavelength of the incident light varies continuously, which is possible by using a spectrometer, there will be many wavelengths for which the constructive interference condition is met. Let's imagine that this occurs at two wavelengths, λ_1 and λ_2 with only one minimum in between and with $\lambda_2 > \lambda_1$. Then $2dn_{\parallel,\perp} = m_1\lambda_1 = m_2\lambda_2$, where m_1 and m_2 are integers and where $m_1 = m_2 + 1$. These two equations show that $n_{\parallel,\perp}$ only depends on the wavelengths at which the transmission intensity is maximum and the thickness of the sample d,

$$n_{\parallel,\perp} = \frac{\lambda_1\lambda_2}{2d(\lambda_2 - \lambda_1)}. \tag{11.6}$$

If only the birefringence Δn is of interest, this can be determined by measuring the optical retardation δ of a properly aligned sample of known thickness d, since $\delta = 2\pi\Delta nd/\lambda_0$, where λ_0 is the vacuum wavelength of the light (see Section 13.4).

Yet another method to measure the birefringence also utilizes a spectrometer. The liquid crystal is placed between crossed polarizers with the director making a 45° angle with the polarizers. In order for the transmission intensity to be a minimum, the optical retardation must be a multiple of 2π. Again, let's imagine that this occurs at two wavelengths, λ_1 and λ_2 with only one maximum in between and with $\lambda_2 > \lambda_1$. This means that $2\pi m_1 = 2\pi\Delta nd/\lambda_1$ and $2\pi m_2 = 2\pi\Delta nd/\lambda_2$, where m_1 and m_2 are integers and where $m_1 = m_2 + 1$. Assuming that the birefringence is the same at the two wavelengths, one obtains an expression for Δn that only depends on the wavelengths at which the transmission intensity is minimum and the thickness of the sample d,

$$\Delta n = \frac{\lambda_1 \lambda_2}{d(\lambda_2 - \lambda_1)}. \tag{11.7}$$

If Δn varies considerably with wavelength (dispersion), then the birefringence result is approximate.

Most liquid crystals absorb light for some range of wavelengths. Therefore, another optical measurement that is often done is to measure the absorption for light polarized parallel to the director $A_\parallel$ and for light polarized perpendicular to the director $A_\perp$. Here the absorption A is just the logarithm (base 10) of the ratio of the incident intensity to the transmitted intensity. $A_\parallel / A_\perp$ is called the dichroic ratio.

Absorption measurements allow the orientational order parameter S to be determined if the indices of refraction are known.

$$S = \frac{n_\parallel A_\parallel - n_\perp A_\perp}{n_\parallel A_\parallel + 2n_\perp A_\perp} \tag{11.8}$$

11.10 MECHANICAL MEASUREMENTS

One of the most important characteristics of a liquid crystal is how it responds to deformation. As discussed in Section 3.7, the equilibrium state of a liquid crystal when there are no outside forces on it is for the director to be uniform everywhere. If the director varies from one location to another, the liquid crystal possesses additional elastic energy. In general there are three types of deformation (splay, twist, and bend), and the energy associated with each type of deformation is characterized by an elastic constant K_1, K_2, or K_3 for splay, twist, or bend, respectively (see Eq. 3.30). Since the behavior of a liquid crystal depends crucially on the values of the elastic constants, measuring them is a routine and important task.

The most common method used to measure the elastic constants is to make use of the Frederiks transition described at length in Section 12.3. The liquid crystal sample forms a uniform planar or homeotropic director configuration between two glass substrates, and an electric field is applied that deforms the director so it is no longer uniform. As discussed in Section 12.3 and shown in Figure 11.17, there are three common geometries involving the directions of the glass surface normal, the uniform director, and the applied electric field. For each geometry there is a critical value for the electric field for which the uniform director starts to deform. For each geometry, the value of the critical field depends on a single elastic constant, so if the thickness of the sample and the anisotropy of the electric susceptibility are known, measuring the critical electric field allows the elastic constant to be calculated.

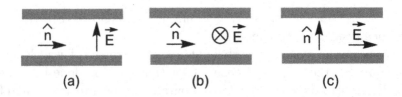

(a)   (b)   (c)

Figure 11.17   Frederiks geometries: (a) splay, (b) twist, and (c) bend. The uniform director before application of an electric field $\vec{E}$ is given by $\hat{n}$. The direction of $\vec{E}$ in (b) is into the page.

While any physical parameter that is sensitive to a change in the uniform director configuration can be utilized to measure the critical electric field, two of the most common methods involve measuring the capacitance or optical retardation of the cell containing the liquid crystal. A magnetic field can be used in place of the electric field, and in this case the anisotropy of the magnetic susceptibility must be known to calculate the elastic constant from the critical value of the magnetic field.

Another important mechanical property of a liquid crystal is its resistance to flow or viscosity. This was discussed in Section 3.1, where it is observed that there are three different viscosities depending on the orientations of the director, the flow, and the direction in which the flow velocity changes (see Fig. 3.1). There are two basic ways for measuring viscosity (defined in Eq. 3.1). The first utilizes flow through a flat capillary tube for which the alignment of the liquid crystal director is achieved by an applied magnetic field. The analysis is complicated because the orientation of the director is affected by the flow, but knowing the pressure gradient, flow rate, and dimensions of the capillary allows the various viscosities to be calculated. The second measures the force on a metal plate as the liquid crystal flows by. Again a magnetic field is used to orient the director properly and the analysis is complicated. But measurement of the force on the plate along with knowledge of the flow rate and dimensions of the plate allow the viscosity to be determined. It should be pointed out that this method has some advantages: the force on the plate can be measured very accurately allowing the flow rate to be small; and, the cross-sectional area of the flowing liquid crystal can be made large enough so there is no influence from the container walls.

11.11 DEVICE TECHNIQUES

Alignment of the liquid crystal in a device is achieved by preparing the glass surfaces containing the liquid crystal in very specific ways. Typically surface layers are placed on the glass to achieve planar or homeotropic alignment. Sometimes a layer of varying thickness is used to align the surface director at a slight angle from planar or perpendicular to the glass surface. Rubbed polymers are typically used to achieve a homogeneous texture and surfactants, hydrophobic polymers, and evaporated SiO are some of the ways to promote a homeotropic texture. Alignment is a crucial factor in how well a liquid crystal device operates, so it is normal to test the alignment of a device by performing some simple measurements. For planar alignment, the intensity of the transmitted light when the two polarizers are parallel to the director divided by the transmitted intensity when the two polarizers are crossed with one parallel to the director and the other perpendicular to the director is a good parameter to use. The better the alignment, the higher the quotient of the two intensities is. Homeotropic alignment can be tested by using two crossed polarizers and rotating the device. The better the homeotropic alignment, the lower the transmitted intensity and the less the transmitted intensity changes as the sample is rotated.

Simple measurements are used to check other aspects of a device. The electrodes are typically thin films of indium-tin oxide. How transparent they are can be determined by measuring the transmitted intensity through an empty cell. The capacitance of the device and the resistance of the electrodes are usually measured with a standard impedance meter. Spacers are used to maintain a constant thickness of the liquid crystal. The variation in thickness is easily probed by shining monochromatic light on the empty device and looking at the interference fringes. Neighboring fringes are located where the thickness changes by half a wavelength. The polarizers and color filters can also be characterized optically. Again, looking at the intensity ratio between parallel and perpendicular polarizers before the device is put together indicates how well the polarizers are working. In a similar way, measuring the absorption of the empty cell as a function of wavelength records exactly what the filters are doing.

Perhaps the most important characteristic of liquid crystal devices is the optical response as a voltage is applied to the device. This is done by monitoring an optical parameter, typically the transmission as a function of applied voltage. A typical response is displayed in Figure 11.18, where two important parameters, the voltage at which the device responds and the voltage range over which the response takes place, can easily be determined.

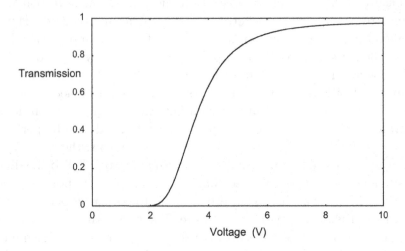

Figure 11.18 The optical transmission of a liquid crystal device as a function of the applied voltage. The applied voltage is usually an alternating voltage with a frequency in the KHz range, so the horizontal axis represents the root-mean-square (rms) voltage value.

Also of paramount importance is how fast the device responds to the application or removal of the applied voltage. Figure 11.19 shows a typical curve following the application of the alternating voltage at time = 0 ms and following the removal of the voltage at time = 10 ms. Notice that both the shape of the response and the time frame of the response are different for the turn-on and turn-off responses. This is because the turn-on response is a reaction to an applied electric field whereas the turn-off response is the natural relaxation of the liquid crystal to its aligned state.

Finally, it may be important to measure other optical characteristics beside the transmission for light propagating normal to the device. For example, a diffuse backlight can be placed behind the device, and the transmission at various angles measured in both the off- and on-

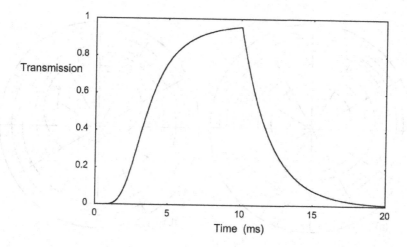

Figure 11.19 The optical transmission of a liquid crystal device when an alternating voltage is applied at time = 0 ms and removed at time = 10 ms. Response times in the range of milliseconds are quite typical.

state. The result of such a measurement might look something like Figure 11.20. The direction of the transmission measurement is given by the radial lines. The angle of the line represents the azimuthal direction and the distance from the center point is a measure of the polar angle. The dark circle on the outside might represent the largest polar angle that can be measured, for example, 85°. The dashed curves are contours of equal transmission, and are labeled by the transmission percentage.

Since almost every characteristic of a liquid crystal depends on temperature (*e.g.*, order parameter, elastic constants), device characterization may require the optical response to be measured as a function of temperature. For such measurements, the device is placed in a temperature-controlled chamber with windows that light can get in and out of. The measurements described above are repeated with the temperature-controlled chamber set to various temperatures.

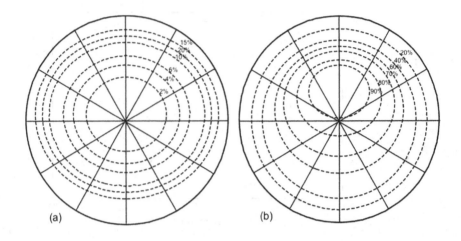

Figure 11.20 The transmission of a liquid crystal device as a function of the viewing angle: (a) off-state, (b) on-state. The radial lines give the azimuthal direction and the distance from the center indicates the polar angle. The thick outside circle represents the maximum polar angle measurable, 85° for example. The dashed lines are contours of equal transmission and are labeled by the percentage transmission.

EXERCISES

11.1 The critical angle for total internal reflection θ_{crit} is given by $\sin\theta_{crit} = n_{lc}/n_p$, where θ_{crit} is defined in Figure 11.10 , n_p is the index of refraction of the prism, and n_{lc} is the index of refraction of the liquid crystal. If the index of the prism is 1.75, what is θ_{crit} for 5CB at room temperature and 546 nm light for the two cases in which the light polarization is parallel and perpendicular to the director. For 5CB under these conditions, $n_\parallel = 1.74$ and $n_\perp = 1.54$.

11.2 What is the approximate tilt angle of the smectic C phase generating the X-ray diffraction pattern shown in Figure 11.12?

11.3 Referring to Figure 11.16, imagine you make a wedge cell and measure the angle γ to be 1.0°. (a) What is the wedge angle α of your cell? You then fill the wedge cell with a nematic liquid crystal, orient it so the director is parallel to the light polarization, and measure the angle β to be 0.3°. (b) Using Snell's law, find $n_\parallel$ of the liquid crystal.

11.4 A nematic liquid crystal is placed between two parallel pieces of glass with the director aligned with the light polarization. The wavelength of the light is varied and two transmission maxima are observed to occur at 472 and 486 nm. (a) If the distance between the two parallel glass pieces is 5.0 μm, what is $n_\parallel$ of the liquid crystal? (b) Estimate for what wavelength $n_\parallel$ is equal to your answer for part (a)?

11.5 The transmission of a nematic liquid crystal is 40% when the light polarization is parallel to the director and 80% when the light polarization is perpendicular to the director. (a) What is the dichroic ratio at this wavelength? (b) Use Eq. 11.8 along with some assumptions to obtain an estimate of the order parameter S of the liquid crystal.

11.6 Two important parameters of a liquid crystal device are V_{10}, the voltage when the transmission reaches 10% of the maximum transmission, and V_{90}, the voltage when the transmission reaches 90% of the maximum transmission. Estimate V_{10} and V_{90} for the device for which data are given in Figure 11.18.

11.7 The rise time for a liquid crystal device is the time it takes for the transmission to rise to $1 - (1/e) = 0.632$ of its maximum transmission after the voltage is turned on (e is the base of natural logarithms). The fall time for a device is the time it takes for the transmission to fall to $1/e = 0.368$ of its transmission just as the voltage is turned off. Estimate the rise and fall times for the device described in Figure 11.19.

CHAPTER 12

Liquid Crystals in Electric and Magnetic Fields – A Delicate Response

12.1 ELECTRIC POLARIZABILITY

The electric polarization in liquid crystals is discussed in Chapter 3 as an example of an anisotropic property. In short, the application of an electric field $\vec{E}$ to a liquid crystal produces a dipole moment per unit volume, called the polarization and given the symbol $\vec{P}$. The polarization depends linearly on the electric field, but the anisotropy of the liquid crystal causes $\vec{P}$ and $\vec{E}$ to have different directions in general. $\vec{P}$ and $\vec{E}$ are thus related by a tensor, $\overleftrightarrow{\chi_e}$, called the electric susceptibility, resulting in the following two identical expressions,

$$\vec{P} = \epsilon_0 \overleftrightarrow{\chi_e} \vec{E} \quad \text{or} \quad \begin{pmatrix} P_x \\ P_y \\ P_z \end{pmatrix} = \epsilon_0 \begin{pmatrix} \chi_\perp & 0 & 0 \\ 0 & \chi_\perp & 0 \\ 0 & 0 & \chi_\parallel \end{pmatrix} \begin{pmatrix} E_x \\ E_y \\ E_z \end{pmatrix}, \qquad (12.1)$$

where the director is oriented along the z-axis and the subscript e has been omitted from χ. The constant ϵ_0 is the permittivity of free space and is equal to 8.85×10^{-12} C^2/Nm^2. The electric field has units of N/C or V/m, so the polarization has units of C/m^2, in accordance with its definition as the dipole moment (Cm) per unit volume (m^3).

The electric field and polarization together define the electric displacement, $\vec{D}$,

$$\vec{D} = \epsilon_0 \vec{E} + \vec{P}, \tag{12.2}$$

which in a material plays the same role as the electric field in a vacuum. It has the same units as the polarization, C/m^2. The linear relation between $\vec{P}$ and $\vec{E}$ results in a linear relationship between $\vec{D}$ and $\vec{E}$,

$$\vec{D} = \overset{\leftrightarrow}{\epsilon} \vec{E}, \quad \text{where} \quad \overset{\leftrightarrow}{\epsilon} = \epsilon_0 (\overset{\leftrightarrow}{1} + \overset{\leftrightarrow}{\chi_e}). \tag{12.3}$$

$\overset{\leftrightarrow}{1}$ is the unit tensor, that is,

$$\overset{\leftrightarrow}{1} = \begin{pmatrix} 1 & 0 & 0 \\ 0 & 1 & 0 \\ 0 & 0 & 1 \end{pmatrix}. \tag{12.4}$$

Operating on a vector with the unit tensor yields the same vector. $\overset{\leftrightarrow}{\epsilon}$ is called the permittivity of the material or, if it is written in units of the permittivity of free space, $\overset{\leftrightarrow}{\epsilon}/\epsilon_0$, is called the relative permittivity or dielectric constant of the material, a quantity that is unitless. Both the permittivity and dielectric constant of liquid crystals are tensor quantities.

Just as in the case of the electric susceptibility, a nematic liquid crystal has a component of the permittivity along the director, $\epsilon_{\parallel}$, and another component perpendicular to the director, $\epsilon_{\perp}$. The anisotropy in the permittivity is just

$$\Delta\epsilon = \epsilon_{\parallel} - \epsilon_{\perp}, \tag{12.5}$$

which can be either positive or negative depending on the permanent dipole moment and polarizability of the molecules. Obviously, the dielectric constant behaves analogously. There are two contributions to the polarization in liquid crystals and therefore two contributions to the permittivity. First, the electric field produces induced dipole moments on the molecules that contribute to the polarization. This is true for solids and liquids, in addition to liquid crystals. Second, the electric field tends to orient permanent dipole moments on the molecules. Orientation in response to an applied electric field is not a strong effect in liquids and solids, but it can be quite large in liquid crystals. For example, if the molecules possess a strong permanent dipole parallel to the long axis of the molecule, as is true for 5CB, the anisotropy of the dielectric constant is very large. The orientational contribution to the dielectric constant dominates, so the anisotropy in the dielectric constant follows the order parameter and increases with decreasing temperature. This is shown in Figure 12.1, where both components of the dielectric constant are shown, and in Figure 12.2, where the anisotropy of the dielectric constant is graphed against temperature. In other cases, the induced polarization and orientational polarization tend to cancel each other out. The result is a much smaller anisotropy and a more complicated temperature dependence. The anisotropy may increase with decreasing temperature just below the nematic-isotropic transition due to the rapid change in the order parameter, and then decrease with decreasing temperature due to the temperature dependence of the induced polarization. An example of this is illustrated in Figure 12.3 for PAA, which has a lateral molecular dipole.

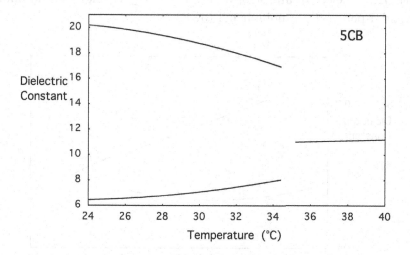

Figure 12.1   Dielectric constants for the nematic and part of the isotropic phases of 4-pentyl-4′-cyanobiphenyl (5CB). In the nematic phase, the upper curve is $\epsilon_{\parallel}$ and the lower curve is $\epsilon_{\perp}$.

One way to understand how liquid crystals respond to electric fields is through energy considerations. Since the energy of a dielectric material depends on $\vec{P}$ and $\vec{E}$, it is useful to express the polarization in terms of $\vec{E}$ and the components of $\overleftrightarrow{\chi}_e$ parallel and perpendicular to the director. If $\hat{n}$ is the director, then the components of the electric field

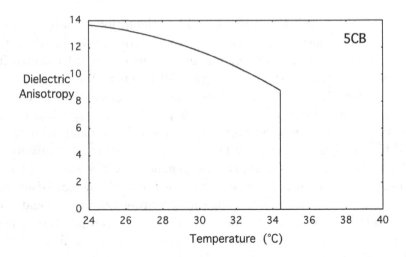

Figure 12.2 Dielectric anisotropy, $\Delta\epsilon = \epsilon_{\parallel} - \epsilon_{\perp}$, of the nematic liquid crystal 4-pentyl-4′-cyanobiphenyl (5CB).

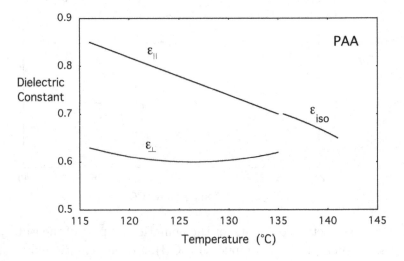

Figure 12.3 Dielectric constants in the nematic and part of the isotropic phase of p-azoxyanisole (PAA).

parallel and perpendicular to the director are

$$E_{\parallel} = \vec{E} \cdot \hat{n} \quad \text{and} \quad E_{\perp} = \left| \vec{E} - (\hat{n} \cdot \vec{E})\hat{n} \right|. \tag{12.6}$$

Each of the components of the electric field parallel and perpendicular to $\hat{n}$ produces a polarization.

$$P_{\|} = \epsilon_0 \chi_{\|} E_{\|} \quad \text{and} \quad P_{\perp} = \epsilon_0 \chi_{\perp} E_{\perp}. \tag{12.7}$$

Substituting in the expressions for the components of the electric field and combining components of $\vec{P}$ parallel to the electric field yield the following expression for the polarization,

$$\vec{P} = \epsilon_0 \left[\chi_{\perp} \vec{E} + \Delta \chi_e (\hat{n} \cdot \vec{E}) \hat{n} \right], \tag{12.8}$$

where $\Delta \chi_e = \chi_{\|} - \chi_{\perp}$.

In many cases of interest, the electric field exists between two parallel conducting plates to which a voltage is applied. The sample of liquid crystal is located between the two parallel conducting plates, and the task is to find the electric energy per unit volume of the liquid crystal. In general, the electric energy per unit volume is given by

$$U_e = \frac{1}{2} \vec{D} \cdot \vec{E}. \tag{12.9}$$

A small change in the director of the liquid crystal $d\hat{n}$ causes a small change in the polarization $d\vec{P}$. If the voltage across the conducting plates remains constant, as is usually the case because they are connected to a power supply, then the electric field also remains constant. This means that $d\vec{D} = d\vec{P}$, so the resulting small change in the electric energy per unit volume can be written

$$dU_e(1) = \frac{1}{2} \vec{E} \cdot d\vec{P}. \tag{12.10}$$

However, the small change in the polarization causes a small amount of additional charge dq to accumulate on the two conducting plates in order to keep the voltage constant. This decrease in the electric energy of the power supply is equal to $-Vdq$, where V is the power supply voltage (and the voltage across the conducting plates). This represents a decrease in electric energy per unit volume of liquid crystal equal to

$$dU_e(2) = -\frac{Vdq}{Ad} = -Ed\sigma, \tag{12.11}$$

where A is the area of the conducting plates, d is the distance between them, and $d\sigma$ is the small increase in charge per unit area on the conducting plates. The small increase in the charge per unit area produces an equal increase in the electric displacement perpendicular to the conducting plates (parallel to $\vec{E}$). This means that

$$d\sigma = \frac{\vec{E}}{E} \cdot d\vec{D} = \frac{\vec{E}}{E} \cdot d\vec{P}, \tag{12.12}$$

so

$$dU_e(2) = -\vec{E} \cdot d\vec{P}. \tag{12.13}$$

The final result is that the total change in electric energy per unit volume of the liquid crystal is the sum of these two contributions

$$dU_e = dU_e(1) + dU_e(2) = -\frac{1}{2}\vec{E} \cdot d\vec{P}. \tag{12.14}$$

But how is this small change in the polarization related to the small change in the director? Since

$$\vec{P} = \epsilon_0 \left[\chi_\perp \vec{E} + \Delta\chi_e(\hat{n} \cdot \vec{E})\hat{n} \right], \tag{12.15}$$

and $\vec{E}$ is constant, then

$$d\vec{P} = \epsilon_0 \Delta\chi_e(\vec{E} \cdot d\hat{n})\hat{n} + \epsilon_0 \Delta\chi_e(\vec{E} \cdot \hat{n})d\hat{n}. \tag{12.16}$$

This means that the change in electric energy per unit volume is given by

$$\begin{aligned} dU_e &= -\frac{1}{2}\epsilon_0 \Delta\chi_e(\vec{E} \cdot d\hat{n})(\vec{E} \cdot \hat{n}) - \frac{1}{2}\epsilon_0 \Delta\chi_e(\vec{E} \cdot \hat{n})(\vec{E} \cdot d\hat{n}) \\ &= -\epsilon_0 \Delta\chi_e(\vec{E} \cdot \hat{n})(\vec{E} \cdot d\hat{n}). \end{aligned} \tag{12.17}$$

If the director is described by the angle θ between it and a plane parallel to the conducting plates (and perpendicular to $\vec{E}$), then

$$\vec{E} \cdot \hat{n} = E \sin\theta \quad \text{and} \quad \vec{E} \cdot d\hat{n} = E \cos\theta d\theta, \tag{12.18}$$

so

$$dU_e = -\epsilon_0 \Delta\chi_e E^2 \sin\theta \cos\theta d\theta. \tag{12.19}$$

If we set $U_e = 0$ at $\theta = 0$, then the electric energy per unit volume of liquid crystal when the director is at an angle θ is given by

$$\begin{aligned} U_e &= -\epsilon_0 \Delta\chi_e E^2 \int_0^\theta \sin\theta' \cos\theta' d\theta' = -\frac{1}{2}\epsilon_0 \Delta\chi_e E^2 \sin^2\theta \\ &= -\frac{1}{2}\epsilon_0 \Delta\chi_e(\hat{n} \cdot \vec{E})^2. \end{aligned} \tag{12.20}$$

12.2 MAGNETIZATION

Understanding the behavior of liquid crystals in a magnetic field is not difficult since it is analogous to what happens in an electric field. Just as an applied electric field causes a polarization or electric dipole moment per unit volume in a liquid crystal, an applied magnetic field $\vec{H}$ produces a magnetization or magnetic dipole moment per unit volume $\vec{M}$. The magnetization is just due to the weak magnetic dipole moments induced on the molecules by the magnetic field. As expected, the strength of the magnetic dipole moments induced on the molecules depends on whether the molecules are oriented with their long axes parallel or perpendicular to the field. Hence there are two coefficients relating $\vec{M}$ and $\vec{H}$.

$$M_{\parallel} = \chi_{\parallel} H_{\parallel} \quad \text{and} \quad M_{\perp} = \chi_{\perp} H_{\perp} \tag{12.21}$$

In analogy with the electric case, $\overset{\leftrightarrow}{\chi}_m$ (or $\overset{\leftrightarrow}{\chi}$ for short) is the magnetic susceptibility tensor. Its components parallel and perpendicular to the director are denoted by $\chi_{\parallel}$ and $\chi_{\perp}$, respectively. $\Delta\chi_m$ is just $\chi_{\parallel} - \chi_{\perp}$. Both $\vec{M}$ and $\vec{H}$ have units of A/m or C/(ms), so $\overset{\leftrightarrow}{\chi}$ is unitless. As is evident from Figure 12.4, values for the magnetic susceptibility of liquid crystals are usually negative (diamagnetic) and on the order of 10^{-6}.

Since there is little magnetic interaction between the molecules, the magnetization is nearly the sum of the molecular magnetic dipole moments. This makes the magnetic anisotropy proportional to the order parameter, as can be seen by the curve in Figure 12.5, which resembles the temperature dependence of S in Figure 1.2.

A second magnetic field, technically called the magnetic induction $\vec{B}$, is usually defined by combining $\vec{M}$ and $\vec{H}$,

$$\vec{B} = \mu_0(\vec{H} + \vec{M}) = \mu_0(\overset{\leftrightarrow}{1} + \overset{\leftrightarrow}{\chi}_m)\vec{H}, \tag{12.22}$$

where μ_0 is a constant called the permeability of free space and equals $4\pi \times 10^{-7}$ N/A^2. $\vec{B}$ therefore has units of N/(mA) or Ns/(mC), which have been given the name tesla. The quantity

$$\overset{\leftrightarrow}{\mu} = \mu_0(\overset{\leftrightarrow}{1} + \overset{\leftrightarrow}{\chi}_m) \tag{12.23}$$

is called the permeability tensor of the material and it has the same units as μ_0. Sometimes a relative permeability is used, μ/μ_0, which is analogous to the dielectric constant and has no units.

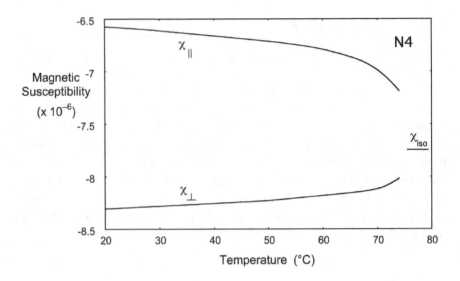

Figure 12.4 Magnetic susceptibilities in the nematic and isotropic liquid phases.

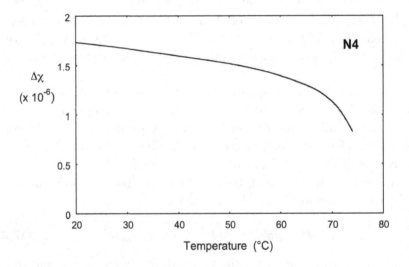

Figure 12.5 Magnetic anisotropy in the nematic phase.

The magnetic energy per unit volume of liquid crystal can be found by exactly the same procedure used to find the electric energy per unit volume. The starting point is the general expression for the magnetic energy per unit volume

$$U_m = \frac{1}{2}\vec{B} \cdot \vec{H},$$ (12.24)

and the result is

$$U_m = -\frac{1}{2}\mu_0 \Delta\chi_m (\hat{n} \cdot \vec{H})^2.$$ (12.25)

12.3 FREDERIKS TRANSITION

To understand how liquid crystals respond to external fields, it must be realized that in most cases interactions between the liquid crystal molecules and boundaries have a large effect. These boundaries can be with a solid material such as glass, but the air-liquid crystal boundary is also both important and interesting. In many cases, the influence of the boundary opposes the response to the electric field, and the result is a threshold phenomenon called the Frederiks transition.

Let us begin with the most simple geometry possible. Imagine that the liquid crystal is contained between two flat pieces of glass, separated by a distance d. A surfactant has been applied to the surfaces of the glass in contact with the liquid crystal, so the director tends to align itself in a single direction parallel to the flat surfaces of the glass. Let this be the x-axis. An electric field is applied perpendicular to the x-axis and parallel to the flat surfaces of the glass. Let this be the y-axis. If the anisotropy of the dielectric susceptibility is positive, then the director tends to align along the electric field, rotating away from the x-axis toward the y-axis. Let us call the angle between the director and the x-axis θ. If we consider the dimensions of the flat pieces of glass to be much larger than the separation, then θ should not be a function of x or y, but should depend on z (an axis normal to the surfaces of the glass). This geometry is illustrated in Figure 12.6.

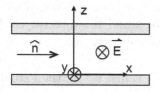 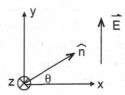

Figure 12.6 "Twist geometry".

If we assume that the director is constrained to point along the x-axis at the surfaces of the glass, then $\theta = 0$ there. θ increases along the z-axis, reaches a maximum at the midpoint ($z = d/2$), and then decreases until it reaches zero again at $z = d$. Since θ varies along the z-axis, the liquid crystal is deformed as discussed in Chapter 3. In general, the energy associated with distortion involves splay, twist, and bend; the terms are easily calculated for this case since

$$n_x = \cos\left[\theta(z)\right] \qquad \text{and} \qquad n_y = \sin\left[\theta(z)\right]. \tag{12.26}$$

The important derivatives are

$$\nabla \cdot \hat{n} = 0$$

$$\hat{n} \cdot (\nabla \times \hat{n}) = -n_x \left(\frac{\partial n_y}{\partial z}\right)_{x,y} + n_y \left(\frac{\partial n_x}{\partial z}\right)_{x,y} = -\frac{d\theta}{dz}$$

$$[\hat{n} \times (\nabla \times \hat{n})]_z = n_x \left(\frac{\partial n_x}{\partial z}\right)_{x,y} + n_y \left(\frac{\partial n_y}{\partial z}\right)_{x,y} = 0. \tag{12.27}$$

As expected, the deformation is pure twist.

The total energy per unit volume is the sum of the distortion and electric field energies.

$$U = \frac{1}{2}K_2 \left(\frac{d\theta}{dz}\right)^2 - \frac{1}{2}\epsilon_0 \Delta\chi_e E^2 \sin^2\theta. \tag{12.28}$$

The actual director configuration is the one that minimizes this energy integrated over the volume of the liquid crystal. Since θ is only a function of z, we need to find the function $\theta(z)$ that minimizes U integrated over z from $z = 0$ to $z = d$. Thus we want to minimize the following free energy per unit area.

$$F_A = \int_0^d \left[\frac{1}{2}K_2 \left(\frac{d\theta}{dz}\right)^2 - \frac{1}{2}\epsilon_0 \Delta\chi_e E^2 \sin^2\theta\right] dz \tag{12.29}$$

This is a problem that the calculus of variations can handle in much the same way as in Chapter 3 when we found the director configuration around a disclination. Since this is a slightly different case, let us quickly review the procedure here.

If the function that minimizes F_A is called $\theta_0(z)$, then small variations from $\theta_0(z)$ cannot change F_A to first-order in the parameter causing the change. To express this in an analytical way, a new function $\theta(z)$ is formed by adding a small amount of another function, $\eta(z)$ to it,

$$\theta(z) = \theta_0(z) + \alpha\eta(z), \tag{12.30}$$

where $\alpha \ll 1$. The change in F_A due to changing $\theta_0(z)$ to $\theta(z)$ cannot depend linearly on α. Since θ must equal zero at $z = 0$ and $z = d$, this means that $\eta(0) = 0$ and $\eta(d) = 0$.

The free energy per unit area for the new path is a function of α. The condition that changes in F_A do not depend linearly on changes in α can be expressed quite simply as

$$\left(\frac{\partial F_A}{\partial \alpha}\right)_{\alpha=0} =$$
$$\int_0^d \left[\left(\frac{\partial U}{\partial \theta}\right)\left(\frac{\partial \theta}{\partial \alpha}\right) + \left(\frac{\partial U}{\partial \left(\frac{d\theta}{dz}\right)}\right)\left(\frac{\partial \left(\frac{d\theta}{dz}\right)}{\partial \alpha}\right)\right] dz = 0. \quad (12.31)$$

But

$$\left(\frac{\partial \theta}{\partial \alpha}\right) = \eta(z) \quad \text{and} \quad \left(\frac{\partial \left(\frac{d\theta}{dz}\right)}{\partial \alpha}\right) = \left(\frac{d\,[\eta(z)]}{dz}\right), \quad (12.32)$$

so

$$\left(\frac{\partial F_A}{\partial \alpha}\right)_{\alpha=0} =$$
$$\int_0^d \left[\left(\frac{\partial U}{\partial \theta}\right)\eta(z) + \left(\frac{\partial U}{\partial \left(\frac{d\theta}{dz}\right)}\right)\left(\frac{d\,[\eta(z)]}{dz}\right)\right] dz = 0. \quad (12.33)$$

The second term can be integrated by parts. When this is done, the integrated part contains $\eta(z)$, thus vanishing when evaluated at $z = 0$ and $z = d$ where $\eta(z) = 0$. The integral can therefore be written as

$$\left(\frac{\partial F_A}{\partial \alpha}\right)_{\alpha=0} = \int_0^d \left[\left(\frac{\partial U}{\partial \theta}\right) - \frac{d}{dz}\left(\frac{\partial U}{\partial \left(\frac{d\theta}{dz}\right)}\right)\right] \eta(z)\,dz = 0. \quad (12.34)$$

Since $\eta(z)$ is an arbitrary function, the only way the integral can always equal zero is if the expression in brackets equals zero. Thus $\theta(z)$ must satisfy the differential equation

$$\left(\frac{\partial U}{\partial \theta}\right) - \frac{d}{dz}\left(\frac{\partial U}{\partial \left(\frac{d\theta}{dz}\right)}\right) = 0, \quad (12.35)$$

which is called the Euler equation.

Substituting the expression for U into this equation yields

$$K_2\left(\frac{d^2\theta}{dz^2}\right) + \epsilon_0 \Delta\chi_e E^2 \sin\theta \cos\theta = 0. \quad (12.36)$$

If this expression is written using dimensionless quantities, then it only needs to be solved once in order to cover all cases. Let us define a new variable ζ, which is just equal to z/d. ζ then goes from 0 to 1 across the liquid crystal sample. This change gives the following equation.

$$\frac{K_2}{d^2}\left(\frac{d^2\theta}{d\zeta^2}\right) + \epsilon_0 \Delta\chi_e E^2 \sin\theta\cos\theta = 0 \tag{12.37}$$

An important length in this situation ξ can be defined through the relation

$$\xi^2 = \frac{K_2}{\epsilon_0 \Delta\chi_e E^2}; \tag{12.38}$$

this length can also be made dimensionless by defining

$$\xi_d = \frac{\xi}{d}. \tag{12.39}$$

With these substitutions, the equation to be solved takes the form

$$\xi_d^2\left(\frac{d^2\theta}{d\zeta^2}\right) + \sin\theta\cos\theta = 0. \tag{12.40}$$

Unfortunately, the solution to this equation, $\zeta(z)$, cannot be expressed in terms of simple functions. It is not difficult to find a numerical solution, however, by using the appropriate value of ξ_d and then guessing a value of $d\theta/d\zeta$ at $\zeta = 0$. Recall that θ equals zero at $\zeta = 0$ and $\zeta = 1$. With this guess, the equation can be integrated to give $\theta(\zeta)$, except that in general it will not return to 0 at $\zeta = 1$. By seeing whether the value of $d\theta/d\zeta$ used is too high or too low, the value that causes $\theta(1)$ to be 0 can be quickly found. This procedure can be repeated for other values of ξ_d (*i.e.*, other values of the electric field). Alternatively, computational software that numerically solves differential equations can be used to solve Eq. (12.40). But such software might need a value of $d\theta/d\zeta$ at $\zeta = 0$ supplied in order to find the solution, especially as the strength of the electric field increases (and ξ_d decreases).

In doing this numerical procedure, it is quickly realized that there is no solution other than $\theta(\zeta) = 0$ everywhere if ξ_d is greater than $1/\pi = 0.3183$. Since ξ_d is inversely proportional to the electric field, this means that below some threshold electric field strength, no distortion takes place. This is the Frederiks transition. Below the threshold electric field strength, the director remains undistorted and aligned parallel to the x-axis. Above the threshold field, the director starts to rotate from its undistorted configuration toward the direction of the electric field. Thus the Frederiks transition is not a phase transition, but a transition from an undistorted to a distorted director configuration.

It is also possible to find the solution to the equation in terms of elliptic integrals. The differential equation must be multiplied by $d\theta/d\zeta$, making each side a perfect differential and thus easily integrated. If the integration constant is written in terms of the value of θ at $\zeta = 1/2$ (let's call it θ_m), the result is, noting that $d\theta/d\zeta = 0$ at $\zeta = 1/2$,

$$\frac{d\theta}{d\zeta} = \frac{1}{\xi_d}\sqrt{\sin^2\theta_m - \sin^2\theta}. \tag{12.41}$$

This equation can be expressed in terms of an elliptic integral if it is integrated from $\zeta = 0$ to $\zeta = 1/2$ with the following substitutions being made

$$m = \sin^2\theta_m \qquad \text{and} \qquad t = \frac{\sin\theta}{\sin\theta_m}. \tag{12.42}$$

The result is

$$\int_0^1 \frac{dt}{\sqrt{1-t^2}\sqrt{1-mt^2}} \equiv K(m) = \frac{1}{2\xi_d}, \tag{12.43}$$

where $K(m)$ is the complete elliptic integral of the first kind.

In order to calculate the threshold value of the electric field, one need only notice that $K(m)$ equals $\pi/2$ at $m = 0$ and increases monotonically to infinity as m approaches 1. This means that $\xi_d = 1/\pi$ at threshold, and that the threshold field E_t is

$$E_t = \frac{\pi}{d}\sqrt{\frac{K_2}{\epsilon_0\Delta\chi_e}}. \tag{12.44}$$

To obtain a graph of θ_m as a function of E/E_t, pick a value for θ_m, then calculate $m = \sin^2\theta_m$, and finally use a mathematical table or computational software to find

$$\frac{E}{E_t} = \frac{1}{\pi\xi_d} = \frac{2}{\pi}K(m). \tag{12.45}$$

A plot of θ_m as a function of E/E_t is shown in Figure 12.7, where the threshold effect is clearly evident. Thus measuring E_t is an excellent method for determining K_2 if $\Delta\chi_e$ is known. Notice that the threshold voltage, $E_d d$, only depends on the liquid crystal. Thus a sample with

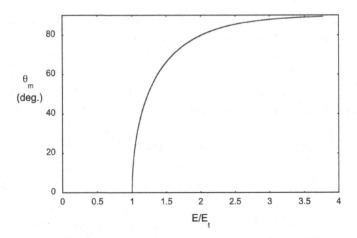

Figure 12.7 Angle of the director at the midpoint as a function of the electric field.

$K_2 = 10^{-11}$ N and $\Delta\chi_e = 11$ requires about 1 volt for distortion to occur.

It is also interesting to see how θ varies as a function of ζ for a particular value of E/E_t. To do this calculation, perform the last integration from $\zeta = 0$ to some unspecified value instead of from 0 to $1/2$ (with θ also going from 0 to some unspecified value). Writing $1/\xi_d$ as $2K(m)$ yields the following expression

$$\int_0^{\sin\phi} \frac{dt}{\sqrt{1-t^2}\sqrt{1-mt^2}} \equiv 2K(m)\zeta, \qquad (12.46)$$

where $\sin\phi = \sin\theta/\sin\theta_m$. The integral is called the elliptic integral of the first kind, $F(\phi|m)$, and the equation can be written

$$\zeta = \frac{F(\phi|m)}{2K(m)}. \qquad (12.47)$$

A simple way to obtain a graph of θ as a function of ζ is first to find $K(m)$ given a value of θ_m (as described previously), and then pick values of θ less than θ_m, in each case calculating

$$\phi = \sin^{-1}\left(\frac{\sin\theta}{\sin\theta_m}\right). \qquad (12.48)$$

At this point a mathematical table or computational software can be used to evaluate $F(\phi|m)$ and thus find the value of ζ corresponding to the chosen value of θ. Figure 12.8 shows how θ varies throughout the liquid crystal for several values of the electric field.

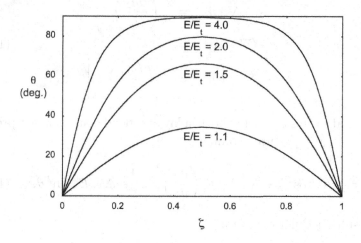

Figure 12.8 Director orientation across the cell for several electric fields.

There are two other geometries besides the "twist" geometry in which a Frederiks transition takes place. Figure 12.9 shows the "splay" geometry, where the boundary conditions again favor a director oriented along the x-axis, but now the electric field is applied in the z direction.

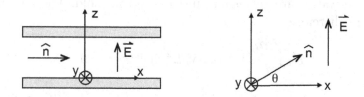

Figure 12.9 "Splay" geometry.

The director now has x and z components and $\theta(z)$ is measured from the x-axis to the director in the xz plane. Calculation of the distortion free energy proceeds as before.

$$n_x = \cos[\theta(z)] \qquad \text{and} \qquad n_z = \sin[\theta(z)]$$

$$\nabla \cdot \hat{n} = \left(\frac{\partial n_z}{\partial z}\right)_{x,y}$$

$$\hat{n} \cdot (\nabla \times \hat{n}) = 0 \qquad\qquad (12.49)$$

$$|\hat{n} \times (\nabla \times \hat{n})| = \left|\left(\frac{\partial n_x}{\partial z}\right)_{x,y}\right|$$

Notice that in this geometry both splay and bend distortion are present. The free energy per unit volume is

$$U = \frac{1}{2}[K_1 \cos^2\theta + K_3 \sin^2\theta]\left(\frac{d\theta}{dz}\right)^2 - \frac{1}{2}\epsilon_0 \Delta\chi_e E^2 \sin^2\theta. \qquad (12.50)$$

The resulting Euler equation is

$$\left[K_1 \cos^2\theta + K_3 \sin^2\theta\right]\left(\frac{d^2\theta}{dz^2}\right)$$
$$+ \left[(K_3 - K_1)\left(\frac{d\theta}{dz}\right)^2 + \epsilon_0 \Delta\chi_e E^2\right]\sin\theta\cos\theta = 0. \quad (12.51)$$

While the solution to this equation is more complicated than in the twist geometry, the threshold field can be found by realizing that $\theta(z)$ and its derivative are very small just above threshold. The important terms are therefore

$$K_1\left(\frac{d^2\theta}{dz^2}\right) + \epsilon_0 \Delta\chi_e E^2 \sin\theta\cos\theta = 0, \qquad (12.52)$$

which with the same substitution as before (but using K_1 instead of K_2), yields

$$\xi_d^2\left(\frac{d^2\theta}{d\zeta^2}\right) + \sin\theta\cos\theta = 0. \qquad (12.53)$$

and a threshold field of

$$E_t = \frac{\pi}{d}\sqrt{\frac{K_1}{\epsilon_0 \Delta\chi_e}}. \qquad (12.54)$$

Thus by this geometry the splay elastic constant K_1 can be measured.

The last geometry also involves both splay and bend. As shown in Figure 12.10, the boundary conditions are such that the undistorted director points along the z-axis and the electric field is applied along the x-axis.

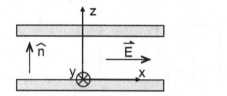
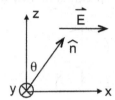

Figure 12.10 "Bend" geometry.

The angle $\theta(z)$ now is measured from the z-axis to the director in the xz plane. The components of the director are

$$n_x = \sin[\theta(z)] \quad \text{and} \quad n_z = \cos[\theta(z)], \qquad (12.55)$$

with the same non-zero derivatives of $\hat{n}$ as in the "splay" geometry (again involving both splay and bend distortion). This time the free energy per unit volume is

$$U = \frac{1}{2}[K_1 \sin^2\theta + K_3 \cos^2\theta]\left(\frac{d\theta}{dz}\right)^2 - \frac{1}{2}\epsilon_0\Delta\chi_e E^2 \sin^2\theta, \qquad (12.56)$$

which looks just like the free energy per unit volume in the "splay" geometry except that the parameters K_1 and K_3 have been switched. With this realization, the threshold field can be written down immediately as

$$E_t = \frac{\pi}{d}\sqrt{\frac{K_3}{\epsilon_0\Delta\chi_e}}, \qquad (12.57)$$

revealing why this is called the "bend" geometry.

12.4   HELIX UNWINDING TRANSITION

An interesting phenomenon takes place when a chiral nematic liquid crystal with positive susceptibility anisotropy (electric or magnetic) finds itself in a field (electric or magnetic) perpendicular to the helical axis. The undistorted helical structure possesses regions with directors making all angles with the field. Obviously, application of the field promotes alignment of the director parallel to the field, and the result is a distorted helical structure. As shown in Figure 12.11, the part of the helix where the director is in the general direction of the field is lengthened, while the other part of the helix where the director is more or less perpendicular to the field is shortened.

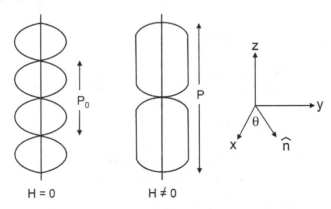

Figure 12.11 Distortion of the helical structure of a chiral nematic liquid crystal by a magnetic field. The magnetic field H is along the y-axis.

The distance between identical parts of the helix (*i.e.*, half the pitch) lengthens as this distortion takes place. As the field increases, the distortion increases until the helix is mostly composed of nematic-like regions oriented along the field with short "walls" in between where the director rotates by 180°. At some critical value of the field, the "walls" become separated by an infinite distance, meaning that the helix has been completely unwound and the chiral nematic phase has been transformed to a nematic phase.

It is instructive to work out the energetics of this unwinding transition. Although it involves the same concepts as were used in understanding the Frederiks transition, the mathematical complexity is higher. The end result is remarkably simple and worth the trouble. Since the theory requires that the liquid crystal be unaffected by boundaries, it applies to the case where the sample is fairly large. Since it is easier to generate a uniform magnetic field over a large area (as opposed to an electric field), experimental work usually employs magnetic fields. It is therefore appropriate to consider the magnetic field case in our discussion, in addition to its being a change of pace from our previous use of the electric field.

Consider a chiral nematic liquid crystal with its helical axis oriented in the z direction. The director lies in the xy plane; let θ be the angle from the x-axis to the director. The magnetic field is applied along the y-axis. This situation is shown in Figure 12.11. Let the pitch and the chirality of the undistorted chiral nematic be P_0 and q_0, respectively. Remember that $q_0 = 2\pi/P_0$. From Chapter 3, it is also known that the pitch and chirality depend on the two elastic constants k_2 and K_2.

$$P_0 = \frac{2\pi K_2}{k_2} \quad \text{and} \quad q_0 = \frac{k_2}{K_2} \tag{12.58}$$

The situation is the same as in the "twist" geometry considered earlier, in that the director has only x and y components, the field is in the y direction, and only twist deformation is present. The only differences are (1) there is an extra term in the elastic energy per unit volume due to the intrinsic chirality of the liquid crystal, and (2) the field energy per unit volume is due to a magnetic field rather than an electric field. With these two changes, the energy per unit volume is

$$U = -k_2\left(\frac{d\theta}{dz}\right) + \frac{1}{2}K_2\left(\frac{d\theta}{dz}\right)^2 - \frac{1}{2}\mu_0\Delta\chi_m H^2 \sin^2\theta. \tag{12.59}$$

It is convenient to combine the elastic energy terms as follows,

$$U_1 = U + \frac{1}{2}K_2 q_0^2 = \frac{1}{2}K_2\left(\frac{d\theta}{dz} - q_0\right)^2 - \frac{1}{2}\mu_0\Delta\chi_m H^2 \sin^2\theta, \tag{12.60}$$

where U_1 is a new energy per unit volume differing from the old one by a constant. The free energy per unit area for a slice of the sample perpendicular to the helical axis, F_A, is just the integral of U_1 along the portion of the z-axis occupied by the slice. The Euler equation resulting from the minimization of F_A is identical to the one obtained in the case of the "twist" geometry.

$$\xi^2\left(\frac{d^2\theta}{dz^2}\right) + \sin\theta\cos\theta = 0 \quad \text{where} \quad \xi^2 = \frac{K_2}{\mu_0\Delta\chi_m H^2} \tag{12.61}$$

If the Euler equation is multiplied by $d\theta/dz$ and integrated, one obtains

$$\frac{1}{2}\xi^2\left(\frac{d\theta}{dz}\right)^2 + \frac{1}{2}\sin^2\theta = \frac{1}{2\kappa^2}, \tag{12.62}$$

where we have used $1/(2\kappa^2)$ as the integration constant. This equation can be solved for $d\theta/dz$

$$\left(\frac{d\theta}{dz}\right) = \frac{1}{\xi\kappa}\sqrt{1 - \kappa^2 \sin^2\theta} \tag{12.63}$$

and integrated as follows,

$$\xi\kappa \int_0^\theta \frac{d\theta'}{\sqrt{1 - \kappa^2 \sin^2 \theta}} = \int_0^z dz' \tag{12.64}$$

or

$$\xi\kappa F(\sin^{-1}\theta|m) = z, \tag{12.65}$$

where $m = \kappa^2$ and $F(\sin^{-1}\theta|m)$ is the elliptic integral of the first kind.

Even in the distorted helix, a full $360°$ revolution of the director can be divided into four equal lengths of $90°$ rotation each. This means that the pitch in the distorted structure, P, is four times the distance for θ to go from 0 to $\pi/2$,

$$\frac{P}{4} = \xi\kappa F(\frac{\pi}{2}|m) = \xi\kappa K(m), \tag{12.66}$$

where $K(m)$ is the complete elliptic integral of the first kind. Our only problem now is that κ is unknown. Its value is the one that minimizes the average free energy for one full pitch of the distorted structure. Our task then is to return to the free energy and write it in a form that lets us find the value of κ that minimizes it.

The average free energy per unit volume is just the reciprocal of the pitch times F_A integrated over one pitch length.

$$\langle F_V \rangle = \frac{\mu_0 \Delta\chi_m H^2}{2P} \int_0^P \left[\xi^2 \left(\frac{d\theta}{dz} - q_0 \right)^2 - \sin^2\theta \right] dz \tag{12.67}$$

Changing the integration variable from z to θ and using the expression for $d\theta/dz$ obtained before, the average free energy per unit volume becomes

$$\langle F_V \rangle = \frac{2\mu_0 \Delta\chi_m H^2}{P} \times$$

$$\int_0^{\pi/2} \left[\frac{\xi^2 \left(\frac{1}{\xi\kappa}\sqrt{1 - \kappa^2 \sin^2\theta} - q_0 \right)^2 - \sin^2\theta}{\frac{1}{\xi\kappa}\sqrt{1 - \kappa^2 \sin^2\theta}} \right] d\theta. \tag{12.68}$$

This expression can be manipulated so it can be written in terms of complete elliptic integrals. The algebra is tedious but straightforward, and the expression we obtain for P must be substituted. The result is

$$\langle F_V \rangle = \frac{\mu_0 \Delta\chi_m H^2}{2} \xi^2 q_0^2 \times$$

$$\left[1 - \frac{\pi}{\xi q_0 \kappa K(m)} - \frac{1}{\xi^2 \kappa^2 q_0^2} \left(1 - 2\frac{E(m)}{K(m)} \right) \right], \tag{12.69}$$

where $m = \kappa^2$ and $E(m)$ is the complete elliptic integral of the second kind.

$$E(m) = \int_0^{\pi/2} \sqrt{1 - m\sin^2\theta}\,d\theta = \int_0^1 \frac{\sqrt{1 - mt^2}}{\sqrt{1 - t^2}}\,dt \qquad (12.70)$$

We now use the standard technique of finding the minimum of a function by taking the derivative of the average free energy per unit volume with respect to κ and setting it equal to zero. This is made easier by realizing that

$$\frac{d}{d\kappa}\left(\frac{E(m)}{\kappa}\right) = -\frac{K(m)}{\kappa^2}. \qquad (12.71)$$

This last relation is not difficult to show; switching the order of differentiation and integration is the key step. One last hint in finding the minimum is to avoid taking the derivative

$$\frac{d}{d\kappa}\left(\frac{1}{\kappa K(m)}\right). \qquad (12.72)$$

This turns out to be a factor in all of the remaining terms of the derivative and thus is not important in finding the value of κ that makes the derivative of $\langle F_V \rangle$ equal to zero. The final result is amazingly simple. The value of κ that minimizes the average free energy per unit volume obeys the following equation

$$\frac{E(m)}{\kappa} = \frac{\pi}{2}\xi q_0. \qquad (12.73)$$

Substituting this in the expression for the pitch, and realizing that $P_0 = 2\pi/q_0$, yields an equation for P in terms of P_0

$$\frac{P}{P_0} = \left(\frac{2}{\pi}\right)^2 E(m)K(m). \qquad (12.74)$$

The zero field case is simply $\kappa^2 = m = 0$, where $E(0)$ and $K(0)$ both equal $\pi/2$. As κ and m increase, $E(m)$ decreases to 1 at $\kappa = m = 1$ and $K(m)$ increases, diverging to infinity as κ and m approach 1. Thus the pitch becomes infinite when $\kappa = 1$, implying a critical value for the length parameter ξ_c.

$$\frac{E(1)}{1} = \frac{\pi}{2}\xi_c q_0 \qquad \text{or} \qquad \xi_c = \frac{2}{\pi}\frac{1}{q_0}. \qquad (12.75)$$

This in turn gives us an expression for the magnetic field strength necessary to unwind the chiral nematic helix.

$$H_c = \frac{\pi}{2}\sqrt{\frac{K_2}{\mu_0 \Delta \chi_m}} q_0 = \frac{\pi^2}{P_0}\sqrt{\frac{K_2}{\mu_0 \Delta \chi_m}} \tag{12.76}$$

Again, an easy way to generate a curve of P versus field is to pick a value for κ (and therefore m) and then use tables or computer software to evaluate $E(m)$ and $K(m)$. The magnetic field and pitch corresponding to this κ value are just

$$\frac{H}{H_c} = \frac{\kappa}{E(m)} \quad \text{and} \quad \frac{P}{P_0} = \left(\frac{2}{\pi}\right)^2 E(m)K(m). \tag{12.77}$$

A graph of the pitch versus magnetic field is contained in Figure 12.12. Notice how little change in the pitch takes place until the magnetic field starts to approach the critical value.

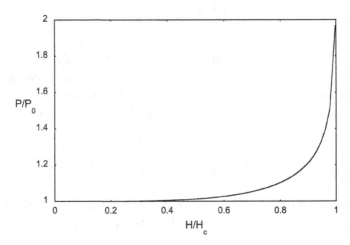

Figure 12.12 Dependence of the pitch on magnetic field.

12.5 BOUNDARIES

The effect of solid boundaries on liquid crystals is very important and is frequently utilized in applications. For the most part, two types of solid boundaries are used. First, a glass substrate is coated with a thin film of a solid polymer and stretched in one direction by rubbing with cloth or paper. This creates anisotropy in the polymer film, which is transferred to the liquid crystal molecules. As a result, the director of the liquid crystal in contact with the polymer film orients parallel to the substrate and in the direction of rubbing in what is referred to as homogeneous or planar boundary conditions. Second, a thin film of amphiphilic molecules is applied to the glass substrate. The polar part of these molecules strongly associates with the glass and the non-polar part extends into the liquid crystal material. This tends to cause the director to orient perpendicular to the glass substrate, a situation called homeotropic boundary conditions.

Theoretically, the effect of such boundaries is described in two general ways. As in this chapter, the boundaries can simply impose a constraint on the director at the boundary. The term for this is "strong anchoring" and it is by far the easiest way to do calculations. However, there is plenty of evidence that this is far from the real situation in many cases, so several methods to describe "weak anchoring" have been developed. One example is to assign a surface elastic energy to the boundary, with the size of this energy determined by the angle θ between the director at the surface and the alignment direction of the substrate. An example is the Rapini function, by which the free energy per unit area of the boundary is just $(1/2)W\sin^2\theta$, where W is called the anchoring strength. This boundary energy term is considered with all the other energy terms when minimizing the energy to find the equilibrium director configuration. Depending on the surface treatment and the liquid crystal, values of W range from 10^{-7} N/m (weak anchoring) to 10^{-3} N/m (strong anchoring).

Let us consider the boundary region of a liquid crystal a little more carefully. Imagine the situation in which the director is constrained to point in the x-direction at a surface in the xy plane at $z = 0$ and a magnetic field is applied in the y direction. This is just the "twist" geometry we discussed before (Figure 12.6). If the liquid crystal sample is extremely thick, however, $\theta = 0$ at the boundary ($z = 0$) and $\theta = \pi/2$ far from the boundary. The Euler equation we obtained in our discussion of the unwinding transition still holds here, because the extra terms due to the chirality in that case do not contribute to the Euler equation. Thus for this situation

$$\xi^2\left(\frac{d^2\theta}{dz^2}\right) + \sin\theta\cos\theta = 0 \qquad \text{where} \qquad \xi^2 = \frac{K_2}{\mu_0\Delta\chi_m H^2}. \quad (12.78)$$

As we did before, this equation can be integrated if we multiply by

$d\theta/dz$ first. If the integration constant is denoted by $C/2$, the result is

$$\xi^2 \left(\frac{d\theta}{dz}\right)^2 + \sin^2\theta = C. \tag{12.79}$$

At $z \to \infty$, $\theta = \pi/2$ and $d\theta/dz = 0$. This implies that $C = 1$ and the equation becomes

$$\xi \left(\frac{d\theta}{dz}\right) = \pm\cos\theta, \tag{12.80}$$

where the plus and minus signs denote either right-handed twist or left-handed twist. Choosing the plus sign and introducing $\phi = \pi/2 - \theta$, this equation can be written

$$\xi \left(\frac{d\phi}{dz}\right) = -\sin\phi. \tag{12.81}$$

Finally, let $u = \tan(\phi/2)$ and rewrite this equation in terms of u. This is easily done considering the triangle in Figure 12.13 as defining the variable u.

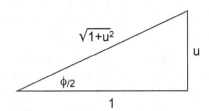

Figure 12.13 Relationship between u and ϕ.

Clearly

$$\sin\phi = 2\sin\left(\frac{\phi}{2}\right)\cos\left(\frac{\phi}{2}\right) = \frac{2u}{1 + u^2} \tag{12.82}$$

and

$$du = \left[1 + \tan^2\left(\frac{\phi}{2}\right)\right] d\left(\frac{\phi}{2}\right) = \frac{1}{2}(1 + u^2)\,d\phi. \tag{12.83}$$

The result is

$$\xi\left(\frac{du}{dz}\right) = -u, \tag{12.84}$$

which can be integrated from $z' = 0$ to $z' = z$ to give

$$u(z) = u(0)e^{-z/\xi}, \tag{12.85}$$

where $u(z)$ is the value of u at an arbitrary value of z and $u(0)$ is the value of u at $z = 0$. Since $\theta = 0$ and $\phi = \pi/2$ at $z = 0$, $u(0) = 1$. Thus the distance over which the angle changes from its constrained value at the surface to its value far from the surface is given by ξ, which depends inversely on the magnetic field strength and also on K_2 and $\Delta\chi_e$. This distance ξ is usually called the nematic coherence length.

12.6 CONVECTIVE INSTABILITIES

Electric fields, especially strong ones, can cause effects in liquid crystals in addition to alignment of the director. In most cases the fact that liquid crystals have non-zero conductivities is important, meaning that the flow of ions in the liquid crystal must be taken into account. This ion flow can in turn cause liquid crystal material to flow, resulting in a situation where the hydrodynamic properties of the liquid crystal are important. The analysis of this type of phenomenon can be extremely complicated. Here we concentrate on the fundamental mechanism behind the more simple effects.

In an isotropic material, the current density $\vec{J}$ is proportional to the electric field $\vec{E}$. The proportionality constant is called the conductivity σ. Since $\vec{J}$ has units of A/m^2 and $\vec{E}$ has units of N/C, σ has units of A/(Vm) or $(\Omega m)^{-1}$. $\vec{J}$ and $\vec{E}$ are collinear in an isotropic material.

Just as the polarization and the electric field do not in general point in the same direction in an anisotropic material, the current density and electric field are not necessarily collinear in an anisotropic substance. σ must therefore be a tensor quantity and

$$\vec{J} = \overset{\leftrightarrow}{\sigma}\vec{E} \quad \text{or} \quad \begin{pmatrix} J_x \\ J_y \\ J_z \end{pmatrix} = \begin{pmatrix} \sigma_\perp & 0 & 0 \\ 0 & \sigma_\perp & 0 \\ 0 & 0 & \sigma_\parallel \end{pmatrix} \begin{pmatrix} E_x \\ E_y \\ E_z \end{pmatrix}, \quad (12.86)$$

where $\sigma_\parallel$ and $\sigma_\perp$ are the conductivities parallel and perpendicular to the director, respectively, and it has been assumed that the director points along the z-axis. Depending on the molecular structure, the anisotropy of the conductivity,

$$\Delta\sigma = \sigma_\parallel - \sigma_\perp, \quad (12.87)$$

can be either positive or negative.

The most dramatic case of a convective instability occurs when the liquid crystal has negative dielectric anisotropy ($\epsilon_\parallel < \epsilon_\perp$) and positive conductivity anisotropy ($\sigma_{\parallel} > \sigma_{\perp}$). Consider the case where the liquid crystal is contained between two surfaces across which a voltage can be applied and on which a surfactant promoting homogeneous alignment has been placed. Application of a small electric field simply stabilizes the director configuration promoted by the surface interactions, producing a uniformly oriented liquid crystal. Since only two directions need be considered, let us call the alignment direction the x-axis and the normal to the surfaces the y-axis. This situation is shown in Figure 12.14.

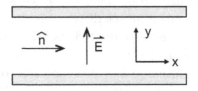

Figure 12.14 Undistorted director configuration.

As the electric field is increased and the liquid crystal is viewed from above through a microscope using light polarized along the alignment direction, at a threshold value of the electric field a striped pattern appears. The stripes run in a direction perpendicular to the alignment direction and are separated by a distance on the order of the thickness of the film. One set of lines is visible if the microscope is focused on the top of the liquid crystal sample and another set of lines, shifted by half the spacing, is seen if the microscope is focused on the bottom of the liquid crystal sample. If light polarized perpendicular to the alignment direction is used, no patterns are observed. These patterns are called Williams domains and the threshold voltage is nearly independent of sample thickness. Dust particles can be seen to go back and forth within these patterns.

If the electric field is increased even more, at some higher threshold value the striped patterns become distorted and their position fluctuates, the flow of material becomes turbulent, and the director gains components parallel to the surfaces and perpendicular to the alignment direction. This condition is known as dynamic scattering and has been considered for possible applications due to the magnitude of the scattering and the low power consumption necessary to achieve the condition.

The basis for the instability associated with these two effects is not difficult to understand. Consider first the undistorted director configuration of Figure 12.14. In this coordinate system the director points along the x-axis and the electric field is along the y-axis. The current density is given by

$$\vec{J} = \begin{pmatrix} J_x \\ J_y \end{pmatrix} = \begin{pmatrix} \sigma_\parallel & 0 \\ 0 & \sigma_\perp \end{pmatrix} \begin{pmatrix} E_x \\ E_y \end{pmatrix}$$

$$= \begin{pmatrix} \sigma_\parallel & 0 \\ 0 & \sigma_\perp \end{pmatrix} \begin{pmatrix} 0 \\ E \end{pmatrix} = \begin{pmatrix} 0 \\ \sigma_\perp E \end{pmatrix}, \qquad (12.88)$$

where for simplicity we have considered only the x and y directions. Notice that the current density is normal to the surfaces and collinear with the electric field. Now imagine that a small periodic distortion spontaneously appears in the uniform director configuration of Figure 12.14. Such a distortion is shown in Figure 12.15(a), where it is clear that at some points the director is not perpendicular to the electric field.

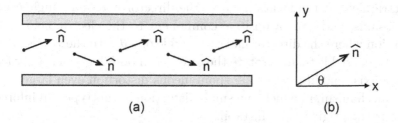

(a)   (b)

Figure 12.15 (a) Distorted director configuration, (b) definition of θ.

Consider a location in the liquid crystal where the director makes an angle θ with the x-axis as shown in Figure 12.15(b). To find $\vec{J}$ at this point, we must write the conductivity tensor in the xy coordinate system, which is rotated from the "director" coordinate system by an angle $-\theta$.

$$\overset{\leftrightarrow}{\sigma} = \overset{\leftrightarrow}{R}_z(-\theta) \begin{pmatrix} \sigma_\parallel & 0 \\ 0 & \sigma_\perp \end{pmatrix} \overset{\leftrightarrow}{R}_z^t(-\theta) =$$

$$\begin{pmatrix} \cos\theta & -\sin\theta \\ \sin\theta & \cos\theta \end{pmatrix} \begin{pmatrix} \sigma_\parallel & 0 \\ 0 & \sigma_\perp \end{pmatrix} \begin{pmatrix} \cos\theta & \sin\theta \\ -\sin\theta & \cos\theta \end{pmatrix} \qquad (12.89)$$

Using this, it quickly becomes obvious that $\vec{J}$ now has components in both the x and y directions.

$$\vec{J} = \begin{pmatrix} \Delta\sigma E \sin\theta\cos\theta \\ (\sigma_\parallel \sin^2\theta + \sigma_\perp \cos^2\theta) E \end{pmatrix} \qquad (12.90)$$

If the conductivity anisotropy is positive, then there is a component of the current density in the positive x direction. If this same calculation is performed where the director makes an angle of $-\theta$ with the x-axis, the current density has a component in the negative x direction. As a result, an accumulation of positive charge builds up between these two locations. This is shown in Figure 12.16.

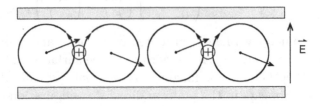

Figure 12.16 Accumulation of charge and convective instabilities.

This localization of charge adds a negative x-component to the electric field at the location where the director makes an angle θ with the x-axis, and adds a positive component to the electric field at the location where the director makes an angle of $-\theta$ with the x-axis. These additional fields tend to rotate the director in both locations away from the x-axis, making the small, spontaneous distortion even larger. It is this mechanism, in which any small distortion of this type is reinforced, that is the origin of the instability.

This distortion is responsible for the stripe patterns. The periodic changes in the director alternately focus and defocus the light passing through the liquid crystal, just as long as the light is polarized parallel to the alignment direction. An observer looking through a microscope can sharpen the image by either focusing above or below the middle of the sample, thus negating the change introduced by the director variation.

This accumulation of charge also sets up the flow pattern shown in Figure 12.16. If the electric field is not too strong, these convection cells are quite stable (Williams domains). At larger electric fields, however, the flow becomes turbulent, with a complex, fluctuating director configuration (dynamic scattering).

12.7 FERROELECTRIC LIQUID CRYSTALS

For the nematic and chiral nematic phases discussed in this chapter, an electric polarization develops in the liquid crystal in response to the application of an external electric field. That is, the electric polarization is zero if no electric field is applied. As discussed in Chapter 7, chiral smectic C liquid crystals are ferroelectric, meaning there is a spontaneous polarization present, $\vec{P}_s$, even when there is no applied electric field $\vec{E}$. As shown in Figure 7.12, this spontaneous polarization is in the plane of the smectic C layers and perpendicular to plane defined by the layer normal and the tilted director. So when an electric field is applied parallel to the layers, there are two effects: (1) the tilt direction rotates around the layer normal so that $\vec{P}_s$ is parallel to $\vec{E}$, and (2) an induced polarization in the direction of the electric field develops just as discussed for nematic liquid crystals. The total electric polarization $\vec{P}$ is therefore the sum of the spontaneous polarization $\vec{P}_s$ and the induced polarization $\vec{P}_{in}$, both of which are parallel to $\vec{E}$.

The energy per unit volume of a material with spontaneous polarization $\vec{P}_s$ in an electric field $\vec{E}$ is given by $-\vec{P}_s \cdot \vec{E}$. Thus when $\vec{P}_s$ and $\vec{E}$ are parallel, the energy is minimized and equals $-P_s E$. The fact that the energy linearly depends on the electric field strength is a significant difference between the case of spontaneous polarization and induced polarization, for which the energy of the latter depends on the electric field strength squared according to Eq. (12.20). The linearity of the energy with electric field and the fact that the spontaneous polarization can be much greater than the induced polarization produces a strong effect in ferroelectric liquid crystals. For this reason, ferroelectric switching using chiral smectic C liquid crystals is an important application, achieving switching speeds much faster than most devices using non-ferroelectric liquid crystals.

All this can be put together in a way very similar to the Landau-de Gennes theory discussed in Chapter 4. This Landau-Ginsburg theory considers a phase transition between a ferroelectric phase ($\vec{P}_s \neq 0$) and a paramagnetic phase ($\vec{P}_s = 0$). The order parameter is the electric polarization, but because the free energy cannot depend on the direction of $\vec{P}_s$ if there is no electric field, only even powers of P_s are allowed. So the following expansion is used, where the term due to an applied field has been added and the polarization is the total polarization $P = P_s + P_{in}$,

$$G(P) = G_0 + \frac{1}{2}AP^2 + \frac{1}{4}BP^4 + \frac{1}{6}P^6 - PE, \qquad (12.91)$$

where G_0 is the free energy of the paramagnetic phase ($P = 0$ if $E = 0$), and A has the same temperature dependence as the first coefficient of the Landau-de Gennes theory. Taking the derivative of $G(P)$ with respect to P and setting it equal to zero to find the minimum free energy, gives the relationship between P and E.

$$AP + BP^3 + CP^5 - E = 0 \qquad (12.92)$$

This equation can be numerically solved for P as a function of E if the values of A, B, and C are given. The result for A = -0.5, B = 1, and C = 10 is shown in Figure 12.17. For a range of electric field strengths, there are three solutions, only two of which are stable. For electric field strengths larger than this, there is one solution. The value of P_s is shown in the figure; notice that P increases from this value as E increases in either direction. In practice, there will be some hysteresis when the direction of the electric field is reversed, which is exaggerated in the figure. The dotted arrows show that to switch the direction of $\vec{P}$, E must be decreased through zero and increase a bit in the other direction to get $\vec{P}$ to change directions.

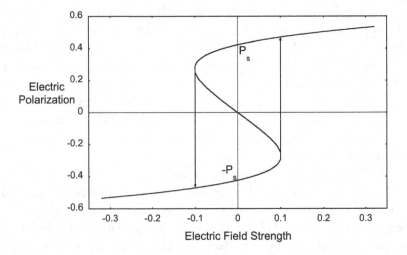

Figure 12.17 Dependence of the electric polarization on the electric field strength according to the Landau-Ginsburg theory. The part of the curve for which the slope is negative is unstable. The vertical arrows indicate the maximum extent of the hysteresis when the direction of the electric field is reversed.

EXERCISES

12.1 The two electric susceptibilities for 5CB in the nematic phase at room temperature are $\chi_\parallel = 20.3$ and $\chi_\perp = 6.4$. Assume the director is oriented along the z-axis. Find the angle between (1) the polarization $\vec{P}$ and the electric field $\vec{E}$ and (2) the electric displacement $\vec{D}$ and the electric field $\vec{E}$ if the electric field is oriented (a) in the xy plane making angles of $45°$ with both the x- and y-axes, and (b) in the yz plane making angles of $45°$ with both the y- and z-axes.

12.2 Since the magnetic susceptibility for nematic liquid crystals is on the order of 10^{-6}, according to Eq. (1.22) the relationship between the magnetic induction B and the magnetic field H is $B = \mu_0 H$ to a good approximation.

(a) Use this to find an expression for the critical magnetic induction B_c necessary to unwind the helix of a chiral nematic liquid crystal with a pitch P_0, a twist elastic constant K_2, and a magnetic anisotropy $\Delta\chi_m$.

(b) A small amount of a chiral dopant is added to the nematic liquid crystal 5CB to produce a chiral nematic phase. At room temperature, the pitch is 20 μm, the twist elastic constant is 4 pN, and the magnetic anisotropy is 1.4×10^{-6}. What is the critical magnetic induction B_c necessary to unwind the chiral nematic helix?

12.3 While SI units are standard, it is not unusual to see cgs (centimeter-gram-second) units being used for magnetic phenomena. Do both parts of the previous exercise using cgs units. If cgs units are used, Eq. (12.76) becomes

$$H_c = \frac{\pi}{2}\sqrt{\frac{K_2}{\Delta\chi_m}}q_0 = \frac{\pi^2}{P_0}\sqrt{\frac{K_2}{\Delta\chi_m}},$$

and Eq. (12.22) becomes $\vec{B} = \vec{H} + \vec{M} = (\overset{\leftrightarrow}{1} + \overset{\leftrightarrow}{\chi}_m)\vec{H}$. Also, the magnetic anisotropy is reduced by a factor of 4π and thus equals 1.1×10^{-7} in cgs units. When using cgs units for magnetic calculations, express each variable in cgs units, substitute them into the correct formula for cgs units, and append the correct cgs units to your answer. In this specific case, your answer for B_c is in gauss. You can check your answer by converting gauss into tesla using 1 tesla = 10,000 gauss.

12.4 What is the size of the coherence length for strong anchoring on a surface with an electric field applied in the plane of the surface but perpendicular to the surface alignment direction? Let the twist elastic constant be 3.5 pN, the electric anisotropy be 12, and the electric field be 0.1 V/μm.

12.5 An early LCD design was called the twisted nematic cell because the alignment directions on the two substrates holding the thin layer of liquid crystal were perpendicular to one another. In the limit of strong anchoring, the directors at the surfaces are perfectly aligned with the alignment directions. This means the director uniformly twists through $\pi/2$ radians in going from one surface to the other. But in all other cases, including less than infinitely strong anchoring and weak anchoring, the director configuration is the one that minimizes the sum of the surface elastic energy per unit area given by $(1/2)W\sin^2\phi$, where ϕ is the angle between the director at the surface and the surface alignment direction, and the volume elastic energy given by Eq. (3.30).

(a) Using the development leading to Eq. (3.36), determine that the bulk elastic energy per unit volume in this case is $(1/2)K_2(\Omega/d)^2$, where Ω is the angle the director twists in going from one surface to the other (in radians), and d is the distance between the two surfaces.

(b) Show that the total elastic energy per unit area of the cell is $(1/2)K_2d(\Omega/d)^2 + W\sin^2\phi$.

(c) The actual value of Ω is the one that minimizes this total elastic energy. Show that this value of Ω can be found by solving the following equation: $a\cos\Omega - \Omega = 0$, where $a = (1/2)Wd/K_2$. Keep in mind that Ω and ϕ are related because $\Omega + 2\phi = \pi/2$.

(d) Use computational software or simply trial and error to determine Ω for a nematic liquid crystal with $K_2 = 4.5$ pN, a cell thickness of 10 μm, and two anchoring conditions: $W = 10^{-4}$ N/m (strong anchoring) and $W = 10^{-6}$ N/m (weak anchoring).

12.6 The spontaneous polarization P_s can be found quite easily in the Landau-Ginsburg theory. Eq. (12.92) relates P and E, so setting $E = 0$ allows P to be solved when there is no electric field. This is simply P_s. Since it is a fifth-order equation, there are five roots.

One is zero, two are imaginary, and two are real and only differ in sign.

(a) Put the values for A, B, and C used in Figure 12.17 in the expression for one of the real roots to check that the P_s value you get is the same as shown in the figure.

(b) If the parameter A is given the same temperature dependence as in the Landau-deGennes theory, $A(T) = A_0(T - T_c)$, what happens to P_s as T approaches T_c from below (remember $A < 0$, so $T < T_c$)?

(c) What is P_s above T_c?

(d) Is the phase transition at T_c discontinuous or continuous?

CHAPTER 13

Light and Liquid Crystals – A Panoply of Color

13.1 POLARIZED ELECTROMAGNETIC WAVES

An electromagnetic wave consists of propagating electric and magnetic fields. In order to propagate as an electromagnetic wave, both the electric and magnetic fields must point in a direction perpendicular to the direction of propagation, and in addition, the electric field must always be perpendicular to the magnetic field. The electric and magnetic fields oscillate in both time and space. That is, if a snapshot of the wave is made at a certain time, both the electric and magnetic fields oscillate as a function of the propagation direction. The distance it takes the fields to repeat themselves is called the wavelength λ_0. Alternately, if the electric or magnetic field at a single location is graphed as a function of time, a similar oscillation takes place. The time it takes the field variation to repeat itself is called the period of the wave T. Since an electromagnetic wave travels one wavelength in one period, the velocity of a wave is equal to its wavelength divided by its period. Electromagnetic waves in a vacuum travel with a velocity equal to 3×10^8 m/s, a constant of nature usually denoted by the symbol c. A sketch of an electromagnetic wave is shown in Figure 13.1.

For mathematical reasons, the wavelength and period are not usually employed when discussing electromagnetic waves. Instead, the wavevector $\vec{k}_0$ is used in place of the wavelength and the angular frequency ω (2π times the frequency) is used in place of the period. The direction associated with the wavevector is simply the direction of propagation of the wave. The definitions of the wavevector and angular

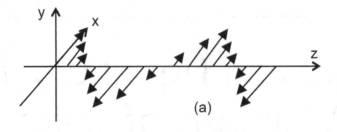

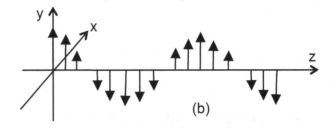

Figure 13.1 Electric (a) and magnetic (b) fields in an electromagnetic wave propagating in the z-direction.

frequency are

$$k_0 = |\vec{k}_0| = \frac{2\pi}{\lambda_0} \quad \text{and} \quad \omega = \frac{2\pi}{T}. \tag{13.1}$$

The relationship between λ_0 and T implies that $\omega/k_0 = c$.

Light is an electromagnetic wave with a wavelength between 0.4 and 0.7 μm. Electromagnetic waves exist with wavelengths ranging from 10^{-14} m up to several kilometers. What makes electromagnetic waves with a wavelength between 0.4 and 0.7 μm unique is that the human retina is sensitive to these wavelengths. So as we discuss light and the optical properties of liquid crystals in this chapter, keep in mind that much of what is described is also true for electromagnetic waves with wavelengths outside the visible part of the electromagnetic spectrum.

Notice that in Figure 13.1 the electric field always points along the x-axis and the magnetic field always points along the y-axis. The orientation of the electric and magnetic fields in the plane perpendicular to the direction of propagation determines the polarization of the wave. Since the magnetic field is perpendicular to the electric field at all times, the convention of specifying the orientation of the electric field is used. Because the electric field is always oriented along the x-axis, the wave shown in Figure 13.1 is referred to as a linearly polarized wave, with the polarization direction being the x-axis. As shown in Figure 13.2, the polarization direction can be in any direction in the plane perpendicular to the propagation direction. Since the electric field is oscillating, the amplitude of the oscillation is used to specify the size of the electric field. To indicate the polarization of a wave, the amplitudes of the x- and y-components of the electric field, E_{0x} and E_{0y}, must be given. The angle the electric field makes with the x-axis is therefore

$$\theta = \tan^{-1}\left(\frac{E_{0y}}{E_{0x}}\right). \tag{13.2}$$

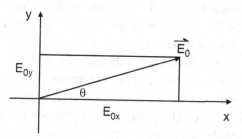

Figure 13.2 Electromagnetic wave linearly polarized in the xy plane.

The fundamental electromagnetic wave possesses electric and magnetic fields that oscillate sinusoidally in both space and time. Other waves exist, but they can be represented as a sum of sinusoidally oscillating waves. A general expression for a sinusoidally varying electromagnetic wave propagating in the z direction must specify both the x- and y-components of the electric field,

$$\begin{aligned} E_x(z,t) &= E_{0x}\cos(kz - \omega t + \epsilon_x) \\ E_y(z,t) &= E_{0y}\cos(kz - \omega t + \epsilon_y). \end{aligned} \tag{13.3}$$

The parameters ϵ_x and ϵ_y allow for the fact that the x- and y-components of the electric field can vary in phase. The possible values for ϵ_x and ϵ_y cover a range of 2π; we use the range from $-\pi$ to $+\pi$. Thus ϵ_x and ϵ_y can be considered angles with units of radians.

The components of sinusoidally oscillating electromagnetic waves are often written in a different form for two reasons. First, for ease of mathematical manipulation, exponential functions with imaginary arguments are used in place of sines and cosines. This introduces real and imaginary parts into the calculations, and one need only take the real or imaginary part of the calculation result to obtain the correct answer. Second, calculating how the components change when traversing optical devices is made simple by writing the components as a two-element column vector.

$$\vec{E}(z,t) \; = \; \begin{pmatrix} E_x(z,t) \\ E_y(z,t) \end{pmatrix} = \begin{pmatrix} E_{0x}e^{i(kz-\omega t+\epsilon_x)} \\ E_{0y}e^{i(kx-\omega t+\epsilon_y)} \end{pmatrix}$$

$$= \; \begin{pmatrix} E_{0x}e^{i\epsilon_x} \\ E_{0y}e^{i\epsilon_y} \end{pmatrix} e^{i(kz-\omega t)}, \qquad (13.4)$$

where it is understood that the upper element of the column vector is the x-component and the lower element is the y-component of the electric field. Because the z and t dependence is the same for both components, it can be factored out, leaving only electric field amplitudes and phases in the column vector.

The column vector involving only amplitudes and phases is called the Jones vector of the electromagnetic wave and is part of what is known as Jones calculus, a mathematical representation used for calculations in optics. Since the part of the expression outside of the column vector is simply a traveling wave of wavevector k and angular frequency ω, it is dropped from calculations involving polarization. In fact, it is often convenient to factor out a constant and write it outside the Jones vector. This is perfectly acceptable because it is only the ratio of amplitudes and difference in the phases that determine the polarization.

For example, if $E_{0y} = 0$, then the Jones vector written quite simply with a zero and a one gives the polarization of the electromagnetic wave.

$$\begin{pmatrix} E_{0x}e^{i\epsilon_x} \\ 0 \end{pmatrix} = E_{0x}e^{i\epsilon_x} \begin{pmatrix} 1 \\ 0 \end{pmatrix} \qquad (13.5)$$

However, it is often necessary to keep track of the scalar factor in front of the simple Jones vector, in this case E_{0x}, since it contains the amplitude of the polarized electromagnetic wave. Sinusoidal terms, in this case $e^{i\epsilon_x}$, can usually be dropped because they represent an overall phase or are not important in determining the intensity of the electromagnetic wave. It is also necessary to define the x- and y-axes properly. According to convention, the electromagnetic wave is propagating along the positive z-axis, the y-axis points vertically upward, and the x-axis is horizontal in a direction that makes the (x, y, z) coordinate system right-handed (see Figure 13.3). For the electromagnetic wave described by Eq. (13.5), the electric field is always along the x-axis, so the electromagnetic wave is horizontally polarized. Vertically polarized electromagnetic waves are likewise represented by a column vector with a 0 as the upper element and a 1 as the lower element.

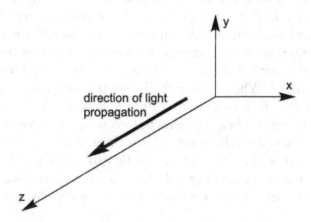

Figure 13.3 Convention used for Jones calculus.

Since it is the difference between ϵ_x and ϵ_y that matters, let us set $\delta = \epsilon_y - \epsilon_x$. Notice that δ can also be thought of as an angle with units of radians. If $\delta = 0$ or $\pm\pi$, then the electromagnetic wave is linearly polarized, independent of the values of E_{0x} and E_{0y}. If $\delta = 0$, then the direction of polarization is given by Eq. (13.2) and shown in Figure 13.2. The Jones vector for this electromagnetic wave is

$$E_0 \begin{pmatrix} \cos\theta \\ \sin\theta \end{pmatrix}. \tag{13.6}$$

If $\delta = \pm\pi/2$, then the fact that the x- and y-components are 90° out of phase makes the electric field rotate around the z-axis in the x-y plane. If $E_{0x} = E_{0y} = E_0$, then the wave is circularly polarized, since the tip of the electric field vector maps out a circle during each period. This situation is shown in Figure 13.4, where the technique of locating the tip of the electric field vector at successive times has been used to visualize the variation of the electric field. These diagrams show the electric field vector at a position $z = 0$ at various times during one period. In Figure 13.4(a), $\delta = \pi/2$ and the Jones vector of the electric field is given by

$$E_0 e^{i\epsilon_x} \begin{pmatrix} 1 \\ e^{i\delta} \end{pmatrix} = E_0 e^{i\epsilon_x} \begin{pmatrix} 1 \\ e^{i\pi/2} \end{pmatrix} = E_0 e^{i\epsilon_x} \begin{pmatrix} 1 \\ i \end{pmatrix}. \tag{13.7}$$

Remembering that the time and space dependence is $e^{i(kz-\omega t)}$ and that only the real parts are important, at $z = 0$, and with $\epsilon_x = 0$ for simplicity, $E_x(0, t) = E_0 \cos(-\omega t) = E_0 \cos(\omega t)$ and $E_y(0, t) = i^2 E_0 \sin(-\omega t) = E_0 \sin(\omega t)$. As time increases, the electric field vector rotates counterclockwise to an observer looking back along the direction of propagation toward the source. Such a wave is referred to as left circularly polarized. When $\delta = -\pi/2$, the electric field vector has components $E_x(0, t) = E_0 \cos(\omega t)$ and $E_y(0, t) = E_0 \sin(-\omega t) = -E_0 \sin(\omega t)$, and therefore rotates in a clockwise fashion to an observer looking back along the propagation direction. This is called a right circularly polarized wave and is illustrated in Figure 13.4(b). Thus the Jones vector for a right (or left) circularly polarized electromagnetic wave has a 1 as the upper element and a $+i$ (or $-i$) as the lower element.

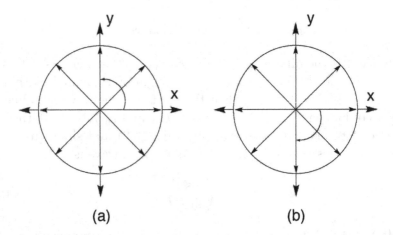

(a)   (b)

Figure 13.4 Left (a) and right (b) circularly polarized electromagnetic waves.

It is important to realize that a circularly polarized electromagnetic wave can be described through its time variation, as we have just done, or through its space variation. If the left circularly polarized wave of Figure 13.4(a) is considered at a specified time, then the tip of the electric field vector maps out a left-handed helix in space. To identify a left-handed helix, imagine the thumb of your left hand pointing in the direction of propagation. The fingers of your left hand then curl around the direction of propagation in the direction the electric field rotates in advancing along the wave in the direction your thumb is pointing. Conversely, the tip of the electric field vector in a right circularly polarized wave maps out a right-handed helix described analogously using your right hand.

If $\delta = \pm\pi/2$ but E_{0x} does not equal E_{0y}, then the electromagnetic wave is elliptically polarized as shown in Figure 13.5(a). The semimajor axis of the ellipse is either along the x- or y-axis, depending on which component of the electric field is larger. If δ does not equal a multiple of $\pm\pi/2$, then the wave is again elliptically polarized, but the angle the semi-major axis makes with the x-axis depends on both the ratio of E_{0x} to E_{0y} and δ. Such a case is illustrated in Figure 13.5(b).

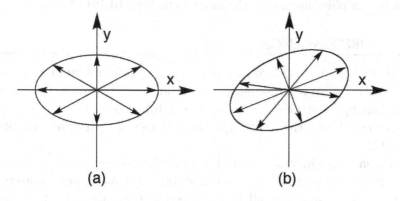

(a)   **(b)**

Figure 13.5 Elliptically polarized electromagnetic waves with (a) δ equal to $\pm\pi/2$ and (b) δ not equal to $\pm\pi/2$.

To see this, the expressions for the components of the electric field must be combined mathematically in a way that eliminates the space and time dependence. Dividing the expressions for $E_x(z,t)$ and $E_y(z,t)$ by E_{0x} and E_{0y}, respectively, the result is (assuming $\epsilon_x = 0$)

$$\frac{E_x(z,t)}{E_{0x}} = \cos(kz - \omega t) \tag{13.8}$$

$$\frac{E_y(z,t)}{E_{0y}} = \cos(kz - \omega t + \delta) = \cos(kz - \omega t)\cos\delta - \sin(kz - \omega t)\sin\delta.$$

After multiplying the first expression by $\cos\delta$, subtracting one from the other, and squaring both sides, one obtains

$$\left(\frac{E_x(z,t)}{E_{0x}}\right)^2 + \left(\frac{E_y(z,t)}{E_{0y}}\right)^2 - 2\left(\frac{E_x(z,t)}{E_{0x}}\right)\left(\frac{E_y(z,t)}{E_{0y}}\right)\cos\delta =$$
$$\sin^2\delta \quad (13.9)$$

Believe it or not, this is the equation for an ellipse oriented at some angle θ (counterclockwise from the positive x-axis). The ellipse is bounded by a box with sides at $\pm E_{0x}$ and $\pm E_{0y}$, and if $E_{0x} > E_{0y}$, θ is given by

$$\theta = \frac{1}{2}\tan^{-1}\left(\frac{2E_{0x}E_{0y}\cos\delta}{E_{0x}^2 - E_{0y}^2}\right). \quad (13.10)$$

If $E_{0x} < E_{0y}$, the proper counterclockwise angle from the positive x-axis is $\pi/2$ plus the angle calculated from Eq. (13.10).

13.2 BIREFRINGENCE

So far we have been talking about electromagnetic waves in a vacuum. In this section we will discuss the propagation of electromagnetic waves in anisotropic materials. For liquid crystals, the optical part of the spectrum is by far the most important, so we will restrict our attention to light.

When light enters a material, its wavelength and velocity decrease by a factor called the index of refraction. In non-magnetic materials, which cover just about all liquid crystals, the index of refraction is simply equal to the square root of the relative permittivity or dielectric constant at optical frequncies. An isotropic material has a single index of refraction, since light polarized in any direction travels at the same velocity in the material. The index of refraction of water is about 1.3 and of glass about 1.5. When light travels from one material to another, in general some of the light is reflected and some is transmitted with a change of direction (refracted). In most materials the index of refraction increases with decreasing wavelength, a phenomenon responsible for the ability of a prism to separate white light into its various colors.

If n is the index of refraction of a material, then the wavelength and velocity of the light in the material are given by the following relations

$$v = \frac{c}{n} \quad \text{and} \quad \lambda = \frac{\lambda_0}{n}, \quad (13.11)$$

where λ_0 is the wavelength of the light in a vacuum. The period and angular frequency of the light are the same in all materials, as can be seen by finding the period by dividing the wavelength by the velocity. The wavevector in the material is given by

$$k = \frac{2\pi}{\lambda} = \frac{2\pi n}{\lambda_0} = nk_0, \qquad (13.12)$$

where k_0 is the wavevector of the light in a vacuum.

Because the relative permittivity of an anisotropic substance is different for electric fields in different directions, the index of refraction for light polarized with its electric field in different directions is also different. This optical phenomenon is called birefringence, and since we are discussing linearly polarized light, it should really be called linear birefringence. In a nematic liquid crystal, this means that light polarized parallel to the director propagates according to one index of refraction, $n_\parallel$, and light polarized perpendicular to the director has another index of refraction, $n_\perp$. Figure 13.6 shows how these two indices of refraction for one wavelength vary with temperature.

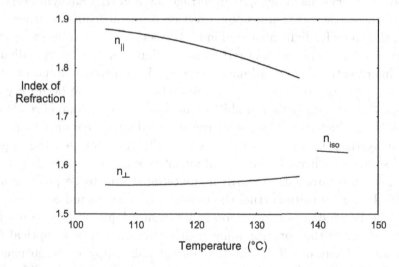

Figure 13.6 Indices of refraction for a typical nematic liquid crystal.

The difference between the two indices of refraction $\Delta n = n_\parallel - n_\perp$, (called its optical anisotropy or birefringence) clearly reflects the fact that the order parameter decreases with increasing temperature. In fact, Δn qualitatively follows the variation of the order parameter as Figure 13.7 demonstrates.

Birefringence is a property of all anisotropic materials, whether uniaxial or biaxial. Nematic liquid crystals fall into the uniaxial category

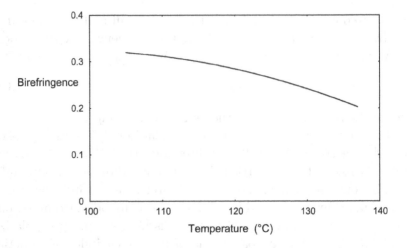

Figure 13.7 Birefringence for a typical nematic liquid crystal.

along with crystals of hexagonal, tetragonal, and trigonal symmetry. In all of these systems there is one direction different from the other two, and the index for light polarized in this direction is called the extraordinary index n_e. The index of refraction for light polarized perpendicular to this direction is the ordinary index n_o. In a nematic liquid crystal, n_e corresponds to $n_{\parallel}$ and n_o corresponds to $n_{\perp}$. Some liquid crystal phases along with orthorhombic, monoclinic, and triclinic crystals are biaxial and have three indices of refraction. The optical anisotropy for these systems is usually defined as the difference between the largest and smallest indices. If the optical anisotropy is positive (n_e is greater than n_o in a uniaxial system), the material is said to be positive uniaxial. If the opposite is true, the material is negative uniaxial.

The two indices of refraction in a nematic liquid crystal equal the square root of the corresponding relative permittivities at optical frequencies. Therefore it is not the optical anisotropy or birefringence that is most directly related to the order parameter but the difference in the squares of the two indices. For this reason the quantity $n_{\parallel}^2 - n_{\perp}^2$ is frequently used as a parameter proportional to the order parameter.

Since the index of refraction is a measure of the velocity of the light, it is worth looking at a uniaxial birefringent medium from the viewpoint of the velocities. The director defines a unique direction in the medium called the optic axis. Light with a polarization parallel to the director is called an extraordinary ray and its velocity is given by the extraordinary index $n_e = n_{\parallel}$. Light polarized perpendicular to the director is called an ordinary ray and its velocity is given by the ordinary index $n_o = n_\perp$. Imagine a plane containing the director. Ordinary rays (with their polarization perpendicular to the plane containing the director) can travel in any direction in this plane. So if we imagine all these waves emanating from some point, at some time later all of them have traveled the same distance. Hence we can draw a circle representing the wavefront surface of all these waves. This circle is labeled the ordinary ray surface in Figure 13.8. Now consider waves emanating from a single point with their polarization in the plane containing the director. If the wave is traveling along the director, the polarization is perpendicular to the director, so the wave travels at the velocity of the ordinary wave. However, if the wave is traveling perpendicular to the director, the polarization is parallel to the director, so the wave travels at the velocity of the extraordinary wave. If the wave travels at some angle between these two directions, the velocity is between the velocities of the ordinary and extraordinary waves. Thus the wavefront surface of all these waves is an ellipse that matches the ordinary wavefront surface along the director but is different from the ordinary wavefront surface for all other directions. This is also shown in Figure 13.8 and is called the extraordinary ray surface. In addition, Figure 13.8 shows the polarizations for the two possibilities of waves traveling perpendicular to the director. In one case the polarization is parallel to the director and the proper wavefront surface is the extraordinary one. In the other case, the polarization is perpendicular to the director and the proper wavefront surface is the ordinary one.

These wavefront surfaces look different depending on which index of refraction is larger. If $n_e = n_\parallel > n_o = n_\perp$, then the medium has positive birefringence and the extraordinary ray surface sits inside the ordinary ray surface. If $n_e = n_\parallel < n_o = n_\perp$, then the medium has negative birefringence and the extraordinary ray surface sits outside the ordinary ray surface. Both of these situations are shown in Figure 13.8.

To calculate the index of refraction for the extraordinary waves when they are traveling at an angle θ to the director other than $0°$ or $90°$ is a very involved calculation. The end result is fairy simple and is given by

$$n_e(\theta) = \frac{n_e n_o}{\sqrt{n_e^2 \cos^2 \theta + n_o^2 \sin^2 \theta}}. \tag{13.13}$$

As a check, let's calculate the index of refraction if extraordinary waves

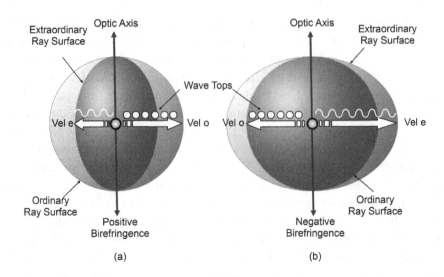

Figure 13.8 Positive and negative birefringence wavefronts. The polarization of the extraordinary ray makes an angle between $0°$ and $90°$ to the director and the polarization of the ordinary wave is perpendicular to the director.

are traveling parallel to the director. This means $\theta = 0°$ and the index equals n_o, which is what we expect. If $\theta = 90°$, then the index is n_e, which is again what we expect.

13.3 OPTICAL RETARDATION

Because liquid crystals are birefringent, light polarized along different directions travels at different velocities. Thus two perpendicular components of light that enter the liquid crystal in phase grow out of phase as they propagate through the liquid crystal. This effect is known as optical retardation and is very important in liquid crystals. Let us examine optical retardation in general and apply it to liquid crystals.

Imagine that linearly polarized light enters a retarder of thickness d. The x- and y-components of the entering light are in phase, but they have different amplitudes, E_{0x} and E_{0y}. As each component of the light travels through the retarder, both have the same frequency ω but they have different wavelengths (or wavevectors, k_x and k_y). Thus the light enters at $z = 0$ and $t = 0$ with a Jones vector of

$$\begin{pmatrix} E_x(0,0) \\ E_y(0,0) \end{pmatrix} = \begin{pmatrix} E_{0x} \\ E_{0y} \end{pmatrix}, \qquad (13.14)$$

and exits at $z = d$ some time t later with a Jones vector of

$$\begin{pmatrix} E_x(d,t) \\ E_y(d,t) \end{pmatrix} = \begin{pmatrix} E_{0x}e^{ik_x d} \\ E_{0y}e^{ik_y d} \end{pmatrix} = e^{ik_x d} \begin{pmatrix} E_{0x} \\ E_{0y}e^{i(k_y - k_x)d} \end{pmatrix}. \qquad (13.15)$$

As usual, the time dependence is not included in the Jones vector. But a wavevector times a distance is simply a phase angle since $kd = 2\pi(d/\lambda)$, which is the phase angle in radians. So the exiting Jones vector can be written

$$\begin{pmatrix} E_{0x}e^{i\epsilon_x} \\ E_{0y}e^{i\epsilon_y} \end{pmatrix} = e^{i\epsilon_x} \begin{pmatrix} E_{0x} \\ E_{0y}e^{i(\epsilon_y - \epsilon_x)} \end{pmatrix} = e^{i\epsilon_x} \begin{pmatrix} E_{0x} \\ E_{0y}e^{i\delta} \end{pmatrix}. \qquad (13.16)$$

In order to put the action of the retarder into the Jones calculus, we just have to realize that the Jones vector of the light is different as it exits the retarder compared to as it enters the retarder. Therefore, the retarder can be represented by something that takes a two-element Jones vector and changes it to a different two-element Jones vector. Our discussion of anisotropy in Chapter 3 showed us that a 2×2 matrix does exactly this.

$$\begin{aligned} \begin{pmatrix} E_x \\ E_y \end{pmatrix}_{after} &= \begin{pmatrix} e^{i\epsilon_x} & 0 \\ 0 & e^{i\epsilon_y} \end{pmatrix} \begin{pmatrix} E_x \\ E_y \end{pmatrix}_{before} \\ &= e^{i\epsilon_x} \begin{pmatrix} 1 & 0 \\ 0 & e^{i\delta} \end{pmatrix} \begin{pmatrix} E_x \\ E_y \end{pmatrix}_{before} \end{aligned} \qquad (13.17)$$

Either of the 2×2 matrices above is called a Jones matrix and can be used to represent a retarder.

If $\epsilon_x > \epsilon_y$, then $n_x > n_y$. Therefore the light travels slower if it is polarized along the x-axis. For this reason, the x-axis of this retarder is referred to as the slow axis (SA) while the y-axis is the fast axis (FA). If $\epsilon_x < \epsilon_y$, then $n_x < n_y$ and the x-axis of this retarder is the fast axis while the y-axis is the slow axis.

To relate this to a nematic liquid crystal sample, imagine the director is parallel to the y-axis and that the wavevector and index of refraction of light polarized parallel to the director are $k_\parallel$ and $n_\parallel$, respectively. Let the wavevector and index of refraction of light polarized perpendicular to the director be $k_\perp$ and $n_\perp$, respectively. Then

$$\epsilon_x = k_\perp d = \frac{2\pi}{\lambda_0} n_\perp d \qquad (13.18)$$

$$\delta = \epsilon_y - \epsilon_x = (k_\parallel - k_\perp)d = \frac{2\pi}{\lambda_0}(n_\parallel - n_\perp)d = \frac{2\pi}{\lambda_0}\Delta n d.$$

Notice that the retardation tends to increase as the wavelength decreases. Measuring the optical retardation of a known thickness of liquid crystal is a convenient method of measuring the birefringence of a liquid crystal at the wavelength used. Finally, if $n_\parallel > n_\perp$, then the director is the slow axis.

Retarders with special values of δ have important effects. If $\delta = \pi/2$ and the SA is vertical, then we can set ϵ_x to $-\pi/4$ and ϵ_y to $+\pi/4$. If the SA is horizontal, then we can set ϵ_x to $+\pi/4$ and ϵ_y to $-\pi/4$. Both of these retarders are called quarter wave plates (QWPs) and have the following Jones matrices.

$$\text{QWP (SA vertical):} \quad e^{-i\frac{\pi}{4}}\begin{pmatrix} 1 & 0 \\ 0 & i \end{pmatrix}$$

$$\text{QWP (SA horizontal):} \quad e^{i\frac{\pi}{4}}\begin{pmatrix} 1 & 0 \\ 0 & -i \end{pmatrix} \qquad (13.19)$$

If light polarized at 45° to the x-axis enters a QWP with its SA horizontal, then the light that emerges is right circularly polarized as can be seen from the following equation,

$$\begin{pmatrix} E_x \\ E_y \end{pmatrix}_{\text{after}} = e^{i\frac{\pi}{4}}\begin{pmatrix} 1 & 0 \\ 0 & -i \end{pmatrix}\begin{pmatrix} E_0 \\ E_0 \end{pmatrix} = E_0 e^{i\frac{\pi}{4}}\begin{pmatrix} 1 \\ -i \end{pmatrix}. \qquad (13.20)$$

If the QWP is rotated by 90° about the z-axis, the indices of refraction for light polarized in the x and y directions are switched and the light emerges left circularly polarized.

If $\delta = \pi$ and the SA is vertical, then we can set ϵ_x to $-\pi/2$ and ϵ_y to $+\pi/2$. If the SA is horizontal, then we can set ϵ_x to $+\pi/2$ and ϵ_y to $-\pi/2$. Both of these retarders are called half wave plates (HWPs) and have the following Jones matrices.

$$\text{HWP (SA vertical):} \quad -i\begin{pmatrix} 1 & 0 \\ 0 & -1 \end{pmatrix}$$

$$\text{HWP (SA horizontal):} \quad i\begin{pmatrix} 1 & 0 \\ 0 & -1 \end{pmatrix} \qquad (13.21)$$

If light polarized 45° to the x-axis enters an HWP with its SA horizontal, then the light that emerges is linearly polarized perpendicular to the original polarization direction:

$$\begin{pmatrix} E_x \\ E_y \end{pmatrix}_{after} = i \begin{pmatrix} 1 & 0 \\ 0 & -1 \end{pmatrix} \begin{pmatrix} E_0 \\ E_0 \end{pmatrix} = iE_0 \begin{pmatrix} 1 \\ -1 \end{pmatrix}. \tag{13.22}$$

Rotating the HWP by 90° about the z-axis again switches the two indices of refraction, but the emerging light is again linearly polarized perpendicular to the incident light.

The birefringence of a liquid crystal is responsible for the patterns seen in a polarizing microscope. In such a microscope, the light traveling towards the observer's eyes first passes through a polarizer, then through the liquid crystal sample, and finally through a second polarizer at 90° to the first polarizer. Because the two polarizers are "crossed", no light gets to the observer's eyes unless the liquid crystal changes the polarization state of the light. If the director in some part of the liquid crystal is parallel to either polarizer, the incident light propagates according to a single index of refraction and suffers no change in polarization. The second polarizer extinguishes this light and that part of the liquid crystal appears dark. If in another part of the sample the director is at an angle to both polarizers, then the light must be thought of as having two in-phase components, one parallel to the director and one perpendicular to the director. These components travel according to two indices of refraction, a phase difference is introduced between them, and the emerging light is elliptically polarized. Some of this light passes through the second polarizer and the liquid crystal appears bright in this location. In addition, since the amount of optical retardation is a function of wavelength (since Δn varies with wavelength), some colors pass through the second polarizer with more intensity than other colors, so the liquid crystal can appear quite colored. One last case is when light propagates along the director. Since the index of refraction for light polarized in all directions perpendicular to the director is the same, there is no optical retardation and the liquid crystal appears dark. Optical retardation produces the many patterns seen in a polarizing microscope that are so useful for identifying the various liquid crystal phases.

13.4 JONES CALCULUS

We are now in a position to fill in some of the details of Jones calculus and demonstrate its usefulness in understanding the optics of liquid crystals. First, let us determine the Jones matrices for linear polarizers. This is easy for horizontal and vertical polarizers because each selects out a single component of the incoming light.

$$\text{horizontal polarizer:} \begin{pmatrix} 1 & 0 \\ 0 & 0 \end{pmatrix} \qquad \text{vertical polarizer:} \begin{pmatrix} 0 & 0 \\ 0 & 1 \end{pmatrix} \quad (13.23)$$

To find the Jones matrix for a linear polarizer at an angle θ (measured counterclockwise from the x-axis), we just have to remember how we used rotation matrices to change the coordinate system in Chapter 3. How does the Jones matrix of a horizontal polarizer change if the x-y coordinate system is rotated by an angle $-\theta$ about the z-axis?

$$\begin{aligned} \theta\text{-polarizer} \quad &= \quad \begin{pmatrix} \cos\theta & -\sin\theta \\ \sin\theta & \cos\theta \end{pmatrix} \begin{pmatrix} 1 & 0 \\ 0 & 0 \end{pmatrix} \begin{pmatrix} \cos\theta & \sin\theta \\ -\sin\theta & \cos\theta \end{pmatrix} \\ &= \quad \begin{pmatrix} \cos^2\theta & \sin\theta\cos\theta \\ \sin\theta\cos\theta & \sin^2\theta \end{pmatrix} \end{aligned} \quad (13.24)$$

Another Jones matrix of interest is one that rotates the components of the polarization by an angle β (counterclockwise in the x-y plane). The derivation is left as an exercise.

$$\beta\text{-rotator} = \begin{pmatrix} \cos\beta & -\sin\beta \\ \sin\beta & \cos\beta \end{pmatrix} \quad (13.25)$$

As an example of how Jones calculus works, let's imagine viewing a nematic liquid crystal between crossed polarizers. The polarizers are fixed, but the liquid crystal sample can be rotated around the z-axis. How does the intensity of light getting through the second polarizer depend on the orientation of the liquid crystal sample? One way to do this calculation is to start with the Jones matrix of the liquid crystal, apply the rotation matrices so it represents a liquid crystal sample at angle θ, and then consider the two fixed polarizers on either side of it. But let us do it the other way, pretending the polarizers are rotated clockwise in the x-y plane and keeping the liquid crystal sample fixed. If the director of the liquid crystal sample is along the x-axis, then $\epsilon_x > \epsilon_y$, and $\epsilon_y - \epsilon_x < 0$. Therefore, if we let $\delta = 2\pi\Delta nd/\lambda_0$, the Jones matrix for the liquid crystal sample is

$$e^{i\epsilon_x} \begin{pmatrix} 1 & 0 \\ 0 & e^{-i\delta} \end{pmatrix}. \tag{13.26}$$

Now we put the Jones calculus to use, starting with the Jones vector for light of amplitude E_0 polarized at $-\theta$ to the x-axis, applying the liquid crystal Jones matrix to it, and then applying the Jones matrix for a polarizer at $\pi/2 - \theta$ to the result. In short, the Jones matrices operate one after the other starting from the right. The Jones matrix for the polarizer at an angle of $\pi/2 - \theta$ is

$$\begin{pmatrix} \sin^2 \theta & \sin \theta \cos \theta \\ \sin \theta \cos \theta & \cos^2 \theta \end{pmatrix}, \tag{13.27}$$

so the calculation goes as follows:

$$\begin{pmatrix} E_x \\ E_y \end{pmatrix} = \begin{pmatrix} \sin^2 \theta & \sin \theta \cos \theta \\ \sin \theta \cos \theta & \cos^2 \theta \end{pmatrix} \times$$
$$e^{i\epsilon_x} \begin{pmatrix} 1 & 0 \\ 0 & e^{-i\delta} \end{pmatrix} E_0 \begin{pmatrix} \cos \theta \\ -\sin \theta \end{pmatrix}. \tag{13.28}$$

After bringing a number of factors out of the Jones vector, the result is

$$\begin{pmatrix} E_x \\ E_y \end{pmatrix} = E_0 e^{i\epsilon_x} (1 - e^{-i\delta}) \sin \theta \cos \theta \begin{pmatrix} \sin \theta \\ \cos \theta \end{pmatrix}. \tag{13.29}$$

Since the last optical device is a polarizer at $\pi/2 - \theta$, a good check is to make sure the polarization is consistent with this. As can be seen from the Jones vector at the end, this is the case. The intensity is the modulus squared of E_x plus the modulus squared of E_y. This simply means finding the modulus squared of the factor before the Jones vector and then multiplying this by the sum of the modulus squared of the two elements of the Jones vector. The intensity is therefore

$$I = E_0^2 [2 - (e^{i\delta} + e^{-i\delta})] \sin^2 \theta \cos^2 \theta = E_0^2 \sin^2 (\frac{\delta}{2}) \sin(2\theta). \tag{13.30}$$

Because of the $\sin(2\theta)$ factor, the intensity is zero whenever the director is parallel to either of the polarizers ($\theta = 0, \theta = \pi/2$, etc.), as it should be. The maximum intensity is when $\theta = \pi/4, 3\pi/4, 5\pi/4$, etc. But the value of the intensity at these maxima depends on the optical retardation of the liquid crystal, $\delta = 2\pi \Delta n d/\lambda_0$. As δ increases, the maximum value of the intensity oscillates between 0 and E_0^2.

It should be noted that one must be careful about amplitudes when working with Jones calculus. If only the polarization state is desired, then there are no overall amplitude concerns, since only the ratios of E_x and E_y are important. However, if overall amplitudes or intensities are important, one must start with a proper Jones vector. This means writing it as the amplitude E_0 times a Jones vector in which the sum of the modulus squared of E_x and E_y is 1. In this way, the amplitude of the starting Jones vector is E_0. After applying Jones matrices, the amplitude of the final Jones vector can be found in the same way. Simply factor out what is necessary to make the sum of the modulus squared of E_x and E_y 1. The modulus of the factor in front of this "normalized" final Jones vector is the amplitude.

An additional example both emphasizes this point and allows the optical retardation (and therefore the birefringence) of a nematic liquid crystal to be measured. Rotating the crossed polarizers together on either side of an aligned liquid crystal sample and producing Eq. (13.29) is not a way to measure δ because in most cases the initial intensity of the light, E_0^2, is not known. So let us imagine the situation that the incident light is polarized along the x-axis, the director of the liquid crystal is oriented at 45°, and the second polarizer is at an arbitrary angle θ.

We first must find the Jones matrix for a liquid crystal with its director at 45°. As done before for a horizontal polarizer, we simply start with the Jones matrix for a liquid crystal with its director at 0°, and transform it to a coordinate system that has been rotated by -45° about the z-axis.

$$\text{LC at } 45° = \begin{pmatrix} \sqrt{1/2} & -\sqrt{1/2} \\ \sqrt{1/2} & \sqrt{1/2} \end{pmatrix} e^{i\epsilon_x} \begin{pmatrix} 1 & 0 \\ 0 & e^{-i\delta} \end{pmatrix} \begin{pmatrix} \sqrt{1/2} & \sqrt{1/2} \\ -\sqrt{1/2} & \sqrt{1/2} \end{pmatrix}$$

$$= e^{i\epsilon_x} e^{-i\delta/2} \begin{pmatrix} \cos(\delta/2) & i\sin(\delta/2) \\ i\sin(\delta/2) & \cos(\delta/2) \end{pmatrix} \quad (13.31)$$

If we now allow this Jones matrix to operate on horizontally polarized light of amplitude E_0 and then operate on the result with a polarizer at angle θ, we obtain the Jones vector

$$E_0 e^{i\epsilon_x} e^{-i\delta/2} \begin{pmatrix} \cos(\delta/2)\cos^2\theta + i\sin(\delta/2)\sin\theta\cos\theta \\ \cos(\delta/2)\sin\theta\cos\theta + i\sin(\delta/2)\sin^2\theta \end{pmatrix}. \quad (13.32)$$

The intensity of this light, after some algebra, is

$$I(\theta) = \frac{I_0}{2}[1 + \cos\delta\cos(2\theta)], \quad (13.33)$$

where $I_0 = E_0^2$. Notice that the term containing δ is no longer multiplied by E_0^2. The $I(\theta)$ function can be fit to the I versus θ data, with δ and I_0 the two fitting constants. Once a value for δ is determined, the birefringence Δn can be calculated as long as the wavelength and sample thickness are known. One drawback to this method of measuring the optical retardation is that δ appears as the argument of a cosine function. Therefore, there are multiple values of δ that produce the same $I(\theta)$. Likewise, since the cosine function does not depend on the sign of its argument, this method cannot discriminate between positive and negative optical retardation.

13.5 CIRCULAR BIREFRINGENCE AND OPTICAL ACTIVITY

Consider now linearly polarized light propagating along the helical axis of a chiral nematic liquid crystal. At every point in the sample the director is at some angle with respect to the initial polarization direction, but this angle is different for each "slice" of the liquid crystal. Thus it makes no difference if the incident light is polarized along the x- or y-axis (assuming the helical axis is along the z-axis), because in each case the distribution of director angles is the same. Thus whatever phase retardation an x-component of the light suffers, the y-component suffers the same retardation. Therefore, chiral nematic liquid crystals are not linearly birefringent for light propagating along the helical axis.

The situation is not the same for circularly polarized light. The helix of the chiral nematic liquid crystal is either right-handed or left-handed. If it is right-handed, then right circularly polarized light experiences a material with a director that rotates in space in the same sense as the electric field of the light rotates in space. Left circularly polarized light experiences a material with a director rotation in the opposite sense of its electric field rotation. The effect on the circularly polarized light is that right- and left-handed circularly polarized light propagate at different velocities. This is called circular birefringence and the anisotropy for circularly polarized light is defined by the difference between n_R and n_L, the indices of refraction for right- and left-handed circularly polarized light, respectively.

We can first understand the consequences of circular birefringence through some of the same types of diagrams we used earlier in the chapter. Figure 13.9 shows the time evolution of right- and left-handed circularly polarized light at some point in space. Notice that the electric field vector rotates in opposite directions for each polarization. The figure also shows that the combination of right- and left-handed circularly polarized light with equal amplitudes produces linearly polarized light.

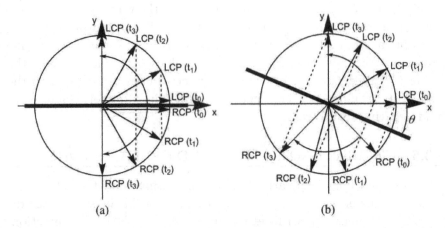

Figure 13.9 Left circularly polarized (LCP) and right circularly polarized (RCP) light combining to produce light linearly polarized along the x-axis in (a) and along a direction at an angle θ to the x-axis in (b). The electric field vectors are shown for four different times, with $t_3 > t_2 > t_1 > t_0$.

At all points in space, the electric field vector rotates clockwise for right circularly polarized light and counterclockwise for left circularly polarized light. If the light incident on the chiral nematic liquid crystal is polarized along the x-axis, then Figure 13.9(a) shows the time development of the electric field of this light. At all times the y-components of the two electric field vectors cancel and the x-components add, producing light linearly polarized along the x-axis. After propagating through the chiral nematic liquid crystal, however, the two circular polarizations are out of phase with one another. Figure 13.9(b) shows the time evolution of the electric field at the location where both polarizations emerge from the liquid crystal. Notice that the left circularly polarized light has suffered retardation relative to the right circularly polarized light (because $n_L > n_R$), so the addition of the two components is no longer linearly polarized along the x-axis. As Figure 13.9(b) shows, the sum of the two components is still linearly polarized light, but the direction of polarization has been rotated in a clockwise direction. This phenomenon is called optical activity, and its magnitude is just the rotation angle divided by the thickness of the sample. If the light is rotated clockwise to an observer looking back along the beam toward the light source, then the optical activity is positive by convention. If the rotation of the polarization is counter-clockwise, the optical activity is negative.

Let's look at this mathematically using Jones calculus. First, it is easy to show that the combination of RCP and LCP light of equal amplitudes produces linearly polarized light.

$$E_0 \begin{pmatrix} 1 \\ -i \end{pmatrix} + E_0 \begin{pmatrix} 1 \\ +i \end{pmatrix} = E_0 \begin{pmatrix} 2 \\ 0 \end{pmatrix} = 2E_0 \begin{pmatrix} 1 \\ 0 \end{pmatrix} \tag{13.34}$$

If the light has traveled through a chiral nematic liquid crystal of thickness d with $n_L > n_R$ ($k_L > k_R$), then the light emerging from the sample is

$$\begin{aligned} \begin{pmatrix} E_x \\ E_y \end{pmatrix} &= E_0 \begin{pmatrix} 1 \\ +i \end{pmatrix} e^{ik_L d} + E_0 \begin{pmatrix} 1 \\ -i \end{pmatrix} e^{ik_R d} \\ &= E_0 e^{i(k_L+k_R)d/2} \begin{pmatrix} e^{i(k_L-k_R)d/2} + e^{-i(k_L-k_R)d/2} \\ i e^{i(k_L-k_R)d/2} - i e^{-i(k_L-k_R)d/2} \end{pmatrix} \\ &= 2E_0 e^{i(k_L+k_R)d/2} \begin{pmatrix} \cos[(k_L - k_R)d/2] \\ -\sin[(k_L - k_R)d/2] \end{pmatrix}. \end{aligned} \tag{13.35}$$

But this is just linearly polarized light rotated clockwise from the x-axis by an angle $\theta = (k_L - k_R)d/2$, with θ having units of radians. Since this rotation is considered positive optical activity, the optical activity β is

$$\beta = \frac{\theta}{d} = \frac{k_L - k_R}{2} = \frac{(n_L - n_R)k_0}{2} = \frac{\pi(n_L - n_R)}{\lambda_0}. \tag{13.36}$$

As with optical retardation, notice that the optical activity tends to increase as the wavelength decreases and is directly proportional to the difference in the indices of refraction for left and right circularly polarized light.

13.6 OPTICS OF CHIRAL NEMATIC LIQUID CRYSTALS

The optics of chiral nematic liquid crystals are very complicated in general. An understanding of how light propagates in a chiral nematic liquid crystal demands that the equations of electromagnetism be solved for a material in which the anisotropy of the indices of refraction form a helix. The scope of such a calculation is beyond what space allows.

Still, it is instructive to describe the results of such a calculation for a few different cases.

In understanding the optics of different materials, it is often useful to ask what polarization passes through the material unchanged. In the case of a retarder with the fast or slow axis oriented along the x-axis, the polarizations that propagate unchanged (sometimes called the normal modes) are the horizontal and vertical polarizations. Another example is a chiral nematic liquid crystal if the pitch P is much less than the wavelength of the light λ_0. When this condition is met, the chirality of the liquid crystal q_0 is much greater than the wavevector k_0 of the light as it propagates along the helix of the liquid crystal.

$$P << \lambda_0 \quad \Rightarrow \quad 2\pi/P >> 2\pi/\lambda_0 \quad \Rightarrow \quad q_0 >> k_0 \qquad (13.37)$$

In this case, the liquid crystal acts like a rotator with the rotation angle equal to the optical activity β times the thickness of the liquid crystal sample d. Its Jones matrix is therefore

$$\begin{pmatrix} \cos(\beta d) & -\sin(\beta d) \\ \sin(\beta d) & \cos(\beta d) \end{pmatrix}. \qquad (13.38)$$

For theses chiral nematics, it is right and left circularly polarized light that propagates unchanged, as can be seen by the following equation.

$$\begin{pmatrix} \cos(\beta d) & -\sin(\beta d) \\ \sin(\beta d) & \cos(\beta d) \end{pmatrix} \begin{pmatrix} 1 \\ \pm i \end{pmatrix} = \begin{pmatrix} \cos\beta \mp i\sin\beta \\ \sin\beta \pm i\cos\beta \end{pmatrix} \qquad (13.39)$$

$$= (\cos\beta \mp i\sin\beta) \begin{pmatrix} 1 \\ \pm i \end{pmatrix}$$

This is exactly the case described in the previous section, so Eq. (13.35) applies. Unfortunately, n_R and n_L are complicated functions of $n_{\parallel}$, $n_{\perp}$, λ_0, and P.

Another case of propagation of light along the helical axis of a chiral nematic liquid crystal is very important for display applications. This special case is when the pitch P of the chiral nematic is much greater than the wavelength of the light λ_0. Now the chirality q_0 is much less than the wavevector of the light k_0. When this condition is met, the normal modes are two perpendicular linear polarizations that rotate with the director as it rotates in helical fashion. Since linearly polarized light along any axis can be formed from the combination of two perpendicular linear polarizations, any linear polarization incident of this liquid crystal rotates through the same angle as the director rotates. This is called the waveguide or Mauguin regime.

In or near the waveguide regime, a useful relation has been derived for the transmission (ratio of the outgoing light intensity to the incident light intensity) for a chiral nematic liquid crystal sandwiched between two polarizers at arbitrary angles. The propagation is along the helical axis of the chiral nematic liquid crystal. The x-axis of the coordinate system is parallel to the director at the front surface of the liquid crystal. The director at the back surface of the liquid crystal is at an angle Ω (counterclockwise from the x-axis). Similarly, the transmission axes of the entrance and exit polarizers are at angles ϕ_{ent} and ϕ_{exit}, respectively. If two variables are defined first,

$$\Gamma = \frac{2\pi \Delta n d}{\lambda_0} \qquad X^2 = \Omega^2 + \left(\frac{\Gamma}{2}\right)^2, \qquad (13.40)$$

where Δn and d are the birefringence and thickness of the liquid crystal sample, respectively, then the transmission T is given by

$$
\begin{aligned}
T = &\ \cos^2(\Omega - \phi_{exit} + \phi_{ent}) + \sin^2 X \sin[2(\Omega - \phi_{exit})]\sin(2\phi_{ent}) \\
&+ \frac{\Omega}{2X}\sin(2X)\sin[2(\Omega - \phi_{exit} + \phi_{ent})] \\
&- \frac{\Omega^2}{X^2}\sin^2 X \cos[2(\Omega - \phi_{exit})]\cos(2\phi_{ent}).
\end{aligned}
\qquad (13.41)
$$

A third and just as important case occurs when the pitch P is of the same order as the wavelength λ_0. This condition produces an effect much like the constructive interference of X-rays discussed in Chapter 3, but in this case constructive interference occurs for light in reflection. This is most easily seen using Figure 13.10, in which light is incident on a chiral nematic liquid crystal at an angle θ from the helical axis, while reflected light makes an angle of ϕ with the helical axis.

If the pitch of the chiral nematic liquid crystal is P, then the structure repeats itself over a distance equal to $P/2$. Reflected waves emanating from parts of the liquid crystal separated by $P/2$ are in phase if the path length difference is equal to a multiple of λ, where λ is the wavelength of light in the liquid crystal. As is evident from Figure 13.10, the path length difference $L_2 - L_1$ can be found by realizing that there are two triangles with L_2 as the hypotenuse. One triangle has an angle equal to θ and the other has an angle equal to $\pi - (\theta + \phi)$. The difference in length between the two paths is then

$$L_2 - L_1 = L_2 - [-L_2\cos(\theta + \phi)] = \frac{P}{2\cos\theta}[1 + \cos(\theta + \phi)]. \quad (13.42)$$

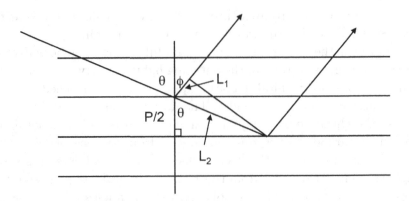

Figure 13.10 Reflection from repeating parts of the chiral nematic structure.

The condition for constructive interference is then

$$m\lambda = \frac{m\lambda_0}{\bar{n}} = \frac{P}{2\cos\theta}[1 + \cos(\theta + \phi)], \qquad (13.43)$$

where m is an integer greater than zero, λ_0 is the wavelength of light outside the liquid crystal, and $\bar{n} = (n_\parallel + n_\perp)/2$. If $\theta = \phi$ (the incident and reflected angles are equal), then this simplifies to

$$m\lambda_0 = \frac{\bar{n}P(2\cos^2\theta)}{2\cos\theta} = \bar{n}P\cos\theta. \qquad (13.44)$$

The implication of this result is apparent in the case of white light and a liquid crystal with a pitch close to the wavelength of the light in the liquid crystal. At any particular angle of reflection, only one wavelength in the visible part of the spectrum is reflected, making the liquid crystal appear quite colored. The color of the reflected light changes as the viewing angle changes. In addition, if the pitch of the chiral nematic liquid crystal changes (due to a change in temperature perhaps), then the color of the reflected light at all viewing angles changes. The phenomenon is called selective reflection, and thermometers utilizing chiral nematic materials in which the pitch is very sensitive to temperature are available commercially.

In a full theoretical development, the selective reflection shows itself as the inability of circularly polarized light with the same handedness as the liquid crystal to propagate along the helix if its wavelength λ_0 is between $n_\perp P$ and $n_\parallel P$, where $n_\parallel$ and $n_\perp$ are the indices of refraction for light polarized parallel and perpendicular to the director, respectively. Thus if linearly polarized light in this wavelength region is incident on the liquid crystal, since it can be considered to consist of equal amounts of right- and left-circularly polarized light, clearly one circular polarization is transmitted and the other is reflected. This is illustrated in Figure 13.11 for a right-handed chiral nematic liquid crystal. A left-handed structure would reflect the opposite circular polarization in the selective reflection regime.

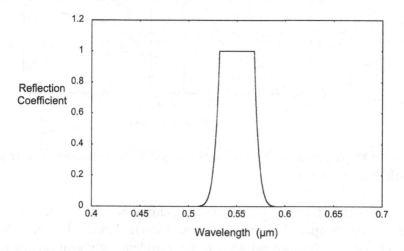

Figure 13.11 Selective reflection of right circularly polarized light from a right-handed chiral nematic liquid crystal.

For wavelengths outside this range, the normal modes are elliptically polarized light of opposite handedness that propagate with different velocities. As discussed previously, this condition leads to optical activity. What is interesting about this, however, is that the index of refraction for the elliptical polarization of the same handedness as the liquid crystal behaves anomalously in the vicinity of the reflection band. On one side of the reflection band it gets much larger than the index for the other elliptical polarization and on the other side it gets much smaller than for the other elliptical polarization. This fact in conjunction with our previous discussion of how optical activity depends on these two indices indicates that the optical activity gets very large and positive on one side of the reflection band and very large and negative on the other side of the reflection band. Near the reflection band, therefore, huge optical activity can result, with values like 10,000°/mm quite typical. This anomalous optical activity is illustrated in Figure 13.12.

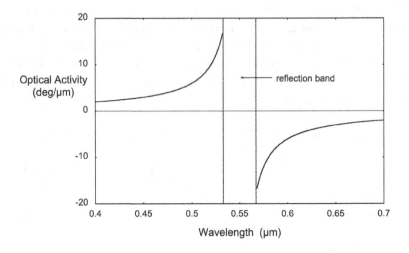

Figure 13.12 Anomalous optical activity for a right-handed chiral nematic liquid crystal.

Figure 13.12 is for a right-handed chiral nematic liquid crystal. Thus positive optical activity for wavelengths below the anomalous region and negative optical activity for wavelengths above the anomalous region is a clear signature of a right-handed structure. The optical activity in a left-handed chiral nematic liquid crystal behaves in just the opposite way. In addition, the exact shapes of the optical activity and selection reflection curves in the anomalous region depend on the thickness of the sample. Thinner samples show more rounded optical activity and reflection coefficient dependences, and the reflection coefficient is not necessarily 1 at any wavelength.

13.7 LIGHT SCATTERING

A nematic liquid crystal sample with the director oriented uniformly throughout the sample causes light to suffer reflection, refraction, and a change in its polarization state. The only light emanating from the sample is either in the reflected or transmitted beam. One look at a nematic liquid crystal demonstrates, however, that this is hardly the case. The liquid crystal sample looks turbid from any viewing angle, indicating that light is being scattered. The origin of the scattering is the random variation of the director at all points in the sample due to thermal effects.

The reason thermal fluctuations of the director are present is the small amount of energy required to distort a liquid crystal slightly. At room temperatures, the normal statistical fluctuations are enough to cause variations in the director that are quite capable of scattering light. This should be compared to fluctuations in the density, which depend on the magnitude of the compressibility. For liquid crystals, the compressibility is high enough so that thermal fluctuations in the density are very small, and thus produce little light scattering.

One characteristic of the scattered light from liquid crystals is that it is highly depolarized. That is, if linearly polarized light is incident on the sample, the scattered light has polarization components perpendicular to the incident polarization direction. The reason for this becomes obvious if we return to our discussion on anisotropy in Chapter 3. If the amplitude of the electric field in the incident light has components E_{0x}, E_{0y}, and E_{0z}, then the polarization this induces in the liquid crystal has components given by

$$\vec{P} = \begin{pmatrix} P_{0x} \\ P_{0y} \\ P_{0z} \end{pmatrix} = \epsilon_0 \overset{\leftrightarrow}{\chi_e} \vec{E} =$$

$$\begin{pmatrix} \chi_\perp & 0 & 0 \\ 0 & \chi_\perp & 0 \\ 0 & 0 & \chi_\parallel \end{pmatrix} \begin{pmatrix} E_{0x} \\ E_{0y} \\ E_{0z} \end{pmatrix} = \begin{pmatrix} \chi_\perp E_{0x} \\ \chi_\perp E_{0y} \\ \chi_\parallel E_{0z} \end{pmatrix}. \quad (13.45)$$

In general $\vec{P}$ is not parallel to $\vec{E}$, so the oscillating induced polarization produces light radiating in all directions and polarized in a different direction from the incident light.

A more detailed discussion can be based on how fluctuations and light scattering were described in Chapters 3 and 4. To simplify the development, let us assume that only the x-component of $\hat{n}$ changes and that this change is independent of y and z. From Eq. (3.31) it is clear that only one term in the splay deformation relation is non-zero. Hence, using Eq. (3.30), the free energy per unit area perpendicular to the x-axis due to this fluctuation, $G - G_0$, is very similar to Eq. (4.51) for a fluctuation in the order parameter,

$$G - G_0 + \frac{1}{2} \int_{-\infty}^{+\infty} \left(K_{11} \left[\frac{dn(x)}{dx} \right]^2 \right) dx, \quad (13.46)$$

where K_{11} is the splay elastic constant. Then following the exact same development as in Section 4.5 in which a wavevector representing a sinusoidal variation along the x-axis, $q_x = 2\pi/\lambda$, is defined, the mean-square value of $n(q_x)$ can be obtained.

$$\langle |n(q_x)|^2 \rangle \propto \frac{k_B T}{K_{11} q_x^2} \tag{13.47}$$

The general case of fluctuations in three dimensions is much more complicated, but follows the same reasoning. The result can be expressed more simply by realizing that if the proper coordinate system is used, the algebra simplifies. This coordinate system has the z-axis along the director, the 2-axis perpendicular to $\hat{n}$ and $\vec{q}$, and the 1-axis perpendicular to both the z- and 2-axes. In this coordinate system, all fluctuations can be described by the mean-square fluctuations in the $\hat{1}$ and $\hat{2}$ directions.

$$\langle |n_1(\vec{q})|^2 \rangle \propto \frac{k_B T}{K_{11} q_\perp^2 + K_{33} q_\parallel^2}$$

$$\langle |n_2(\vec{q})|^2 \rangle \propto \frac{k_B T}{K_{22} q_\perp^2 + K_{33} q_\parallel^2} \tag{13.48}$$

K_{22} and K_{33} are the twist and bend elastic constants, respectively, $q_\parallel$ is the component of $\vec{q}$ parallel to $\hat{n}$, and $q_\perp$ is the component of $\vec{q}$ perpendicular to $\hat{n}$, where $\vec{q}$ is the scattering wavevector. The first equation describes fluctuations in splay and bend, while the second equation describes fluctuations in twist and bend.

The intensity of light scattering is directly proportional to these mean-square fluctuations of the director, but the polarization of the incident and scattered light must be considered. The derivation relies on concepts that have not been introduced, but the result turns out to be extremely general and useful. If the incident polarization is denoted by $\hat{i} = (i_1, i_2, i_z)$ and the scattered polarization by $\hat{f} = (f_1, f_2, f_z)$, then the intensity of scattered light, I_s is given by

$$I_s \propto \langle |n_1(\vec{q})|^2 \rangle (i_1 f_z + i_z f_1)^2 + \langle |n_2(\vec{q})|^2 \rangle (i_2 f_z + i_z f_2)^2 \tag{13.49}$$

$$I_s \propto \frac{k_B T}{K_{11} q_\perp^2 + K_{33} q_\parallel^2} (i_1 f_z + i_z f_1)^2 + \frac{k_B T}{K_{22} q_\perp^2 + K_{33} q_\parallel^2} (i_2 f_z + i_z f_2)^2.$$

To demonstrate how useful this result is, let us consider a specific light scattering experiment. Imagine the incident light is polarized parallel to the director and only light polarized perpendicular to the director is detected at a scattering angle θ. Figure 13.13 shows this arrangement along with the $(\hat{1}, \hat{2}, \hat{z})$ coordinate system and $\vec{q}$. Section 3.5 gives the definition of $\vec{q}$ in a scattering experiment, showing that $q = |\vec{q}| = 2k\sin(\theta/2)$, where k is the magnitude of the incident and scattered wavevectors. From the figure it is clear that $\hat{i} = (0, 0, 1)$ and $\hat{f} = (\cos(\theta/2), \sin(\theta/2), 0)$. Note also that $q_{\parallel} = 0$ and $q_{\perp} = q$. With these assignments, the scattered light depends on the scattering angle according to the following relation,

$$I_s \propto \frac{\cos^2(\theta/2)}{K_{11}q^2} + \frac{\sin^2(\theta/2)}{K_{22}q^2}. \tag{13.50}$$

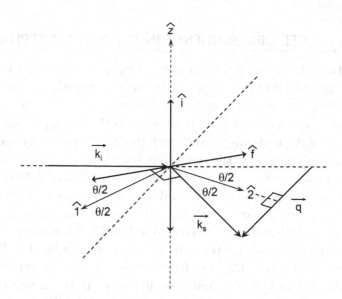

Figure 13.13 Directions in a light scattering experiment in which the scattering angle is θ and the director is along the z-axis. The coordinate system is given by $(\hat{1}, \hat{2}, \hat{z})$, the incident wavevector by $\vec{k}_i$, the scattered wavevector by $\vec{k}_s$, and the scattering wavevector by $\vec{q}$. $\hat{i}$ and $\hat{f}$ are the incident and scattered light polarizations, respectively.

Finally, if q is substituted in this equation, the full angular dependence of the scattered light is revealed.

$$I_s \propto \frac{\cot^2(\theta/2)}{K_{11}} + \frac{1}{K_{22}} \tag{13.51}$$

Therefore, if the scattered light intensity is plotted versus the cotangent squared of half the scattering angle, the slope divided by the intensity intercept equals K_{22}/K_{11}. In this way, light scattering is a very convenient way to measure the ratio between elastic constants.

It should be pointed out that in this analysis, θ is the scattering angle in the sample as opposed to the scattering angle measured in the laboratory. Therefore, in such an experiment, the indices of refraction of the sample, $n_\parallel$ and $n_\perp$, and the index of refraction of the surrounding medium must be known to convert one scattering angle into the other. The evaluation of q also must take the birefringence of the sample into account. In this example, the incident wavevector is $2\pi n_\parallel/\lambda_0$ and the scattered wavevector is $2\pi n_\perp/\lambda_0$, where λ_0 is the vacuum wavelength of the light.

13.8   INDUCED BIREFRINGENCE IN THE ISOTROPIC PHASE

Fluctuations in the isotropic phase were briefly discussed in Chapter 4, where it was seen that the average of the order parameter is zero but the average of the square of the order parameter is non-zero. The situation is different, however, if a field is applied to the isotropic phase. The anisotropy of the molecules and the interactions between the molecules cause slight alignment of the molecules and therefore a non-zero value of the order parameter. This in turn results in a slightly different index of refraction for light polarized parallel to the field compared to light polarized perpendicular to the field. Since this birefringence is only present if a field is applied, it is called induced birefringence.

Since the order parameter S is non-zero, we can use the Landau-de Gennes expression for the free energy of the isotropic phase, Eq. (4.1). But since S is so much smaller than in the nematic phase, we only have to keep the first term quadratic in S. We also saw that for a magnetic field, the free energy of an aligned phase is given by Eq. (12.25). This free energy depends on the anisotropy of the magnetic susceptibility $\Delta\chi_m$, which in turn depends on S. To bring this out explicitly, let us set $\Delta\chi_m = \Delta\chi_{max}S$, where $\Delta\chi_{max}$ is the magnetic susceptibility when the molecules are perfectly aligned. Therefore, $\Delta\chi_{max}$ depends on the molecular structure and does not depend on the order parameter and depends very weakly on the temperature. Since the director is parallel to the magnetic field $\vec{H}$, the free energy per unit volume can be described by

$$G = G_{iso} + \frac{1}{2}A_0(T - T^*)S^2 - \frac{1}{2}\mu_0\Delta\chi_{max}SH^2. \tag{13.52}$$

To find the value of S that minimizes G, simply take the derivative of Eq. (13.52) with respect to S, set it equal to zero, and solve for S, where

$$S = \frac{\mu_0\Delta\chi_{max}H^2}{2A_0(T - T^*)}. \tag{13.53}$$

But earlier in this chapter we discussed that $n_\parallel^2 - n_\perp^2 = (n_\parallel + n_\perp)\Delta n$ is proportional to S. Therefore, since $n_\parallel + n_\perp$ is almost independent of temperature and magnetic field, the induced birefringence is inversely proportional to $T - T^*$.

$$\Delta n \propto \frac{H^2}{(T - T^*)} \tag{13.54}$$

Figure 13.14 shows plots of the induced birefringence for several values of the applied magnetic field according to this relationship. Notice how small the induced birefringence is.

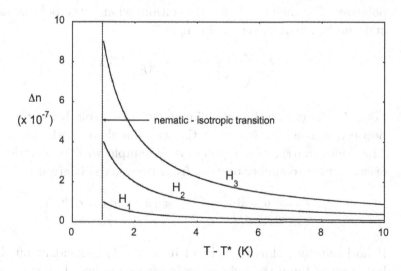

Figure 13.14 Variation of the induced birefringence Δn as a function of temperature T for fixed values of the magnetic field H ($H_1 < H_2 < H_3$). T^* is the temperature at which the isotropic phase is unstable according to Landau-de Gennes theory.

EXERCISES

13.1 Light from a laser has a wavelength λ_0 of 514 nm. What is the value of the wavevector k_0, frequency $\nu = 1/T$, and angular frequency ω of this light?

13.2 For the following Jones vectors, determine if the light is linearly, circularly, or elliptically polarized. If circularly or elliptically polarized light, is it right- or left-handed? What is its amplitude, and what angle does the linearly polarized or elliptically polarized light make with the x-axis?

$$\begin{pmatrix} 0 \\ 4 \end{pmatrix} \qquad 3\begin{pmatrix} 1 \\ -i \end{pmatrix} \qquad \begin{pmatrix} 2 \\ 5 \end{pmatrix} \qquad \begin{pmatrix} 1 \\ 3 + 4i \end{pmatrix}$$

13.3 Verify that the Jones matrix for a β rotator is correct by operating the Jones matrix given in Eq. (13.25) on linearly polarized light at an angle θ. Is the result linearly polarized light rotated by an angle β?

13.4 Each of the polarizations of light below passes through a vertical polarizer. For each incident polarization, what is the polarization and amplitude of the emergent light?

$$\frac{E_0}{2}\begin{pmatrix} -1 \\ \sqrt{2} \end{pmatrix} \qquad 3E_0\begin{pmatrix} 3 \\ 5i \end{pmatrix}$$

13.5 Let's do the calculation of a liquid crystal sample being rotated between crossed polarizers in the way not done in the chapter. The Jones matrix of a liquid crystal sample with retardation δ oriented at θ (counterclockwise from the x-axis) is given by

$$e^{i\epsilon_x}\begin{pmatrix} \cos^2\theta + \sin^2\theta e^{-i\delta} & \sin\theta\cos\theta(1 - e^{-i\delta}) \\ \sin\theta\cos\theta(1 - e^{-i\delta}) & \sin^2\theta + \cos^2\theta e^{-i\delta} \end{pmatrix}.$$

If horizontally polarized light of intensity I_0 is incident on this liquid crystal and the light emerging from the liquid crystal then passes through a vertical polarizer, show that the intensity of light emerging from the vertical polarizer is the same as given in Eq. (13.30), namely $I_0 \sin^2(\delta/2)\sin(2\theta)$.

13.6 A 0.1 g/mL solution of glucose in water has an optical activity of $+52.7\,°/dm$ at $20°C$ for light with a wavelength of 589 nm (1 dm $= 0.1$ m). What is the difference between the indices of refraction for left and right circularly polarized light in this solution? Which index is larger?

13.7 Eq. (13.41) simplifies if the input polarization is aligned with the director at the front surface of the liquid crystal sample, *i.e.*, $\phi_{ent} = 0$. If the waveguide condition holds, then $\Omega \ll X$ in Eq. (13.41). With this approximation, what is the expression for the transmission T? Explain that this is what you expect if the polarization of the light rotates with the director as the director rotates in helical fashion.

13.8 White light is incident on a chiral nematic liquid crystal as illustrated in Figure 13.10. The pitch is equal to 450 nm and the average index of refraction $\bar{n}$ equals 1.4. (a) If the white light is incident with $\theta = 0$, what wavelength light is reflected backwards? (b) If the white light is incident with $\theta = 25°$, what wavelength of light is reflected with $\phi = 25°$?

13.9 Instead of the arrangement given in Figure 13.13, the director is perpendicular to the polarization of the incident light and in the plane defined by $\vec{k}_i$ and $\vec{k}_s$. (a) Find the appropriate coordinate system $(\hat{1}, \hat{2}, \hat{z})$. (b) Find the two components of $\vec{q}$, namely $q_{\parallel}$ and $q_{\perp}$, in terms of q. (c) Use Eq. (13.49) to find how the intensity varies with the scattering angle θ. Be sure to take the angular dependence of q into account.

13.10 After measuring the induced birefringence Δn at a fixed value of the magnetic field H at many temperatures in the isotropic phase, a student makes a graph of $1/\Delta n$ versus temperature. (a) Describe what the data should look like if graphed this way, and (b) at what value should an extrapolation of the data cross the temperature axis?

CHAPTER 14

Liquid Crystal Displays – Here, There, and Everywhere

14.1 BASIC STRUCTURE OF LIQUID CRYSTAL DISPLAYS

Chapter 1 contains a short description of liquid crystal displays (LCDs) and Figure 1.17 illustrates a vertically aligned LCD. Before we discuss some of the different types of LCDs, let us examine components of a display that are common to nearly all LCDs. Figure 14.1 shows a cross-section of a typical LCD. There are almost always polarizing films on the outside of the top and bottom glass substrates. The inside of the top and bottom glass substrates are coated with a thin layer of indium-tin oxide (ITO), which acts as a transparent conducting film, and then coated with a thin film that promotes alignment of the liquid crystal. In some displays the conducting film is coated only on the inside of one of the glass substrates. The liquid crystal itself is between the glass substrates and completely sealed from the atmosphere. If the display is to use ambient light, there is a reflector behind one of the glass substrates. If the display is to use a light source, then there is a source of light (often an LED) in back of the display that provides illumination.

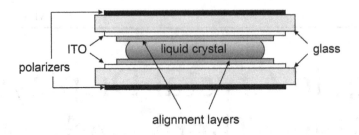

Figure 14.1 Components of a typical liquid crystal display. A display using ambient light has a reflector behind the display, but most LCDs utilize a light source behind the display.

14.2 TWISTED NEMATIC DISPLAYS

The twisted nematic (TN) display was the first successful LCD that demonstrated that high quality displays could be fabricated using liquid crystals. In this type of display, the alignment layers cause the director of the liquid crystal next to the glass to align parallel to the glass. In addition, the alignment direction on each piece of glass is perpendicular to one another. The transmission axis of the polarizer on the outer surface of each glass substrate is parallel to the alignment direction on the inside surface. Therefore, the alignment layers cause the director of the liquid crystal between the glass substrates to rotate by 90° in going from one glass surface to the other. If the cell thickness is on the order of 10 μm and the birefringence of the liquid crystal is not too large, the light passes through the liquid crystal with its direction of polarization rotated by 90° (see Section 13.6). Since the polarization direction of the polarizing film on one glass substrate is perpendicular to the polarizing film on the other glass substrate, light passes through the display. If a reflector is placed behind the display, the light passing through the display is reflected back through the display, rotating once again by 90° and passes again through the polarizer on the other glass substrate. In ambient light, the fact that the display reflects gives it a silver appearance.

If the liquid crystal has positive dielectric anisotropy, as discussed in Chapter 12 the director tends to align itself with the electric field. A voltage applied to the ITO films therefore causes the director of the liquid crystal to orient perpendicular to the glass surfaces where the electrodes are if it is above the threshold value. There is a layer of liquid crystal approximately one coherence length thick next to each surface in which the director remains parallel to the glass surfaces, but otherwise the director is nearly perpendicular to the glass surfaces and therefore nearly parallel to the propagation direction of the light throughout most of the display. Figure 14.2 depicts the zero voltage "off" state and the above threshold voltage "on" state. This almost uniform director configuration causes the light polarized by the first polarizer to pass through the liquid crystal without any change to its polarization state. It is therefore extinguished by the polarizer on the other glass substrate, and as a result no light is transmitted. In this state, the cell appears dark.

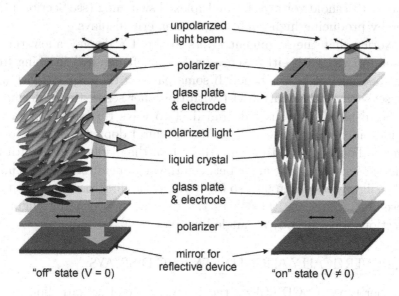

Figure 14.2 "Off" and "on" states for a twisted nematic liquid crystal display.

The LCD just described produces black numbers, letters, or figures on a silver background in ambient light. Since little current flows from one ITO electrode to the other, this type of display consumes extremely little power, making it ideal for battery-powered applications. It is called a passive display, since no light is generated by the display. It can be made to work as an active display by placing a diffuse light source behind the display. The light passes once through the display, forming black characters on a bright background. Power consumption is increased a great deal due to the use of a light source behind the display, but it does allow the display to function when the ambient light is too dim.

There are many variations to this basic design. If the polarization directions of the two polarizing films are parallel to each other rather than perpendicular, then the display appears dark in the "off" state and bright in the "on" state. The addition of colored filters can make the bright state highly colored instead of silver in either configuration of the polarizers. In the super-twist nematic (STN) LCD, the director twists by 270° instead of 90° in going from one glass surface to the other. Such a display possesses several superior qualities, such as sharper threshold voltages for multiplexed switching (see Section 14.6), thereby producing high-quality alpha-numeric displays.

Notice that the alignment layers in the TN display allow the director to twist with either sense (right- or left-handed) in going from one glass surface to the other. If some parts of the display twist with one sense and other parts twist with the other sense, the performance of the display is seriously degraded. Two ways this situation can be avoided include: (1) a small amount of chiral dopant is added to the nematic liquid crystal to ensure that all of the display twists with the same sense, and (2) the angle between the alignment directions is made to be slightly less than 90°, so everywhere the director twists with one sense, *i.e.*, the sense that takes the director between the alignment directions separated by an angle less than 90°.

14.3 VERTICALLY ALIGNED NEMATIC DISPLAYS

Another type of LCD utilizes the birefringence of nematic liquid crystals to change the polarization state of light. In Chapter 13, the optical retardation introduced by a liquid crystal film is described at length. The important point is that optical retardation changes linearly polarized light into elliptically polarized light. If the retardation is exactly 90°, circularly polarized light emerges. If the retardation is 180°, then the light emerges linearly polarized but along the direction perpendicular to the original polarization direction. Imagine a display with polarizing films perpendicular to each other on the outside of each glass substrate but now with films that promote alignment of the director perpendicular to the glass surfaces. This is called a homeotropic or vertical texture (as opposed to a uniform director configuration parallel to the glass surfaces everywhere — called a homogeneous or planar texture). Such a display is called a vertically aligned nematic (VAN) display. Since light propagation along the director does not change its polarization state, the display appears black as no light emerges through the crossed polarizers. This "off" state is shown in Figure 14.3.

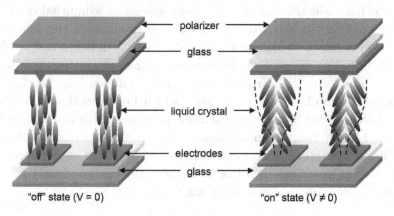

Figure 14.3 "Off" and "on" states for a vertically aligned nematic (VAN) liquid crystal display.

If the liquid crystal has negative dielectric anisotropy, application of a voltage to the ITO films causes the director to tilt away from the normal to the glass surfaces as shown in Figure 14.3. As long as the direction of this tilt is not parallel to either of the polarizers, this introduces optical retardation because the index of refraction for light polarized parallel to the director is different from the index of refraction for light polarized perpendicular to the director. Some of the resultant elliptically polarized light (all of it if the retardation is 180°) passes through the second polarizer and the display appears bright. In fact, since the amount of retardation (and therefore the brightness of the display in its "on" state) depends on the magnitude of the voltage applied to the display, this type of display can be used to produce intensities between its brightest state and its dark state. This is called a grayscale.

If the dielectric anisotropy is negative, the director tilts to be closer to perpendicular to the electric field, but in principle it can do this along any azimuthal direction. Additional steps must be taken to avoid the possibility that in some regions the tilt is along one azimuthal direction and in another region the tilt is along another azimuthal direction. One way to do this to construct an alignment layer that causes the director to be not exactly perpendicular to the glass surface, but tilted by a slight angle in the same azimuthal direction. This slight deviation from the perpendicular direction (or from the direction parallel to the glass surface) is called pretilt. If there is some pretilt on both surfaces, the electric field will tilt the director along the same azimuthal direction throughout the display. Another technique to engineer the azimuthal direction of tilt is to fabricate a surface coating that has a very shallow incline. The director aligns perpendicular to the incline, producing pretilt. The azimuthal direction of the pretilt can also be varied so the director tilts uniformly around 360°, which improves the characteristics of the display as the viewing angle increases. This is illustrated in Figure 14.3 as due to a protuberance on the upper substrate. Pretilt is utilized in displays with planar alignment too, such as TN and STN displays. In these cases the director at the surface it not exactly parallel to the surface, but makes a sight angle with it.

14.4   IN-PLANE SWITCHING NEMATIC DISPLAYS

As mentioned previously, the electrodes of a liquid crystal display need not be on the opposite glass substrates. An important example is a display that utilizes a nematic liquid crystal with the positive and negative electrodes on the same glass substrate. Known as an in-plane-switching (IPS) display, this type of LCD has some superior characteristics and is widely used in large, high-quality displays. Let us imagine a liquid crystal sample with positive dielectric anisotropy between two glass substrates with films on each glass substrate producing parallel alignment in the same direction. The electrodes are thin strips parallel to each other, so when a voltage is applied to them, the resulting electric field is mostly parallel to the piece of glass and perpendicular to the electrodes (see Figure 14.4). If the alignment layer causes the director to be parallel to the electrodes, the electric field causes the director to rotate toward being perpendicular to the electrodes. In principle the rotation could be either clockwise or counterclockwise, so to keep the director uniform, the alignment direction on both glass substrates makes a slight angle with the electrodes. As shown in Figure 14.4, the polarizer on the glass substrate with the electrodes is parallel to the alignment direction, while the polarizer on the glass substrate without the electrodes is perpendicular to the alignment direction.

The electric field is not uniform everywhere between the electrodes, and the rotation of the director also is not uniform, being very small

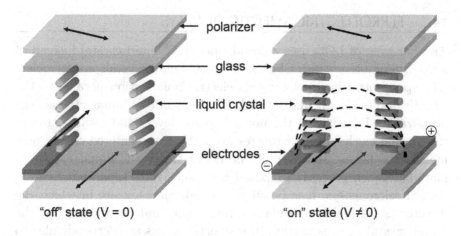

"off" state (V = 0)   "on" state (V ≠ 0)

Figure 14.4 "Off" and "on" states for an in-place switching (IPS) nematic liquid crystal display. The double arrows represent the transmission axes of the polarizers on each glass substrate. In the "off" state,
the director of the liquid crystal is uniform throughout the display and
parallel to the polarizer on the glass substrate with the electrodes. In
the "on" state, the director makes an angle with the polarizers.

near the glass substrates and maximum halfway between the glass substrates. But if we assume that the director rotates uniformly throughout the cell, then we can use some results from Chapter 13 to determine
what happens to light passing through the cell. Eq. (13.30) describes
the intensity transmitted by a liquid crystal sample with its director
making an angle θ with crossed polarizers on either side of the sample. When the director is parallel to one of the polarizers, as is true
when no voltage is applied to the electrodes, $\theta = 0$ and the transmitted intensity is zero. When a voltage is applied to the electrodes,
θ is no longer zero and the transmitted intensity depends on the initial intensity, the optical retardation of the liquid crystal sample, and
$\sin^2(2\theta)$. The maximum transmittance when $V \neq 0$ can be achieved
by designing the birefringence of the liquid crystal, the value of the
applied voltage, and the thickness of the sample so that the optical
retardation is 180°. Since θ depends on the magnitude of the voltage
applied to the electrodes, IPS displays can produce a grayscale.

14.5 FERROELECTRIC SMECTIC DISPLAYS

Another type of LCD uses a chiral smectic C liquid crystal instead of a nematic liquid crystal. Chiral smectic C liquid crystals are ferroelectric, spontaneously developing an electric polarization parallel to the smectic layers. In an undistorted chiral smectic C liquid crystal, the polarization is at 90° to the normal to the layers and rotates around the normal as the director rotates around a cone centered on the normal to the layers. However, if the chiral smectic C liquid crystal is placed between properly prepared glass substrates separated by only several micrometers, it is possible to establish a texture in which the director is parallel to the glass surfaces and uniform throughout the liquid crystal. In this texture, the smectic planes are perpendicular to the glass surfaces and the spontaneous polarization is along the normal to the glass surfaces. This is shown in the "off" state of Figure 14.5.

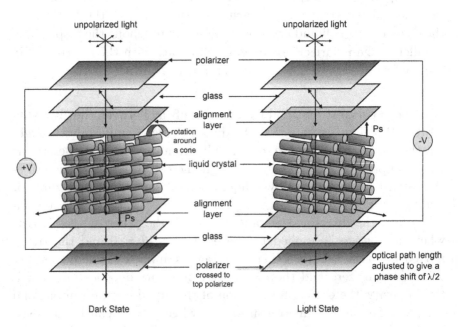

Figure 14.5 "Off" and "on" states for a ferroelectric smectic liquid crystal display.

The director is tilted by some angle (the tilt angle) away from the normal to the smectic planes and the spontaneous polarization is directed from the top glass substrate toward the bottom glass substrate. It must be pointed out that to obtain this texture, a DC electric field directed from top to bottom to orient the spontaneous polarization must be present. If the electric field is switched so it is directed from the bottom glass substrate towards the top glass substrate, the polarization rotates 180° as the director maintains the same tilt angle but rotates around the cone centered on the normal to the smectic planes. This situation is depicted in the "on" state of Figure 14.5.

All that is necessary to make this device into a display is to add polarizers to the outside of the two glass substrates. If the bottom polarizer is parallel to the director when the cell is in the configuration shown in the "off" state of Figure 14.5 and the top polarizer is perpendicular to the bottom polarizer, then light entering the cell from below is polarized parallel to the director, therefore suffering no change in its polarization state. It is extinguished by the top polarizer so the display appears dark. If the tilt angle of the chiral smectic C liquid crystal is 22.5°, then light entering the display shown in the "on" state of Figure 14.5 from below is polarized at an angle of 45° to the director. If the thickness and optical anisotropy of the liquid crystal are just right, the light suffers a phase retardation of 180° and exits with a polarization oriented at 90° relative to the director. Thus it passes through the top polarizer and the display appears bright. Because the switching in both directions is driven by the interaction of the spontaneous polarization with the electric field (rather than the interaction between an induced polarization and the electric field or a relaxation process when the electric field is removed), these surface stabilized ferroelectric liquid crystal (SSFLC) displays are orders of magnitude faster than twisted nematic and vertically aligned displays.

There is another difference between SSFLC displays and most other LCDs that is as important as the faster switching. Although the electric field is necessary to obtain each of the states shown in Figure 14.5, no electric field is necessary to maintain either state. In most LCDs, one state is stable with no field present and the other state is stable with an electric field present. In an SSFLC display, both states are stable with no electric field present, making it a bistable display. This feature is especially advantageous when a display is used in one state for a long time, such as is true for an information sign, price tag, or advertisement. In such a case, power to the display itself, i.e., not to a backlight, is only supplied to switch the display. Figure 14.6 shows an SSFLC display that was switched off six months before the picture was taken, yet still is in the same state.

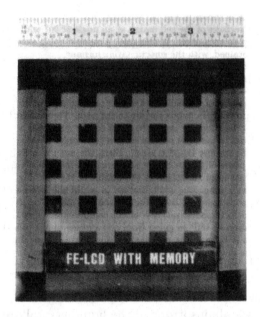

Figure 14.6 A ferroelectric liquid crystal display that was written using an electrical field, but with the electrical connections removed to demonstrate that the image remains in a bistable state. The picture was taken six months after the removal of the electrical connections. Image courtesy AT&T Bell Laboratories.

14.6 LCD TECHNOLOGY

So far the discussion has been on the basic working mechanism for a number of LCDs. In order to make a commercially successful LCD, many more factors must be considered and a huge number of problems must be solved. To discuss all of the considerations that go into the design and construction of these LCDs would take volumes, since an enormous amount of effort has gone into creating displays with superior characteristics. This section contains a discussion of some of the most important design requirements and the techniques employed to produce LCDs that meet these requirements.

The first consideration is the liquid crystal material itself. It must be stable over long periods of time and many switching cycles; it must have a liquid crystal temperature range that spans all of the temperatures at which the device is likely to be used; it must have dielectric and mechanical properties that allow it to be switched at the required speed; and its optical properties must be just right for it to function well as a display. These various requirements are achieved by using carefully designed mixtures of many different liquid crystal compounds. It should be remembered, however, that these requirements are contradictory, in that optimizing one property usually means not optimizing another property. Thus selecting the final mixture calls for a complicated balancing of these requirements against each other.

The use of eutectic mixtures often accomplishes the requirement of wide liquid crystal range. Such mixtures are discussed in Section 2.3, and a typical phase diagram showing the eutectic point is shown in Figure 2.13. The situation is much more complicated when many components are mixed together, but even in this case it is possible to obtain a wide nematic range by proper choice of the various concentrations.

Often it is desirable to employ a mixture with a high value of the electric susceptibility anisotropy $\Delta\chi_e$ in order to keep the threshold voltage low. Compounds with permanent dipoles at one end, such as the cyano group, tend to increase $\Delta\chi_e$, but in some cases such molecules tend to pair with anti-parallel dipole moments, effectively reducing $\Delta\chi_e$. For this reason, non-polar compounds are sometimes mixed with polar compounds to achieve the desired properties. Depending on the type of display, the desired ratio of the bend elastic constant to the splay elastic constant, K_3/K_1, should be large or small. Both of these conditions can be achieved by utilizing compounds with a smectic phase below the nematic phase, which tends to increase this ratio, and by using highly aromatic compounds with long terminal chains, which tends to have small values for this ratio.

The birefringence of a mixture is closely related to its electric susceptibility anisotropy at optical frequencies. Often the value of the birefringence is not crucial, although in many cases it must not be too large. In general, highly aromatic compounds have higher birefringence. To achieve fast switching times, a low viscosity is usually required. In general, shorter molecules are less viscous than longer molecules, but the nature of the molecular interactions is also extremely important. Viscosity is extremely temperature dependent, and can change by a factor of ten if the temperature is changed by 50°C. Mixtures have a viscosity intermediate between that of the individual compounds, so often some components are present in the final mixture to keep the viscosity low.

The director in a bulk liquid crystal sample tends to be randomly oriented with disclinations separating regions of different orientation. On the other hand, the liquid crystal in a display must be well aligned in order for the display to function effectively. This necessary alignment is almost always achieved by making one dimension of the liquid crystal sample small, perhaps 3 - 50 μm, by confining it between two glass substrates with a coating that causes the director next to the coating to align in a certain direction. Although the coating only interacts with molecules next to it, since the liquid crystal director tends to be correlated over tens of micrometers, the alignment at the glass surfaces is maintained across the small thickness of the sample. A coating made from a thin film of some polymers, usually a cross-linked polyimide, produces planar or homogeneous alignment of the director parallel to the surface. A coating of some surfactants, hydrophobic polymers, or evaporated SiO produces vertical or homeotropic alignment of the director perpendicular to the surface. Such alignment layers are usually applied using the spin-coating technique. In order to specify a direction of alignment parallel to the glass surface, the coating must be treated in some fashion, usually by rubbing, but photo-alignment, micro-scratching, and imprinting techniques are also utilized. As discussed previously, LCDs usually require that the alignment not be perfectly planar or vertical. Therefore, the coating is treated in a way that produces a pretilt angle of less than 5° for planar alignment and greater than 85° for vertical alignment.

Almost all LCDs require polarizers on the outside surface of the glass substrates and color LCDs must have color filters on the inside surfaces of one of the glass substrates. The polarizers are plastic films containing a high concentration of aligned iodine dyes. To protect the polarizers, passivation layers are present on both sides of the polarizing films. The color filter layer is usually about 2 to 3 μm thick and consists of an organic polymer with red, green, or blue organic dye incorporated in it. To prevent degradation of the color filter layer, a passivation layer of a polyimide is usually applied over the color filter layer. An important part of some LCDs is a solid optical element added to the outside of one of the glass substrates. Usually called a compensator, this layer has a fixed optical response to the light passing through the display, and it is this response plus the variable response of the liquid crystal that determines the characteristics of the LCD. LCD compensators are discussed in more detail in the following section.

Another issue in the design of high-quality LCDs concerns the "sharpness" of the display's response as it is turned on and off. This is illustrated in Figure 14.7, where the "sharpness" is defined as the difference between the 90% transmittance voltage and the 10% transmittance voltage.

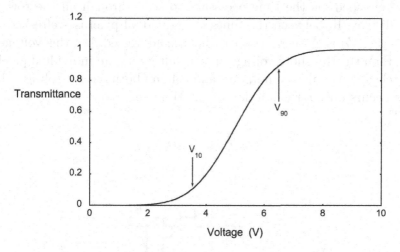

Figure 14.7 Transmittance as a function of applied voltage for an LCD.

The "sharpness" depends strongly on the type of display and less on the material used in the display. It is of paramount importance in displays with many pixels, such as large computer and television screens. The reason for this is that it is impossible to fabricate individual connections to each pixel in such displays. After all, the display may consist of over 2 million individual pixels, each controlling the amount of red, green, or blue light passing though a square area less than 100 μm on a side. Instead of addressing each individual electrode, electrodes on one piece of glass are electrically connected together in rows and the electrodes on the other piece of glass are electrically connected together in columns. While this enormously reduces the number of connections required, some form of multiplexing is necessary in order to turn individual pixels on and off.

A simple multiplexing scheme for a matrix display is shown in Figure 14.8. Each row receives a voltage V_R one at a time from top to bottom, a cycle which is repeated many times per second. Between the times a particular row receives the voltage V_R, the voltage applied to it is zero. As each row receives a voltage V_R, each column receives either a positive or negative voltage, $\pm V_C$, so that the voltage across the "on" pixels in the row is greater than the threshold voltage and the voltage across the "off" pixels in the row is less than the threshold voltage. In other words, the voltages being applied to the columns change each time a new row receives a voltage V_R, so that the individual pixels in each row are addressed in the desired way. The time for each pixel to be "refreshed" is the time necessary to scan through all the rows in the display. In between the times an individual pixel is "refreshed", a voltage $\pm V_C$ is applied as other rows are addressed, but this voltage is less than the threshold voltage. The result is that an individual pixel is switched "on" only when its row and column both receive voltages, but this occurs only for a fraction of the time necessary to cycle through all the rows.

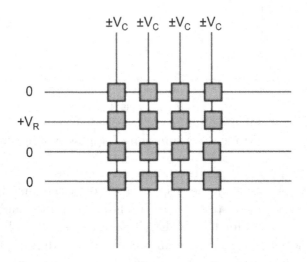

Figure 14.8 Typical multiplexing scheme for an LCD.

In actuality, the situation is slightly more complex than this. If the time to cycle between all the rows is short enough, the liquid crystal display responds to the mean squared voltage applied to it. Therefore, the mean squared voltage over one cycle of all the rows for an "on" pixel must be above the threshold, while the mean squared voltage over one cycle of all the rows for an "off" pixel must be below the threshold. Since the voltage necessary to turn a pixel on is applied for a fraction $1/N$ of the cycle time, where N is the number of rows, it gets more and more difficult to achieve the above threshold and below threshold mean squared voltages as N gets larger and larger. Clearly the "sharpness" of the response curve is crucial, in that the "sharper" the response curve, the closer the below threshold mean squared voltage and the above threshold mean squared voltage can be, and therefore, the larger N can be.

If it is desired to multiplex N rows and N columns, how close together must the above threshold mean squared voltage $\langle V_{ON}^2 \rangle$ and the below threshold mean squared voltage $\langle V_{OFF}^2 \rangle$ be? Since we are free to choose V_R and V_C, we must determine what values of these two voltages maximize the number of rows and columns we can multiplex with the same values of the mean squared "on" and "off" voltages. With N maximized, we are putting the least restriction on the mean squared "on" and "off" voltages. We first calculate the mean squared voltage applied to an "on" and "off" pixel, given values for V_R and V_C. An "on" pixel receives a voltage $V_R + V_C$ when its row is addressed and $\pm V_C$ when other rows are addressed. An "off" pixel receives a voltage $V_R - V_C$ when its row is addressed and $\pm V_C$ when other rows are addressed. The time dependence of the voltage across both an "on" and an "off" pixel is shown in Figure 14.9.

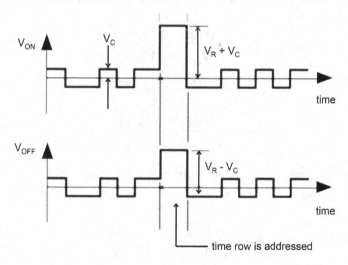

Figure 14.9 Time dependence of the voltages for a multiplexed LCD.

Notice that the mean squared voltage for each pixel when its row is not being addressed is V_C^2. Thus the mean squared voltage for an "on" pixel and an "off" pixel can be written as follows.

$$\langle V_{ON}^2 \rangle = \frac{(V_R + V_C)^2 + (N-1)V_C^2}{N}$$

$$\langle V_{OFF}^2 \rangle = \frac{(V_R - V_C)^2 + (N-1)V_C^2}{N} \tag{14.1}$$

Subtracting these two equations produces an expression for V_R.

$$V_R = \frac{N}{4V_C}\left[\langle V_{ON}^2\rangle - \langle V_{OFF}^2\rangle\right] \tag{14.2}$$

Adding the same two equations we just subtracted, and then substituting the expression for V_R, allows us to solve for N in terms of V_C and the mean squared "on" and "off" voltages.

$$N = \frac{8V_C^2\left[\langle V_{ON}^2\rangle + \langle V_{OFF}^2\rangle - 2V_C^2\right]}{\left[\langle V_{ON}^2\rangle - \langle V_{OFF}^2\rangle\right]} \tag{14.3}$$

To find the value of V_C that maximizes N, we take the derivative of N with respect to V_C^2 and set it equal to zero. This gives us the values of V_R and V_C that maximize N, along with the maximized value of N, $N_{\max}$, for what is known as the Alt-Pleshko optimization condition.

$$V_R^2 = \frac{\left[\langle V_{ON}^2\rangle + \langle V_{OFF}^2\rangle\right]^3}{\left[\langle V_{ON}^2\rangle - \langle V_{OFF}^2\rangle\right]^2}$$

$$V_C^2 = \frac{\langle V_{ON}^2\rangle + \langle V_{OFF}^2\rangle}{4} \tag{14.4}$$

$$N_{\max} = \frac{\left[\langle V_{ON}^2\rangle + \langle V_{OFF}^2\rangle\right]^2}{\left[\langle V_{ON}^2\rangle - \langle V_{OFF}^2\rangle\right]^2}$$

This last equation can be solved for the ratio of the root mean squared voltages, V_{rms}^{ON} and V_{rms}^{OFF}, for the "on" and "off" pixels, respectively.

$$\frac{V_{rms}^{ON}}{V_{rms}^{OFF}} = \frac{\sqrt{\langle V_{ON}^2\rangle}}{\sqrt{\langle V_{OFF}^2\rangle}} = \sqrt{\frac{\sqrt{N_{\max}}+1}{\sqrt{N_{\max}}-1}} \tag{14.5}$$

The need for a "sharp" threshold response in highly multiplexed displays is clear from this relationship. For a display of 300 rows, V_{rms}^{ON} cannot be more than 6% higher than V_{rms}^{OFF}. If the threshold is about 1 volt, then V_R must be about 25 volts and V_C about 0.75 volts. Figure 14.10 shows how the maximum number of rows depends on the ratio of V_{rms}^{ON} to V_{rms}^{OFF}.

One way in which the "sharpness" of an LCD can be increased is to fabricate the display with an active device such as a thin film transistor at each pixel. This is shown in Figure 14.11, where it can be seen that the voltages are applied to the electrodes and transistor in such a way that the response is determined mostly by the transistor rather

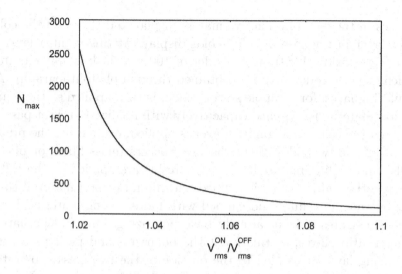

Figure 14.10 Maximum number of rows in a multiplexed display as a function of the ratio of the "on" to "off" voltages.

than the display itself. A properly designed transistor can increase the "sharpness" of an LCD a great deal, allowing for highly multiplexed displays. Such an active matrix LCD is extremely common.

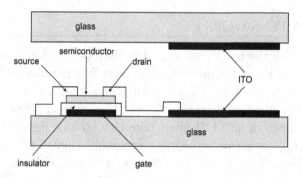

Figure 14.11 Thin film transistor active matrix display pixel.

The actual fabrication of an LCD is quite complex. Although ordinary soda-lime glass can be used in some displays, the high concentration of alkali ions in this type of glass causes problems for active matrix displays. As a result, aluminosilicate alkali-free glass must be used, and even in this case a passivation layer of SiO_2 is often deposited on the glass surface. The next step in the process is to fabricate the active devices in the case of an active matrix display and then add the color filter layer in the case of a full-color display. At this point a layer of indium-tin-oxide (ITO) on the order of 100 nm thick is applied, after which the electrode pattern is obtained through photolithography. An insulating layer, for example silicon oxide, is then applied so the liquid crystal material is switched capacitively with no direct current passing through the cell. An alignment layer is applied, completing the preparation of the two sides of the display. Spacers (glass fibers or plastic balls) and sealing material (an epoxy resin) are applied as the cell is put together, after which it is filled with liquid crystal material (usually by vacuum filling) and capped with more sealing material. If the glass substrates did not already have polarizing films and perhaps a compensating layer, at this point the polarizers and perhaps a compensating layer are added to the outside of the two glass substrates. The last step in the process is to make all of the necessary electrical connections between the display and the driving electronics.

The actual design of an LCD must take into account even more parameters than discussed here. Some of these are the switching time, the contrast ratio (ratio of the brightness in the "on" and "off" states), and the variation with viewing angle (how the display parameters depend on the angle from which the display is viewed).

14.7 COMPENSATORS

In almost all of the examples of LCDs discussed so far, analysis of the optical properties is for on-axis light traveling parallel to the normal to the cell. But an LCD utilizes a diffuse backlight, so there is a good deal of off-axis light seen by the viewer. Therefore, a proper analysis of the performance of an LCD must take this off-axis light into account.

As a simple example, imagine a vertically aligned nematic LCD as shown in Figure 14.3. For on-axis light and crossed polarizers, the "off" state appears dark. But consider off-axis light as shown in Figure 14.12. If the incident direction of the light is given by the angle θ with the normal to the display, then depending on the polarizer direction, the electric field of the light in general has two in-phase components. One is in the plane defined by the propagation direction and the director, and the other is perpendicular to that plane. The index of refraction for the former is given by Eq. (13.13),

$$n_e(\theta) = \frac{n_e n_o}{\sqrt{n_e^2 \cos^2 \theta + n_o^2 \sin^2 \theta}}, \tag{14.6}$$

and the index of refraction of the latter is just n_o. Since the velocities of the two components are not the same, a phase difference is introduced and the light striking the second (crossed) polarizer is elliptically polarized. Therefore, the second polarizer does not extinguish this light and the display does not appear dark. This is called the light leakage effect.

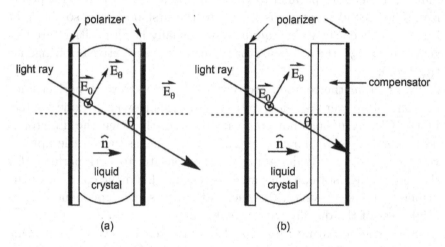

Figure 14.12 Off-axis light propagating through a vertically aligned nematic LCD. The light makes an angle θ with the director $\hat{n}$ and the components of the electric field in and out of the plane containing the director and light propagation direction are given by E_θ and E_0, respectively. (a) Without a compensator, the elliptically polarized light exiting the liquid crystal is not extinguished by the second (crossed) polarizer. (b) With a compensator that introduces an equal and opposite phase difference compared to the LCD, the two components are back in phase when they leave the compensator and thus are extinguished by the second (crossed) polarizer.

Since the path length of the light in the liquid crystal depends on θ, the phase difference introduced by the liquid crystal $\Delta\phi$ depends on the angle of the incident light in a complicated way,

$$\Delta\phi = \left(\frac{2\pi}{\lambda_0}\right)\left(\frac{d}{\cos\theta}\right)\left[\frac{n_e n_o}{\sqrt{n_e^2 \cos^2\theta + n_o^2 \sin^2\theta}} - n_o\right], \qquad (14.7)$$

where d is the thickness of the liquid crystal and λ_0 is the vacuum wavelength of the light. However, a uniaxial birefringent plate that introduces the opposite phase difference, i.e., $-\Delta\phi$, compensates for the LCD phase difference for all θ. The two components of the light are in phase as they exit the compensator (also called a compensating film) and therefore parallel to the transmission axis of the first polarizer. The second polarizer is at $90°$ to the first polarizer, so the light is extinguished. Since the liquid crystal usually has a positive birefringence ($n_e > n_o$), the compensator typically has a negative birefringence ($n_e < n_o$).

Notice that the compensator can be designed to work perfectly for the dark state (zero applied voltage), and this is what is done for real LCDs. Clearly it does not compensate perfectly when the director is no longer perpendicular to the LCD, as is the case when the applied voltage is not zero. By designing the compensator to work perfectly for the zero voltage state, a dark state is produced and the contrast of the display is large. However, this means there is some dependence of the brightness of the display with viewing angle.

Such compensating films can be fabricated in several ways. One is to mechanically stretch an extruded or casted polymer. The drawback of this technique is that only a small amount of birefringence can be achieved. A second method is to use cross-linkable nematic liquid crystals. The liquid crystal is first aligned and then the cross-linking reactions are induced, producing a solid film (see Section 10.6). A third method uses lyotropic liquid crystals that are applied to an orienting substrate and then allowed to dry. Because discotic liquid crystals usually have negative birefringence, they find use as compensating films. When used in this way, the films are passive, meaning that no voltage is ever applied to them. Only the LCD itself that uses calamitic liquid crystals is switched. In fact, by creating a film of a discotic liquid crystal with different boundary conditions on the top and bottom, good compensation can be achieved for both the $V = 0$ and $V \neq 0$ states of the LCD. The discotic liquid crystal is cross-linked while in this configuration resulting in a solid film with the correct optical properties.

EXERCISES

14.1 A TN display with a thickness of 10 μm uses 5CB and a very small amount of a chiral dopant that induces a right-handed twist. The specifications for 5CB at 25°C, 546.1 nm, and 1 KHz are: $K_1 = 1.2$ pN, $K_2 = 0.6$ pN, $K_3 = 1.5$ pN, $n_\parallel = 1.7257$, $n_\perp = 1.5392$, $\chi_\parallel = 19.78$, and $\chi_\perp = 6.59$. Assuming that the threshold voltage necessary to deform the director configuration is the same as for an untwisted cell, calculate the threshold voltage of this TN LCD for an applied electric field with a frequency of 1 KHz.

14.2 Assume that the surface directors of the TN display described in the previous exercise are oriented parallel to the polarizer on the same glass substrate and that the polarizers are oriented perpendicular to one another. Also assume that the director twists with the same sense throughout the display. Use Eq. (13.41) to find the transmission of the cell for 546.1 nm light if no electric field is applied. Does the transmission change if a small amount of chiral dopant that induces a left-handed twist is used?

14.3 (a) Use the discussion of the Frederiks transition in Chapter 12 to determine a formula for the threshold electric field of a VAN display. (b) The specifications for a liquid crystal mixture suitable for a VAN display, Merck MJ961213, for 550 nm light are: $K_1 = 13$ pN, $K_2 = 5.8$ pN, $K_3 = 15$ pN, $n_\parallel = 1.5622$, $n_\perp = 1.4791$, $\chi_\parallel = 3.6$, and $\chi_\perp = 7.4$. Use these specifications to calculate the threshold voltage for a VAN display with a thickness of 3.0 μm.

14.4 Assume that the director of the VAN display described in the previous exercise in the "on" state is uniformly oriented parallel to the glass surfaces and at 45° to the crossed polarizers. What is the transmission of the display for 550 nm light?

14.5 A plot of N_{max} versus $V_{rms}^{ON}/V_{rms}^{OFF}$ is shown in Figure 14.9 for a display with the same number of rows and columns. Derive the equation describing how N_{max} depends on $V_{rms}^{ON}/V_{rms}^{OFF}$ and use it to verify a few points in the plot.

14.6 Let $V_{rms}^{OFF} = 0.5$ volts and $V_{rms}^{ON}/V_{rms}^{OFF} = 1.05$ for a display with the same number of rows and columns. This means that the "below threshold" and "above threshold" voltages differ by only 5%. According to the Alt-Pleshko optimization condition, what must the values of V_R and V_C be, and what is the maximum number of rows this display can have?

14.7 The liquid crystal in a vertically aligned nematic LCD has a birefringence for 550 nm light of 0.083, and the thickness of the liquid crystal layer is 4.0 μm. If the compensator has a birefringence of -0.035 for 550 nm light, how thick should the compensator be?

CHAPTER 15

Other Applications — Optical Fiber Networks to Shampoo

15.1 LIQUID CRYSTAL ON SILICON TECHNOLOGY

Along with the development of the types of LCDs discussed in Chapter 14, another technology exists that has several advantages the others do not. It relies on the mature state of CMOS (complementary-metal-oxide-semiconductor) integrated circuitry, which can be processed to be extremely complex, very small scale, and intrinsically high performance, all on a silicon wafer. The idea is for a silicon wafer with the CMOS circuitry to be the backplane of a reflective LCD. This places the reflector between the silicon substrate and the liquid crystal rather than behind the glass substrate. The new technology, called LCoS (liquid-crystal-on-silicon) allows all the electronic circuitry to be behind the liquid crystal. No longer does space between the pixels have to be reserved for the electronic circuitry.

The basic structure of an LCoS device is shown in Figure 15.1. The electric field is produced by the CMOS circuitry between the reflective metal layer and the ITO film and is directed perpendicular to the substrates. Thus any of the types of LCDs with the applied electric field perpendicular to the substrates can be used with LCoS (*e.g.*, twisted nematic, vertical aligned nematic). Such devices can be put to good use in projection systems, where a small LCoS cell can reflect light that is then projected on a screen. Such small LCoS devices are possible because higher pixel densities can be achieved, a real advantage of the LCoS technology. In the case of such displays, there has to be a polarizer on the outside of the glass substrate (not shown in the figure). A color LCoS display creates red, blue, and green images and combines them optically, or the three monochrome images are displayed sequentially at a fast enough rate so the human eye interprets the correct color image. Projection displays are less popular now because of the huge advances in large LCD screens. However, a small projector casting its image on a screen or wall is much more portable than a large LCD display, so such projectors are still used widely.

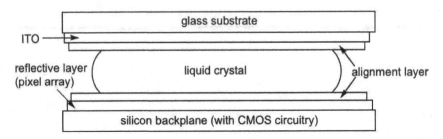

Figure 15.1 Structure of a typical LCoS device. If the device is to be used as a reflective display, in addition to what is shown, a polarizer must be added to the outside of the glass substrate.

When an LCoS device is used as a display, it controls the intensity of light reflected from the device. But without a polarizer, an LCoS device can control the angle of reflection (with little change in intensity) in what is known as a phase-only LCoS device. Such non-display devices have become extremely important in several types of optical systems, with optical communication systems perhaps the best example.

One of the methods used to allow a single optical fiber to carry huge amounts of data is through wavelength division multiplexing (WDM). Such systems take the 5000 GHz wide spectrum centered at 1550 nm used in optical communications and divide it into channels perhaps only 50 GHz wide. In this way 100 data channels can be carried over a single optical fiber simultaneously. If one thinks about each channel being a slightly different "color", then each of the 100 "colors" of the light in the fiber carries independent data. At different points in the communication system, some "colors" must be separated from the other "colors" so the data in the different channels can be directed in different ways. For example, a device that takes an optical signal of many "colors" as its input and outputs differently directed signals each of a single "color" is a component of a wavelength selective switch (WSS) and often relies on LCoS technology.

How can an LCoS device vary the direction of the reflected beam? Imagine a twisted nematic display that has been altered in two ways. First, there are no polarizers. Second, the alignment directions of the films on the two substrates are parallel to each other instead of being perpendicular. Both the "off" and "on" states of such an electrically controlled birefringence (ECB) device, with pretilt, are depicted in Figure 15.2. If the light entering the device in its "off" state is polarized as shown in the figure, it propagates with its polarization nearly parallel to the director. If the light entering the device in its "on" state is polarized as shown in the figure, it propagates with its polarization nearly perpendicular to the director. According to Eq. (13.18), light reflected from an "on" pixel is optically retarded compared to light reflected from an "off" pixel by $\phi = (4\pi/\lambda_0)\Delta nd$, where $\Delta n = n_\parallel - n_\perp$, d is the thickness of the liquid crystal, and λ_0 is the vacuum wavelength of the light. Actually, since the director everywhere is not perfectly parallel and perpendicular to the light polarization in the "off" and "on" states, respectively, the relative retardation is slightly less than $(4\pi/\lambda_0)\Delta nd$. More precisely, depending on the strength of the electric field applied to the liquid crystal, the relative optical retardation can be controlled between 0 and slightly less than $(4\pi/\lambda_0)\Delta nd$. Thus any amount of relative optical retardation in this range can be set for each pixel of the device.

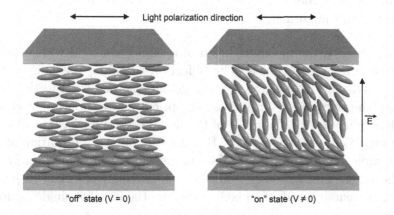

Figure 15.2 "Off" and "on" states of a reflective ECB device. The black film on the lower substrate represents a reflective coating.

Such a device can control the direction of the reflected light. For example, imagine light normally incident on the device strikes two pixels (see Figure 15.3). The light is in phase as it enters both pixels, but if the right pixel introduces some optical retardation ϕ relative to the left pixel, then the light emitted from the pixels is no longer in phase. This means that the wavefront of the emitted beam, *i.e.*, the line of constant phase, is not parallel to the substrate (as it would be for simple reflection). To determine the reflected wavefront, one only has to find the point on the ray emanating from the left pixel that has traveled a distance $L = \lambda_0\phi/(2\pi) = \phi/k_0$ so that its phase is the same as the light emanating from the right pixel (k_0 is the vacuum wavenumber). As can be seen from the figure, this distance is along a ray that makes an angle $\theta = \sin^{-1}[(1/k_0)(\phi/s)] = \sin^{-1}[(\lambda_0/2\pi)(\phi/s)]$ with the normal to the substrate, where s is the distance between pixels.

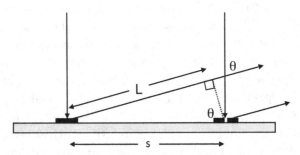

Figure 15.3 The incident light is normal to the device, but because there is more phase delay in the right pixel, the reflected beam makes an angle θ with the normal.

In actuality, the incoming beam of a single "color" is wide enough to strike a number of pixels. The reflection is still at θ as long as the optical retardation increases by ϕ in going from one pixel to the next toward the right. Since an increasing optical retardation might easily reach a value outside the range of the liquid crystal device, the optical retardation only increases to 2π. But since an optical retardation of 2π is the same as zero optical retardation, instead of increasing past 2π, it resets to zero and increases from there in moving from pixel to pixel. This is repeated for all the pixels struck by the incident single "color" beam.

A schematic showing the function of an LCoS device in a WSS is presented in Figure 15.4. Using the previous example, imagine that the incident light containing many "colors" is reflected by a diffraction grating and then directed toward an LCoS device. The diffraction grating separates the different "colors" slightly so they strike different columns of the LCoS device. The device is programmed to reflect the different "colors" in different directions so they end up taking different routes. In this way, the data channels that were combined so they could be carried by the same optical fiber are separated and routed to different optical fibers.

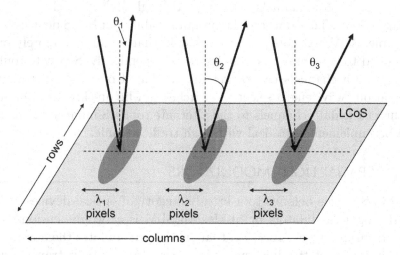

Figure 15.4 Operation of an LCoS device. The incident light contains many data channels, each being carried by a slightly different wavelength and striking different columns of the device. The gray areas show where the light of each channel strikes the LCoS device. As explained in the text, the pixels in each of these areas are programmed to reflect the light in a specific direction, given by the angles θ_1, θ_2, and θ_3.

Notice that the LCoS device only works for light polarized in a certain direction. The light in an optical fiber, however, is not polarized. To overcome this problem, the light from the fiber is separated into two perpendicular linear polarizations, and one of the polarizations is rotated by 90° so it is parallel to the other polarization. Each polarization is directed to different areas of an LCoS device, one of the redirected beams is rotated by 90°, and finally both are combined before being fed into a fiber.

While the discussion has concerned the use of an LCoS device to separate multiple wavelength channels, it should be pointed out that it can be used in reverse. The individual wavelength channels are created separately and then must be combined into the same optical fiber. This can be done by programming an LCoS device to direct multiple input beams coming from different directions into a single output beam. In fact, these two basic LCoS functions make multiple wavelength selective switches the heart of optical communication networks. For example, they are critical to the performance of network nodes, where some data channels are passed through and other channels are separated out. This is where the programmability of LCoS devices is so valuable. A WSS is completely flexible, in that it can be remotely programmed to control each data channel independently. Separate routes for the data channels can be set up, so that if one suffers a failure, the various wavelength selective switches can be quickly programmed to direct the data channels to the alternate route. Similar adjustments can be implemented to deal with high traffic volume.

15.2 SPATIAL LIGHT MODULATORS

An LCoS device belongs to a broad category of optical devices called spatial light modulators or SLMs. An SLM is a two-dimensional device, transmissive or reflective, that spatially modulates the amplitude and/or phase of the incident optical wavefront. Many transmission SLMs are based on the LCD technology described in the previous chapter, while many reflection SLMs rely on the LCoS devices described in the previous section. The two-dimensional array of pixels of a liquid crystal SLM can be controlled optically in an optically addressed SLM (OASLM) or electrically in an electrically addressed SLM (EASLM).

An LCoS device can be considered an electrically addressed, parallel aligned nematic (PAN), phase-only spatial light modulator. The same basic structure can be optically addressed if the CMOS circuitry is removed and replaced with a photosensitive material. In this way, an optical control beam incident on the back of the device produces an electric field pattern in the liquid crystal that corresponds to the amplitude pattern of the control beam wavefront. This electric field pattern in the liquid crystal controls how the light beam incident on the front of the device is reflected. A simplified diagram of such a device is shown in Figure 15.5.

An example of a transmission EASLM that modulates the amplitude and phase of the wavefront is simply a twisted nematic display

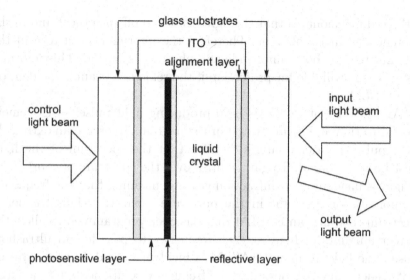

Figure 15.5 Optically addressed liquid crystal spatial light modulator.

with a two-dimensional grid of electrically controlled pixels. Unlike a phase-only SLM and just like a TN LCD, this device has perpendicular polarizers on the outside of each glass substrate.

Liquid crystal SLMs have a broad range of applications. Their ability to manipulate light pixel by pixel allows them to perform operations in parallel, an important process in optical computing. There are some real advantages of computing with optical light beams. For example, if a plane wavefront is incident on an SLM programmed with a transmission pattern, some distance away from the SLM the emerging wavefront is the Fourier transform of the SLM transmission pattern.

This example illustrates a more general use of EASLMs, namely manipulating the shape of a wavefront. By programming the intensity pattern of light emanating from the EASLM, one can produce light with specific wavefront shapes. One commercial use of this technology is in laser surgery. With a slight modification, EASLMs can be used for shaping pulses in the time domain.

An important use of OASLMs is to record holograms. Typically light scattered off an object is made to interfere with a reference beam (usually a plane wave) with the OASLM recording the interference pattern. If the reference beam is then incident on the OASLM with the interference pattern, the light reflected from the OASLM creates a three-dimensional image of the original object. The reason the image is three-dimensional is that the light scattered off the object and in the reference beam is coherent. Therefore the interference pattern of the OASLM records both amplitude and phase information. This information is then available for projection if the reference beam is incident on the OASLM.

Another use of EOSLMs is in producing light pulses of extremely short duration and is illustrated in Figure 15.6. To create an extremely short pulse, Fourier analysis dictates that the spectrum of the light must be quite broad. To gain control over the spectrum, a short pulse of light is incident on a diffraction grating, meaning that the frequency components of the light in the pulse are separated. This frequency spectrum then hits an EASLM that has been programmed to allow the correct amounts of light at each frequency to pass for an ultrashort pulse. The light from the EASLM is incident on a diffraction grating so its reflection combines all of the frequency components into a single beam. The result is an ultrashort pulse.

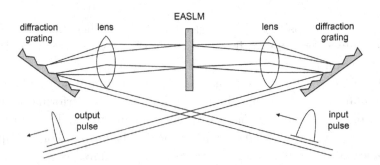

Figure 15.6 Electrically addressed spatial light modulator (EASLM) being used to shape ultrashort light pulses.

15.3 CIRCULAR POLARIZERS

One problem that arises in most types of displays (for example, electroluminescent displays, field emissive displays, LEDs, and OLEDs) is that some light is reflected by surfaces inside the display. These include, for example, glass, metal, conductive coating, and acrylic surfaces. This reflected light emerges from the display and decreases important characteristics of the display, for instance, contrast. In order to keep reflected light from exiting displays, a circular polarizer is used. In many cases, a liquid crystal is the starting material used in fabrication of the circular polarizer.

We saw in Chapter 13 that if light linearly polarized at 45° passes through a quarter-wave plate (QWP) with its slow axis horizontal, then right circularly polarized (RCP) light emerges (see Eq. 13.20). This means that a linear polarizer and a QWP can be combined to produce a circular polarizer. The Jones matrix for a right circular polarizer is just

$$\begin{pmatrix} 1 & 0 \\ 0 & -i \end{pmatrix} \begin{pmatrix} 1/2 & 1/2 \\ 1/2 & 1/2 \end{pmatrix} = \frac{1}{2} \begin{pmatrix} 1 & 1 \\ -i & -i \end{pmatrix}. \tag{15.1}$$

This can be verified by applying the right circular polarizer Jones matrix to an arbitrary input polarization. The amplitude of the resulting light varies, but the light is always RCP. The Jones matrix for a left circular polarizer can be found in a similar way by combining a linear polarizer at 45° with a QWP with its slow axis vertical.

$$\begin{pmatrix} 1 & 0 \\ 0 & i \end{pmatrix} \begin{pmatrix} 1/2 & 1/2 \\ 1/2 & 1/2 \end{pmatrix} = \frac{1}{2} \begin{pmatrix} 1 & 1 \\ i & i \end{pmatrix}. \tag{15.2}$$

The usefulness of circular polarizers in eliminating reflections stems from the fact that RCP light is changed to LCP and *vice versa* when reflected. This can be understood by reminding ourselves of the coordinate system we are using (z-axis is along the propagation direction of the light, y-axis is vertical, and the x-axis is horizontal in the direction that forms a right-handed (x, y, z) coordinate system). As shown in Figure 15.7, when the z-axis is reversed due to reflection, the y-axis remains vertical but the x-axis reverses.

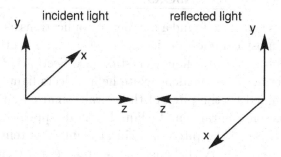

Figure 15.7 Conventional coordinate systems for incident and reflected light.

Thus reflection can be described by the following Jones matrix

$$\begin{pmatrix} -1 & 0 \\ 0 & 1 \end{pmatrix}, \tag{15.3}$$

which can be used to show that RCP light is changed into LCP light upon reflection.

$$\begin{pmatrix} -1 & 0 \\ 0 & 1 \end{pmatrix} \begin{pmatrix} 1 \\ -i \end{pmatrix} = \begin{pmatrix} -1 \\ -i \end{pmatrix} = - \begin{pmatrix} 1 \\ i \end{pmatrix} \tag{15.4}$$

So the light that is circularly polarized with one handedness is reflected with a change in the handedness of its circular polarization and then encounters the same circular polarizer from the other direction. What happens? The QWP looks the same to the reflected light, but the linear polarizer now sits in a coordinate system in which the x-axis is reversed. If this linear polarizer was at 45° for the incident light, it is at -45° for the reflected light. Therefore, the Jones matrix calculation for what emerges is

$$\begin{pmatrix} 1/2 & -1/2 \\ -1/2 & 1/2 \end{pmatrix} \begin{pmatrix} 1 & 0 \\ 0 & -i \end{pmatrix} \begin{pmatrix} 1 \\ i \end{pmatrix} = \begin{pmatrix} 0 \\ 0 \end{pmatrix}, \tag{15.5}$$

meaning that the reflected light is extinguished.

Similar to the compensators described in Section 14.7, the QWP of the circular polarizer can be fabricated in a number of ways, including stretching a polymer sheet, cross-linking a liquid crystal, and allowing the solvent to evaporate from an aligned lyotropic liquid crystal.

15.4 TEMPERATURE SENSORS

Liquid crystals have also found a number of applications as temperature sensors. The most common device is based on the selective reflection of light by chiral nematic liquid crystals (see Section 13.6). A chiral nematic material under incident white light reflects light with a wavelength determined by the pitch of the chiral nematic and the angle of viewing. Thus some chiral nematic liquid crystals appear highly colored in white light. Since the pitch of a chiral nematic is temperature dependent, the color observed changes as the temperature changes. Thus the color of the display can be used to determine the temperature.

Although a thin film of chiral nematic liquid crystal material on a black background, especially if it is properly aligned, provides plenty of reflected light, such an arrangement is prone to contamination and subsequent degradation. To provide some protection to the liquid crystal material and to allow it to be applied more easily, the chiral nematic material is micro-encapsulated into droplets with dimensions of tens of micrometers. These microcapsules are then mixed with a binding material for use in a device. Since spherical microcapsules do not produce the proper alignment of the chiral nematic material, a binding material that contracts upon curing is used to flatten the microcapsules, partially align the liquid crystal, and produce much brighter colors.

Proper use of various mixtures allows the production of temperature sensors with a wide variety of responses. For example, the pitch of many chiral nematic liquid crystals has a weak temperature dependence, running though the visible wavelengths over tens of Celsius degrees. Materials with a visible response in different temperature regions are produced by simply mixing several such compounds together. On the other hand, the pitch of some chiral nematic liquid crystals passes through the entire visible range within a few tenths of a Celsius degree just above the smectic-chiral nematic phase transition temperature. This effect is shown in Figure 15.8. The reason for this divergence is that the twist elastic constant K_2 is very large in the smectic phase, since the smectic layers must deform considerably in order for the director to twist. As we saw in Section 3.7, the pitch is directly proportional to K_2, so both K_2 and the pitch increase as fluctuating regions of smectic-like ordering begin to form near the transition to the smectic phase. All that is necessary to produce extremely sensitive temperature sensors is to mix together the proper compounds so that the smectic-chiral nematic phase transition occurs at the appropriate temperature.

Such devices have found use in a wide variety of applications. These include fever thermometers, hot warning indicators, monitoring devices for the packaging of chilled food, and battery testers, as well as novelties such as "stress" or "mood" sensors, and color-changing jewelry, clothing, decorative wall coverings, and tiles. In the case of some temperature sensors on beer cans to indicate if the beer is cold, the chiral nematic is right in the printing ink used to make the design. Such a device has also been used in medical thermography, in which application of a color sensitive device to a part of the body produces a visual image of the temperature variations of the skin, thus providing an aid for the diagnosis of circulation problems and cancerous growths. Other applications include the use of these temperature sensitive films to detect local heating due to (1) the presence of radiation or certain vapors, (2)

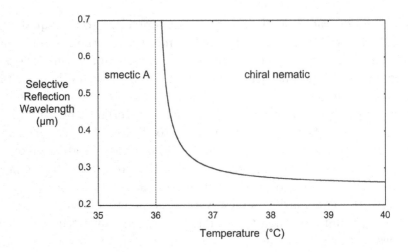

Figure 15.8 Temperature dependence of the selective reflection wavelength of a chiral nematic liquid crystal near the transition to a smectic phase.

plastic flow of structural materials, (3) poor electrical connections on circuit boards, and (4) aerodynamic and fluid flow patterns.

15.5 LUBRICANTS

Liquid crystals often form an oriented boundary layer next to solid surfaces. This property has the potential of allowing the liquid crystal to support high loads and maintain its integrity over time, attributes that are necessary for high-quality lubricants and additives to conventional lubricants. Both calamitic and discotic liquid crystals have this property and are therefore candidates for lubricants. But recently, a good deal of attention is being paid to discotic liquid crystals because the planes of the molecules in the boundary layer are parallel to the surface. This provides the liquid crystal with the lowest viscosity for motion parallel to the surface and allows for stronger bonding of the molecules to the surface. In addition, these discotic liquid crystal lubricants work even better under pressure, where they outdo conventional lubricants, even when the conventional lubricants have a lower viscosity. In fact, the performance of discotic liquid crystals when under high pressure depends weakly on both molecular structure and whether the phase is columnar or nematic. Finally, it is interesting to note that the performance of a non-liquid crystal material with a molecular structure similar to molecules that form discotic liquid crystal phases can be just as good as materials in the liquid crystal phase.

One interesting application of a lubricant based on discotic liquid crystals is in computer hard drives. Discotic molecules with fluorinated side chains form a single molecular layer on the surfaces, with the cores densely packed and parallel to the surfaces, and the side-chains oriented perpendicular to the surfaces. This single layer lubricant provides a stable and low-friction environment, allowing the mechanical parts of the drive to last longer.

15.6 HIGH YIELD STRENGTH POLYMERS

Aromatic main-chain polymers are a class of fairly rigid materials that can form thermotropic or lyotropic liquid crystals. They tend to have high melting temperatures and low solubility in common organic solvents. Perhaps the most important examples of this class of polymers are the aromatic polyamides. Because hydrogen bonding bridges tend to form between the polymer molecules, these materials often remain solid even at high temperatures, sometimes high enough to cause decomposition. On the other hand, these polymers form lyotropic liquid crystals in certain solvents (see Section 10.2). When a fiber is spun from such a solution, the orientational order and strong bonds between the polymer chains in the fiber cause them to be both high strength and heat resistant. The most famous of these polymers is poly(1,4-phenylene-terephthalamide), which has been commercialized by Dupont under the trade name Kevlar®, the structure of which is shown in Figure 15.9. The solvent normally used in the processing of this polymer is concentrated sulfuric acid.

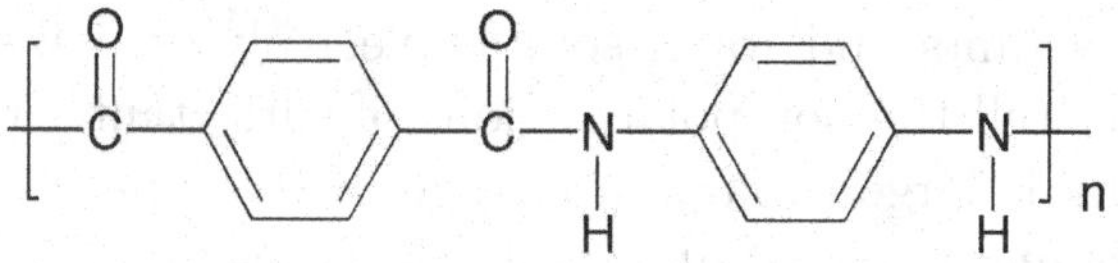

Figure 15.9 Structure of poly(1,4-phenylene-terephthalamide), better known by its trade name Kevlar®.

The basic structure of poly(1,4-phenylene-terephthalamide) has been modified to produce several high-performance materials. Since chain stiffness and solubility are crucial factors in achieving lyotropic liquid crystals, modifications that increase the solubility while maintaining the rod-like structure, high glass transition, and stability have been developed. For example, structural modifications have produced materials soluble in nonpolar aprotic solvents with inorganic salts, in nonpolar aprotic solvents themselves, and even in common organic solvents. For some of these materials, the concentration of polymer must be as high as 40 or 50 wt% in order to form a lyotropic liquid crystal phase.

15.7 DETERGENTS, COSMETICS, FOOD, AND DRUGS

Chapter 9 discussed at length how surfactants form liquid crystal phases at certain concentrations and temperatures. Since surfactants are an extremely important ingredient in many commercial products, it is not surprising that their liquid crystalline properties play a role in their effectiveness. Recall that surfactants are amphiphilic molecules with both hydrophobic and hydrophilic parts. In water, these molecules spontaneously form structures with the polar or hydrophilic parts of the molecules in contact with the water, and with the non-polar or hydrophobic parts of the molecules not in contact with the water. Depending on the molecular structure (relative size of the hydrophilic and hydrophobic parts, charge on the polar part of the molecule, *etc.*), concentration, temperature, and presence of additional compounds, individual micelles (spherical, rod-like, or disk-like) may form, or extended cubic, hexagonal, or lamellar structures may be present. The presence of additives is important in most commercial products. If the additive is water soluble, it adds to the volume of the water solution, making the system behave as if the surfactant concentration were lower. At high enough concentrations, water-soluble additives can change the shape or nature of the surfactant structures. This is especially true for ionic surfactants and water-soluble electrolytes. Oil-soluble additives tend to swell the non-polar regions of surfactant structures, and if this swelling is large enough, a change in the type of structure occurs. Lastly, amphiphilic additives join the surfactant molecules at the polar to non-polar interface and can cause dramatic changes to the surfactant structures. Most important is how the additive changes the average size of the polar portion of the two compounds.

Modern soap bars contain additives that promote the lamellar structure over the hexagonal structures. The lamellar phase has a lower viscosity than the hexagonal phase, which makes for easier processing and superior foaming properties. Liquid detergents sometimes come in the lamellar phase. When diluted during washing, the lamellar phase breaks up into micelles, which are the most effective for cleaning. The lamellar phase also allows abrasive particles to be suspended in the solution. This helps to enhance the cleaning power, and can be safe if the particles are less hard than the surfaces being washed.

Conditioners, both hair and fabric, are made of surfactants in the gel or L_β phase. The idea is that these surfactants completely coat the surface of the fibers or hairs with the tightly packed alkyl chains of the ordered gel phase. This makes fabrics feel "softer" and hair more manageable, both when slightly wet or completely dry. Since the surfactants must remain on the fibers or hair after rinsing, cationic surfactants are used.

Similarly, bread and cake dough must contain surfactants that form lamellar phases. The surfactants constitute the membranes that keep the gas bubbles that form intact, allowing the dough to "rise" and produce a light, textured structure.

Finally, the importance of liquid crystal phases in the processing of food and pharmaceuticals must be mentioned. The final product is a mixture of many substances, and during processing these substances must be brought together into the proper form. By including surfactants in the mixture, and by carefully controlling the liquid crystal phases that form during the processing steps, the desired final formulation is achieved. The result is a wide variety of spreads and frozen desserts, along with successful drug delivery, *e.g.*, in deep heat creams for muscle relief. The use of a liquid crystal to solubilize vitamin C is discussed in Section 9.9.

EXERCISES

15.1 The input beam to an LCoS device has a frequency spread of 5000 GHz. (a) If this beam is diffracted and spread over the 1980 columns of the LCoS device, what is the frequency spread per column? (b) If the data channels are 50 GHz wide, how many columns of the LCoS device receive light from the same channel and therefore should be programmed to reflect the light at the same angle?

15.2 The input beam to the LCoS device of the previous exercise is normal to the LCoS plane. The channel of data with a center wavelength of 1550 nm needs to be reflected by the LCoS device at an angle of 10° to the normal. What should the optical retardation per distance, ϕ/s, of the columns reflecting this channel be?

15.3 The calculation in the text showing that a circular polarizer can be used to extinguish light reflected by the display utilized a right circular polarizer. Perform the same calculation, but instead use a left circular polarizer to show that light reflected by the display is extinguished.

15.4 For the liquid crystal with a pitch described by Figure 15.7, estimate how much the temperature must increase for the selective reflection color to go from red (650 nm) to blue (450 nm).

Bibliography

[1] L. M. Blinov. *Structure and Properties of Liquid Crystals.* Springer. Dordrecht, Germany, 2011.

[2] S. Chandrasekhar. *Liquid Crystals.* Cambridge University Press. Cambridge, UK, second edition, 1992.

[3] R. H. Chen. *Liquid Crystal Displays: Fundamental Physics and Technology.* John Wiley & Sons, Inc. Hoboken, NJ, USA, 2011.

[4] P. J. Collings. *Liquid Crystals: Nature's Delicate Phase of Matter.* Princeton University Press. Princeton, NJ, USA, second edition, 2002.

[5] P. G. de Gennes and J. Prost. *The Physics of Liquid Crystals.* Oxford University Press. Oxford, UK, second edition, 1993.

[6] I. Dierking. *Textures of Liquid Crystals.* Wiley-VCH. Weinheim, Germany, 2003.

[7] D. Dunmur and T. Sluckin. *Soap, Science and Flat Screen TVs.* Oxford University Press, UK, 2011.

[8] J. W. Goodby, P. J. Collings, T. Kato, C. Tschierske, H. F. Gleeson, and P. Raynes, editors. *Handbook of Liquid Crystals*, volume 1-8. Wiley-VCH. Weinheim, Germany, second edition, 2014.

[9] L. S. Hirst. *Fundamentals of Soft Matter Science.* CRC Press. Boca Raton, FL, USA, 2013.

[10] I. C. Khoo. *Liquid Crystals.* John Wiley & Sons, Inc. Hoboken, NJ, USA, second edition, 2007.

[11] D. G. Morris. *Stereochemistry.* Royal Society of Chemistry, Cambridge, UK, 2001.

[12] O. G. Mouritsen. *Life - As a Matter of Fat: The Emerging Science of Lipidomics.* Springer-Verlag, Heidelberg, Germany, 2005.

[13] P. Oswald and P. Pieranski. *Nematic and Cholesteric Liquid Crystals.* CRC Press. Boca Raton, FL, USA, 2005.

[14] P. Oswald and P. Pieranski. *Smectic and Columnar Liquid Crystals.* CRC Press. Boca Raton, FL, USA, 2005.

[15] J. V. Selinger. *Introduction to the Theory of Soft Matter.* Springer International Publishing. Cham, Switzerland, 2016.

[16] S. Singh. *Liquid Crystals Fundamentals.* World Scientific Publishing. Singapore, 2002.

[17] E. A. Wood. *Crystals and Light: An Introduction to Optical Crystallography.* Dover Publications Inc., New York, USA, second edition, 1977.

Index

Printed in the United States
by Baker & Taylor Publisher Services